YOUNG STARS AND PLANETS NEAR THE SUN
# IAU SYMPOSIUM 314

*COVER ILLUSTRATION*

The V4046 Sgr disk as imaged in CO line emission by the Submillimeter Array (left; from Rosenfeld et al. 2013, ApJ, 775, 136), in 427 $\mu$m continuum emission by ALMA during Early Science (Cycle 0) operations (center; from Andrews et al., in prep.), and in polarized intensity at 1.25 $\mu$m by the Gemini Planet Imager in its coronagraphic/polarimetric mode (right; from Rapson et al. 2015, ApJ, 803, L10). The proximity and age of systems like V4046 Sgr — a member of the $\sim$20 Myr-old $\beta$ Pic Moving Group that lies just $\sim$73 pc from Earth — enable unique studies of early stellar evolution and the origins of planetary systems.

IAU SYMPOSIUM PROCEEDINGS SERIES

*Chief Editor*
THIERRY MONTMERLE, IAU General Secretary
*Institut d'Astrophysique de Paris,*
*98bis, Bd Arago, 75014 Paris, France*
*montmerle@iap.fr*

*Editor*
PIERO BENVENUTI, IAU Assistant General Secretary
*University of Padua, Dept of Physics and Astronomy,*
*Vicolo dell'Osservatorio, 3, 35122 Padova, Italy*
*piero.benvenuti@unipd.it*

INTERNATIONAL ASTRONOMICAL UNION

UNION ASTRONOMIQUE INTERNATIONALE

# YOUNG STARS AND PLANETS NEAR THE SUN

## PROCEEDINGS OF THE 314th SYMPOSIUM OF THE INTERNATIONAL ASTRONOMICAL UNION HELD IN ATLANTA, GEORGIA, USA MAY 11–15, 2015

Edited by

**JOEL H. KASTNER**
*Chester F. Carlson Center for Imaging Science and School of Physics &
Astronomy, Rochester Institute of Technology*

**BEATE STELZER**
*Istituto Nazionale di Astrofisica / Osservatorio Astronomico di Palermo*

and

**STANIMIR METCHEV**
*Department of Physics & Astronomy, Centre for Planetary Science and
Exploration, The University of Western Ontario
Department of Physics & Astronomy, Stony Brook University*

CAMBRIDGE UNIVERSITY PRESS
University Printing House, Cambridge CB2 8BS, United Kingdom
40 West 20th Street, New York, NY 10011-4211, USA
10 Stamford Road, Oakleigh, Melbourne 3166, Australia



First published 2016

Printed in the UK by Bell & Bain, Glasgow, UK

Typeset in System LaTeX $2_\varepsilon$

*A catalogue record for this book is available from the British Library Library of Congress
Cataloguing in Publication data*

ISBN 9781107138162 hardback
ISSN 1743-9213

# Table of Contents



## Preliminaries



## 1. The Identification, Ages, and Origins of Nearby Young Stars and Moving Groups

## 2. The Early Evolution of Low- to Intermediate-mass Stars

## 3. The Dispersal of Protoplanetary Disks and The Origins of Debris Disks

## 4. The Early Evolution of Planetary Systems

## 5. Prospects for Advances in The Study of Nearby Young Stars and Planets

## 6. Conference Summary

# Preface

The motivation for a 2015 IAU Symposium dedicated to the study of young stars and planets near the Sun is captured perfectly in the following summary from our original meeting proposal, written by SOC member Ben Zuckerman: *The region surrounding the Sun out to a distance of $\sim$100 pc is often described as the "local bubble" due to the relatively low density of the interstellar medium and an accompanying lack of regions of star formation. In the past two decades, research by many astronomers has revealed an abundance of post T Tauri stars and early type stars of comparable age inside of the bubble. Many of these stars have been classified as members of kinematic moving groups, whose ages range from $\sim$8 Myr up to $\sim$200 Myr. Because these stellar groups are so close to Earth, they provide some of the best samples available to astronomy to investigate the early evolution of low- to intermediate-mass stars. While these nearby, youthful stars are themselves of great interest to stellar astronomy, they also represent the most readily accessible targets for direct imaging (and other measurements) of dusty circumstellar debris disks and young, substellar objects — i.e., newly formed brown dwarfs and, especially, planets. Indeed, <200 Myr-old stars within $\sim$100 pc represent the best laboratories to study the conditions and timescales associated with protoplanetary disk evolution and the formation and early physical and dynamical evolution of planetary systems.*

Our Symposium was intended to highlight the major advances in our understanding of the early evolution of stars and planetary systems, and the potential for further progress, flowing from investigations of nearby young stars. Our aim was to gather scientists approaching such studies from a wide variety of directions: the identification, ages, and origins of local young moving groups; early stellar evolution from theoretical and observational perspectives; the signatures of nascent or recently formed exoplanet systems, including the dispersal of protoplanetary disks, the nature of debris disks, and star-disk, planet-disk, and planet-planet interactions; and the properties of newborn planets.

To draw out the latest results in (and connections between) these diverse topics, the meeting was organized into five, interrelated themes, which also represent the basis for the presentation of the papers in these Proceedings: 1) the identification, ages, and origins of nearby young stars and moving groups; 2) the early evolution of low- to intermediate-mass stars; 3) the dispersal of protoplanetary disks and the origins of debris disks; 4) the early evolution of planetary systems; and 5) the prospects for advances in the study of nearby young stars and planets resulting from new and future observing facilities.

IAU Symposium 314 was the product of a dedicated and thoughtful SOC and LOC whose expertise runs wide and deep through all of the science topics just listed. I am particularly indebted to my co-editors, Beate Stelzer and Stan Metchev, for their time and care in editing these Proceedings.

Our colleague Ben Zuckerman again deserves special mention. Of the many researchers who have driven the field of nearby young star and planet research so far forward over the past two decades, no one has contributed more than Ben. Fittingly, it was Ben who initiated the discussions between a few of us that led to the concept for this Symposium, and it was his steady guidance that resulted in its realization. So, while IAU Symposium 314 may not have been formally dedicated to Ben, our meeting was clearly a testament to that remarkable Zuckermanian vision and insight.

*Joel Kastner, SOC co-Chair*
*Rochester, NY, USA, August 2015*

# IAU Symposium 314:
## Organizing Committees and Acknowledgements

### Scientific Organizing Committee

Isabelle Baraffe (UK)  
Rene Doyon (Canada)  
Michael Ireland (Australia)  
Rob Jeffries (UK)  
Anne-Marie Lagrange (co-Chair; France)  
Stan Metchev (Canada)  
Beate Stelzer (co-Chair; Italy)  
Carlos Torres (Brazil)  

Mike Bessell (Australia)  
Greg Herczeg (China)  
Moira Jardine (UK)  
Joel Kastner (co-Chair; USA)  
Michael Liu (USA)  
David Rodriguez (Chile)  
Motohide Tamura (Japan)  
Ben Zuckerman (USA)  

### Local Organizing Committee

Sebastien Lepine (Chair, USA)  
Russel White (USA)  

Inseok Song (USA)  

### Acknowledgements

IAU Symposium 314 was made possible via the endorsements and support of IAU Divisions G (Stars & Stellar Physics; Coordinating Division), H (Interstellar Matter & the Local Universe), and F (Planetary Systems & Bioastronomy), and Commissions 37 (Star Clusters & Associations), 53 (Extrasolar Planets), and 54 (Optical & Infrared Interferometry). The SOC and LOC thank the IAU for providing travel grants for 17 Symposium participants, and thank Georgia State University and Rochester Institute of Technology's Laboratory for Multiwavelength Astrophysics for their generous financial support and logistical assistance.

The organizers are grateful to Ben Zuckerman for providing a thought-provoking public lecture on the search for higher forms of life in the universe, and to Ben, expert panelists Virginia Trimble and Karin Obërg, and moderator Sebastien Lepine for leading an insightful, dynamic, highly audience-interactive discussion on the same topic following Ben's lecture.

The success of this symposium both professionally and publicly relied heavily on the efforts of the eager, hard working and on-task LOC volunteer staff. Special thanks to the LOC's public outreach coordinator, Nicole Cabrera, for her creative effort in advertising the IAUS 314 public lecture and panel discussion, and for organizing the IAUS 314 Creative Lessons in Astronomy and Space Science (CLASS) Contest. We thank German Chaparro, Virginie Faramaz, Daniel Horenstein, Jeremy Jones, and Laura Vican for presenting at the CLASS Contest, which enabled 40 ninth graders from Cristo Rey Atlanta to learn astronomy from real astronomers. We extend a special thanks to the graduate students who took leading roles in the symposium arrangements: Cassy Davison and Jeremy

Jones (t-shirt design), Mitchell Revalksi (volunteer staffing), Dicy Saylor (badges and social media), and Tara Cotten, Ryan Norris, and Jennifer Winters (program assembly). The lively mid-week excursions were kindly organized and hosted by Nicole Cabrera, Tara Cotten, Cassy Davison, Jinhee Lee, Ryan Norris, Sam Quinn, and Jennifer Winters. Finally, we are grateful to Matt Anderson, Sanam Chaudhary, David Davis, Karen Garcia, Dawn Graninger, Daniel Horenstein, Sushant Mahajan, Rachel Matson, Rahul Patel, Kristina Punzi, Valerie Rapson, Caroline Roberts, Michele Silverstein, and Luqian Wang for logistical support.

The editors wish to thank RIT's Makayla Roof and Cheryl Merrell for assistance with editing and compiling data for the Proceedings, and Elizabeth Woodhouse (Cambridge University Press) and Madeleine Smith (Institut d'Astrophysique de Paris) for timely assistance with and careful attention to the production of the Proceedings.

# CONFERENCE PHOTOGRAPH

# IAU Symposium 314: Participants

| | |
|---|---|
| Kimberly Aller | Institute for Astronomy, University of Hawaii, USA |
| Katelyn Allers | Bucknell University, USA |
| Louis Amard | Geneva Observatory, Switzerland & Montpellier University, France |
| Rebecca Azulay | University of Valencia, Spain |
| Isabelle Baraffe | University of Exeter, UK |
| Cameron Bell | University of Rochester, USA |
| Michael Bessell | Australian National University, Australia |
| Beth Biller | University of Edinburgh, UK |
| Alex Binks | Keele University, UK |
| Til Birnstiel | Max-Planck Institute for Astronomy, Germany |
| Mark Booth | Instituto de Astrofísica Pontificia Universidad Católica de Chile |
| Brendan Bowler | California Institute of Technology, USA |
| Katie Boyce | Rockhurst University, USA |
| Timothy Brandt | Institute for Advanced Study, USA |
| Nicole Cabrera | Georgia State University, USA |
| Justin Cantrell | Georgia State University, USA |
| Gilles Chabrier | University of Exeter, UK |
| German Chaparro | Universidad ECCI, Colombia |
| Gael Chauvin | Institut de Planétologie et d'Astrophysique de Grenoble / CNRS, France |
| Lucas Cieza | Universidad Diego Portales, Chile |
| Ann Marie Cody | NASA Ames Research Center, USA |
| Tara Cotten | University of Georgia, USA |
| Kevin Covey | Western Washington University, USA |
| Sebastian Daemgen | University of Toronto, Canada |
| Cassy Davison | Georgia State University, USA |
| Emmanuel Di Folco | Laboratoire d'Astrophysique de Bordeaux / CNRS, France |
| Jason Dittman | Harvard University, USA |
| Rene Doyon | University of Montreal, Canada |
| Trent Dupuy | University of Texas at Austin, USA |
| Paul Elliott | ESO, University of Exeter, UK |
| Jackie Faherty | Carnegie Institution of Washington, USA |
| Virginie Faramaz | Instituto de Astrofísica Pontificia Universidad Católica de Chile |
| Gregory Feiden | Uppsala University, Sweden |
| Will Fischer | NASA Goddard Space Flight Center, USA |
| Laura Flagg | Northern Arizona University, USA |
| Colin Folsom | Institut de Planétologie et d'Astrophysique de Grenoble / CNRS, France |
| Adam Frank | University of Rochester, USA |
| Jonathan Gagne | Carnegie Institute of Washington, USA |
| Phillip Galli | Instituto de Astronomia, Geofisica e Ciencias Atmosféricas, Brazil |
| Jack Gallimore | Bucknell University, USA |
| Victor (Eugenio) Garcia | Lowell Observatory, USA |
| Bartek (Bartosz) Gauza | Instituto de Astrofísica de Canarias, Spain |
| Doug Gies | Georgia State University, USA |
| Uma Gorti | NASA Ames research center, SETI Institute, USA |
| Dawn Graninger | Harvard University, USA |
| Jose Carlos Guirado | Universidad de Valencia, Spain |
| Gregory Herczeg | Kavli Institute, China |

| | |
|---|---|
| Daniel Horenstein | Georgia State University, USA |
| Gaitee Hussain | European Southern Observatory, Germany |
| Subkhonjon Ibadov | Sternberg Astronomical Institute, Russia and Tajik Academy of Sciences, Tajikistan |
| Michael Ireland | Macquarie University, Australia |
| Jeremy Jones | Georgia State University, USA |
| Joel Kastner | Rochester Institute of Technology, USA |
| Grant Kennedy | University of Cambridge, UK |
| Matthew Kenworthy | Leiden Observatory, The Netherlands |
| Taisiya Kopytova | Max Planck Institute for Astronomy, Germany |
| Agnes Kospal | Konkoly Observatory, Hungary |
| Adam Kraus | University of Texas at Austin, USA |
| Charles Lada | Smithsonian Astrophysical Observatory, USA |
| Jinhe Lee | University of Georgia, USA |
| Sebastien Lepine | Georgia State University, USA |
| Min-Kai Lin | Canadian Institute for Theoretical Astrophysics, Canada |
| Michael Liu | Institute for Astronomy, University of Hawaii, USA |
| Nicholas Lodieu | Instituto de Astrofísica de Canarias, Spain |
| John MacNeil | University of Georgia, USA |
| Thomas Maindl | University of Vienna, Austria |
| Lison Malo | Canada-France-Hawaii Telescope, USA |
| Eric Mamajek | University of Rochester, USA |
| Mark Marley | NASA Ames Research Center, USA |
| Christian Marois | Herzberg Institute of Astrophysics, Canada |
| Jonathan Marshall | University of New South Wales, Australia |
| Sarah Martell | University of New South Wales, Australia |
| Sean Matt | University of Exeter, UK |
| Ben McGimsey | Georgia State University, USA |
| Carl Melis | University of California, San Diego, USA |
| Stanimir Metchev | University of Western Ontario, Canada |
| David Montes | Universidad Complutense de Madrid, Spain |
| Christoph Mordasini | University of Bern, Switzerland |
| Simon Murphy | Australian National University, Australia |
| Elisabeth Newton | Harvard-Smithsonian Center for Astrophysics, USA |
| Eric Nielsen | Institute for Astronomy, University of Hawaii, USA |
| Ryan Norris | Georgia State University, USA |
| Karin Öberg | Harvard University, USA |
| Ryan Oelkers | Texas A&M University, USA |
| Deborah Padgett | NASA Goddard Space Flight Center, USA |
| J. Robert Parks | Georgia State University, USA |
| Rahul Patel | State University of NY at Stony Brook, USA |
| Mark Pecaut | Rockhurst University, USA |
| Jessica Pericaud | Laboratoire d'Astrophysique de Bordeaux, France |
| Ignazio Pillitteri | Harvard-Smithsonian Center for Astrophysics, USA |
| Marc Pinsonneault | Ohio State University, USA |
| Peter Plavchan | Missouri State University, USA |
| David Principe | Universidad Diego Portales, Chile |
| Kristina Punzi | Rochester Institute of Technology, USA |
| Samuel Quinn | Georgia State University, USA |
| Valerie Rapson | Rochester Institute of Technology, USA |
| Jon Rees | University of Exeter, UK |

| | |
|---|---|
| Adric Riedel | CUNY - The College of Staten Island, USA |
| Stephen Rinehart | NASA Goddard Space Flight Center, USA |
| Aaron Rizzuto | University of Texas at Austin, USA |
| Jan Robrade | Hamburger Sternwarte, Germany |
| Joseph Rodriguez | Vanderbilt University, USA |
| David Rodriguez | Universidad de Chile, Puerto Rico |
| Barbara Rojas Ayala | Universidade do Porto, Portugal |
| Dicy Saylor | Georgia State University, USA |
| Joshua Schlieder | NASA Ames Research Center, USA |
| Tobias Schmidt | Astrophysikalisches Institut und Universitäts-Sternwarte Jena, Germany |
| Christian Schneider | European Space Agency, The Netherlands |
| Adam Schneider | University of Toledo, USA |
| Evgenia Shkolnik | Lowell Observatory, USA |
| Michele Silverstein | Georgia State University, USA |
| Michal Simon | Stony Brook University, USA |
| Garrett Somers | Ohio State University, USA |
| Inseok Song | University of Georgia, USA |
| Alessandro Sozzetti | INAF - Osservatorio Astrofisico di Torino, Italy |
| Beate Stelzer | INAF - Osservatorio Astronomico di Palermo, Italy |
| David Stooksbury | University of Georgia, USA |
| Angelle Tanner | Mississippi State University, USA |
| Susan Terebey | California State University, LA, USA |
| Chris Tinney | University of New South Wales Australia |
| Virginia Trimble | University of California, Irvine, USA |
| Adriana Valio | CRAAM - Mackenzie University, Brazil |
| Nienke van der Marel | University of Hawaii, USA & Leiden Observatory, The Netherlands |
| Laura Vican | University of California, Los Angeles, USA |
| Luqian Wang | Georgia State University, USA |
| Russel White | Georgia State University, USA |
| David Wilner | Harvard-Smithsonian Center for Astrophysics, USA |
| Jennifer Winters | Georgia State University, USA |
| Jiwei Xie | Nanjing University, China |
| Andrew Youdin | University of Arizona, USA |
| Olga Zakhozhay | Main Astronomical Obs. of the National Academy of Sciences, Ukraine |
| Ben Zuckerman | University of California, Los Angeles, USA |

*Young Stars & Planets Near the Sun*
*Proceedings IAU Symposium No. 314, 2015*
*J. H. Kastner, B. Stelzer, & S. A. Metchev, eds.*

© International Astronomical Union 2016
doi:10.1017/S1743921315006511

# Fourth Day of Creation:
# The Proto-history of Young Stars,
# Star Streams, and Exoplanets Near the Sun

## Virginia Trimble[1]

[1]Department of Physics and Astronomy, University of California, Irvine, CA, USA

**Abstract.** Items of scientific knowledge at any moment in time have pre-histories when they were debated, doubted, or absolutely denied. The examples considered here are the admitted facts that star formation is an on-going process in the Milky Way, that there are young moving groups (the products of young star clusters in the process of dissolution and perhaps more complex processes), and that planets orbiting other stars are common. It is hard to imagine any of these ceasing to be part of core astronomical knowledge, but you are advised not to place large bets on this.

**Keywords.** Star Formation, Moving Groups, Star-Streams, Exoplanets

## 1. Introduction

The after-dinner version of this approach to the topic of the conference included a large number of images of ancient cosmologies and people, early published texts, and handwritten letters. Not one of them is here, because I do not own copy-rights to the first two classes, and the latter two classes are almost unreadable. However, some attempt is made to convey the flavor of the presentation, with references to where much of the information can be found.

## 2. Young Stars and On-Going Star Formation

*Vaya'ash Elohim et shney hamorot hag'dolim...And G.d made two great lights, the greater light to rule the day and the lesser light to rule the night. V'et hakochabim. And also the stars...Vayechi erev, vayechi boker, yom rivi'i. And there was evening and there was morning. A fourth day.* This version makes clear that the stars are somewhat incidental to the solar system and are all the same age. The writers of Genesis got the time scale wrong by a factor of a million or so, but the idea that everything is the same age lasted long.

In contrast, for the ancient Egyptians, the Chinese, and the Greeks, star formation continued into their present. The shafts extending from the king's burial chamber in Cheops' pyramid were intended for his ka to rise and join the stars; you have undoubtedly heard of the Chinese recording what we now call supernovae as "guest stars;" and the Greeks even provide a tentative ordering of what formed first. By this I mean Greek mythology, not the much less picturesque equants and deferents of Ptolemy and his followers. Careful reading of these leads us to the conclusion that the gods first put the Hyades (the daughters of Atta) in the sky and then the Pleiades, apparently to save them from being chased by Orion, who, however, followed them into the sky. He was joined later by his dogs Canis Major and Canis Minor. This does not correspond to our modern ordering, though the Pleiades are younger than the Hyades, and the Orion Nebula stars younger than either.

Through the 18th, 19th, and into the 20th century there were at least half a dozen swings back and forth between these two views of "all at once, long ago" and "on-going process" (Trimble 1997). The last adherents of "all at once" were probably de Vaucouleurs, who wanted to form galaxies and stars in a coasting phase of a universe with a cosmological constant as late as 1958, and David Layzer with his 1964 cold big bang. Herschel over his lifetime rode both horses in succession, eventually settling on spiral nebulae as systems in formation. Russell, Dugan, and Stewart had "ongoing" in the first, 1927 edition of their two volume classic text and "all at once" in the 1938 second edition. Swings to "all at once" came from the first understanding of thermodynamics (if everything is running down it must all have started in the past) and later from estimates of stellar lifespans from the recognition of hydrogen fusion as the primary energy source. The mistake there was the feeling that, if most things are the same age (stars like the sun; the Hubble time; the oldest earth rocks and meteorites), then probably everything is the same age.

Swings toward "on-going" had come from Herschel's nebulae and Russell's giant and dwarf theory of stellar evolution in 1913. You might have expected Huggins's discovery of gas clouds, an obvious source of material for new stars, to have settled the issue in the 1860's, but he said not. Huggins's own view was that the proto-stellar material advocated by Herschel would have to contain all the elements found in the sun, while he recognized only hydrogen, nebulium, and something he thought was nitrogen, but was not, in the gaseous nebulae.

Thus the issue was settled observationally when V.A. Ambartsumyan and Adriaan Blaauw pointed to expanding groups of stars that included OB stars and associated gas. The discoveries of T Tauri stars and Herbig-Haro objects aided the process. In 1948, Lyman Spitzer was still putting much star formation in the early universe (with incomprehensible and perhaps different physics), but said that the supergiants must be young since hydrogen fusion could not have kept them that bright for long. Cecilia Payne Gaposchkin, writing for general readers in 1952, said, "We can even toy with the idea that stars are being born continuously; and the places where the young stars are found suggest very definitely that they are born of the interstellar dust."

Because papers in 1939 by Spitzer and by Fred Hoyle had come so close to suggesting on-going star formation as the dominant process, I wrote to them in 1996 asking for their reflections on that time and subject. Both answered. Hoyle reminded me that, even now, the origin of spectroscopic binaries from clouds is difficult to understand and also that, in 1939, the astronomical world did not believe in an interstellar medium containing hydrogen. What we now call Bondi-Hoyle accretion (though Lyttleton was co-author on one of the papers) was intended to bulk up old, low mass stars into OB's and supergiants.

Spitzer's response told of a talk he had given at Harvard in 1939 and a paper submitted to ApJ later that year incorporating "certain dynamical processes in the interstellar medium" that would have led to star formation. These were objected to as "too speculative," and the paper was published without that material. He also answered my next letter, in which I had, of course, asked who the objectors were. Spitzer declined to answer "since one of these is still very much alive." But he signed as Lyman, to show we were still friends. I was never able to think of anyone this could have been except Jesse Greenstein, but at lunch the next spring Jesse said no. Only some massive unsealing of ApJ records from 1939 (which may not even still exist) could now tell us. Spitzer has himself told the non-secret parts of the story in his chapter in volume 7 of the Kuiper Compendium, *Stars and Stellar Systems.*

The heroes whose pictures appeared at this point were, of course, Ambartsumyan, Blaauw, and Spitzer.

## 3. Moving Groups and Star Streams

Blaauw's (1952) study of the stars of the Zeta Persei group and of surrounding O to B5 stars provides a natural transition to this second topic. His figure shows proper motions after the subtraction of the mean of the group, thus revealing the expansion and a dynamical time scale of only about 10 million years, so there are both moving groups and ongoing star formation. But we can look much farther back.

Relative proper motions of stars were discovered by Halley, comparing his positions with those of Ptolemy. By 1783 enough numbers had accumulated for William Herschel to declare that there was a component from the reflex of solar motion, meaning that we are moving more or less toward Hercules, a result which stands today (North 2008 pp. 423, 436). But it was Richard Proctor (1869, 1870) who concluded that published catalogues were concealing another effect. Stars were generally listed in declination zones, but, he said, take a star map and put little arrows to represent proper motions and you will find what he called star drift.

Proctor drew attention to the drift in Taurus (the Pleiades and its neighbors) toward the north-west, in Gemini and Cancer toward the south-east, and Leo toward Cancer. These are still more or less, he said, in the direction of solar motion (and there have been frequent claims for a large solar association). Five of the seven bright stars in Ursa Major and three in Arietes drift in the opposite direction. He concluded that it was at least possible that the local stellar system was in rotation around the rich cluster h and $\chi$ Persei, and he looked forward to the "stellar proper motions of recess or approach" to be expected from Mr. Huggins, now called radial velocities. For me, the symbol will always be $V_r$, but most speakers at IAU Symposium 314 used RV (well, recreational vehicles also display proper motions).

Angular proper motions and indeed radial velocities accumulated and by 1927 Russell, Dugan, and Stewart recognized moving clusters Ursa Majoris, the Hyades (with distance determination), Sco-Cen, Perseus, Pleiades, Coma Berenices, and Praesepe in Cancer (some of these in common with Proctor's systems).

Meanwhile, however, there had arisen a confusorium, with Kapteyn in 1905 describing the large system of peculiar motions as two interlaced streams moving in opposite directions. Karl Schwarzschild soon after put forward his velocity ellipsoid. The two pictures are essentially equivalent said RDS, though the latter retains some currency as a way of characterizing the asymmetries of stellar motions perpendicular to the galactic disk toward and away from the galactic center, and in the direction of galactic rotation. What Kapteyn and Schwarzschild were seeing, through a telescope darkly, was galactic rotation of the disk, but not the halo traced by globular clusters and other old, high-proper motion stars. Schwarzschild and Hertzsprung (1913) pointing out that Preasepe moves with the Hyades and Wilson's (1932) discovery that 2% of stars near the sun share the motion of the Hyades and constitute the Taurus group to be found all over the sky, are also part of this story. Plots of star streaming from RDS that I described as "fairly incomprehensible" gave their caption if not their plot to a couple of diagrams shown on Friday by others.

Analyses of stellar peculiar motions were often described as measuring the solar motion relative to some specific population right down to the time when Hubble used the population of external galaxies in 1929 and found something else.

But we leave it to Oort (1926, 1932) to sort out the large scale Milky Way patterns so as to be able to hurry on to Olin Jeuck Eggen. His life (see Eggen 1993) counts as nearly indescribable. The picture of him I showed was his own choice for that memoir (the other three pictured heroes were Kapteyn, K. Schwarzschild, and Oort), though Mike Bessell has one he likes much better. Bessell also updated the "two carburetors" story (Trimble *et al.* 2001) with the topper that, when other astronomers acquired cars with two, Eggen claimed three.

Starting in 1957, Eggen indicated that the Taurus Group included 1% of FK5 stars and was part of a group moving through the solar neighborhood at 44 km/sec. Later studies found superclusters associated with Sirius, Hyades, Pleiades, IC 2391, and NGC 2516 and 2287 (both part of the Pleiades group). Intermediate summaries are Eggen (1965, 1993) up to the final Eggen (1998) in which he concluded that physically meaningful groups might share only motion, not the ages of the stars, compositions, or directions in the sky. Should much of this be classified with his "trend lines" in HR diagrams from earlier work? I have certainly thought so in the past (Trimble (2000). But, in 2004, when SDSS data began to recognize disrupted dwarf satellites as star streams in the Milky Way (Navarro *et al.* 2004) my instantaneous reaction was "Egad! Eggen was right!" The possibility deserves a closer look by some young, kindly person!

Looking ahead, there will, of course, be more star streams coming out of surveys from SDSS and GAIA. GAIA also finds Exoplanets, which brings us to the next section. The fifth person in the photograph I showed was Jim Gunn (in 1971), the founding spirit of SDSS.

## 4. Exoplanets, or, All My Errors are Systematic

A systematic error is the scientific equivalent of doing the same thing over and over again and expecting a different result (one of psychology's definitions of insanity). This is no longer a problem for Exoplanets. The right path was foreshadowed in Zuckerman and Song (2004) whose abstract ends by saying, "These young groups (ages 8–50 Myr) along with other nearby, young stars will enable imaging and spectroscopic studies of the origin and early evolution of planetary systems," and there is no doubt that IAUS 314 marks a major step forward on that trajectory. But let's look back, in order, at direct imaging, radio and optical radial velocities, transits, and finally astrometry, the most interesting, heroic, and tragic of the tales.

### 4.1. *Direct Imaging*

William Herschel is said several places to have suspected a third light in 70 Oph, when the stars were about 4.5″ apart and each of naked eye magnitude. By the last day of IAUS 314 there were "about 10" directly-imaged Exoplanets, not all from HST, but nearly all involving great cleverness in adaptive optics, coronagraphy, and other ways of cheating mother nature. A few retractions have not, I think, cost anybody life, limb or reputation.

### 4.2. *Radio Radial Velocities*

The trick is to time pulsars very, very accurately. Bailes *et al.* (1991) in this way thought briefly that they had found evidence for a six-month-period oscillation in the radial velocity of PSR 1829-10. Well, yes, but it was the earth moving in and out of the sun's gravitational field They had allowed for special relativistic (Doppler) but not general relativistic (gravitational redshift) effects.

Triumph was, however, waiting just around the calendric corner for Wolszczan and Frail (1992) who found first one, then two (and now I think there are three) planets orbiting PSR 1957+12. Masses and orbits are both small compared either to Jupiter or later "hot Jupiters," and they are probably bits of broken-off neutron star material, now mostly iron, rather than habitable rocks and volatiles. Pulsar timing carries on at several observatories, but has not ever quite struck the same kind of iron again.

### 4.3. *Optical Radial Velocities (winners old and new)*

Black (1995), who called his review "Completing the Copernican Revolution," had the misfortune to write just before the Geneva and Lick groups announced 51 Peg b. Published studies at the time included Campbell *et al.* (1988) who had found a 1–5 Jupiter-mass planet orbiting $\epsilon$ Eridani (an early Drake candidate for SETI searches), now I think generally not counted, and Latham *et al.* (1989) who had certainly found something, but not a planet or even a brown dwarf, unless the system is much more nearly edge-on than other data suggested.

The winners (Mayor and Queloz 1995 and Marcy and Butler 1996) won big, with multiple prizes to their names and, within a decade, enough planets for statistical analysis, a large fraction found by their methods and groups (Jones 2004). I had the good fortune to address "The Quest for Other Worlds" at that same conference (Trimble 2004), providing a list of 23 possible detection methods. Of those I then thought dubious, gaps in YSO accretion disks now count as quite reliable, and, according to a comment by Ben Zuckerman, pollution of host star atmospheres has definitely been recognized among white dwarfs (Zuckerman 2015).

### 4.4. *Transits*

At peak, there were more than 20 groups looking for transits by Exoplanets (dimming of a star when the planet crosses its face, Charbonneau *et al.* 2004). Many had come up empty, simultaneously revealing a puzzling brown dwarf desert around normal stars. And the "detections" included a frightening cohort of false positives, 136 of 137 in one OGLE search (Torres e al. 2004). Eclipsing binaries, with grazing incidence or blended on the sky with a bright, non-varying star were the main sources. Meanwhile, the answer was lurking in the literature in a journal much less cited than Nature or ApJ Letters (Borucki and Summers 1984). "Go up, young man," was the answer, and though Kepler has also had its false positives, spot checking by radial velocity methods shows they are very few among the 3000 candidates to date (Borucki 2015). None of the discoveries is quite an earth-mass planet in the liquid-water zone of a sun-like star, but some are not far off, and Kepler has also been the source of most of the multi-planet systems. The record to date is probably 7, organized in a size-period combination nothing like our solar system.

### 4.5. *Astrometry*

Relying on the same physics as radial velocity searches but in the plane of the sky instead of perpendicular to it, this is the oldest method, connected with the saddest stories and very few successes. The method works and led to the discoveries of Sirius B and Procyon B, both later imaged. That it might be extended into the planetary range was the thought of Capt. William Steven Jacobs (Jacobs 1855), who suspected he might have seen a small wiggle in the binary orbit of 70 Oph. Jacob is not obscure, in the sense of being represented in the (British) *Dictionary of National Biography.* He received however, little credit for his idea, though he was director at Madras (now Chennai) for many years, where Pogson (of the magnitude scale and also unfairly obscure) was a staff member and later Jacobs's successor. The saga of his last return to India, via 100 days under sail and leading to his death a few weeks later in Poonah, is found in a diary kept by his daughters. The hunt and peck method of using Google (etc) will find this via a

Jacobs family genealogy. The meals described for the voyage leave one feeling that scurvy or some similar deficiency must have contributed to his death.

Next comes Thomas Jefferson Jackson See, who also looked at 70 Oph and thought (or anyhow claimed) that he had seen astrometric evidence for a planet around the secondary (See 1896). Chamberlin (of the Chamberlin-Moulton hypothesis) showed the triad would be unstable, and so obstreperous did See become (see See 1899, sorry!) that he was forbidden ever again to submit a paper there. He lived to a ripe old, perhaps even happy old, age, mostly on Mare Island, and received a rather nice obituary in *Sky & Telescope* (Ashbrook 1962) as well as in the *New York Times* and *Physics Today*. He belonged, at peak, to at least 25 professional scientific societies in at least 5 countries.

Later players in the astrometric tournament are less colorful, but generally not less unsuccessful. Wrong orbits were reported for 70 Oph by Dirk Reuyl and Eric Holmberg (famous for other things) and for 61 Cygni by K. A. Strand, who became in due course director of the US Naval Observatory. The mistake is an easy one to make. If you had before you the orbit of $\xi$ UMa that appears on page 682 of RDS, you might well be tempted to take a red pencil and connect up the dots from 1880 to 1899 and claim to have seen three cycles of a 5.9 year perturbation of the main 59.8 year period, with an amplitude that would imply a perturbing mass of about 0.05 times the mass of the secondary star.

The last in this lugubrious list is Peter (Piet) van de Kamp, who was director of Sproul Observatory (Swarthmore College) from 1937 to 1972, though he continued research there for some years, returning to Holland only very late in life. He sought through most of that time companions to nearby low-mass stars, and indeed greatly increased our knowledge of visual binaries, but now is almost entirely remembered for the claim of two planets orbiting Barnard's star. This was at one time a major topic of discussion, and I remember a boy of about 10 coming up to me after a very non-technical talk for a 3rd grade class somewhere around 1983 and asking, verbatim, "What is the current status of the companions of Barnard's star?" For van de Kamp to nearly the end of his life (van de Kamp 1986) they had masses of 0.7 and 0.5 $M_J$ and orbit periods of 12 and 20 years. But a long data set from Leander McCormick Observatory found no trace of either, and the astronomical community came to blame a shift in the declination axis of the Sproul telescope and other mechanical and observational problems for the apparent wiggles in the sky.

The two "astrometry heroes" whose pictures I showed were Strand and van de Kamp, for obvious reasons. But, in addition, Strand had offered me a job at USNO in 1968 to work on the observatory's 2-micron survey (which proved not very informative). And at high school graduation in 1961, Ruth Staff Halliday, an unforgettable teacher of US history at Hollywood High School, had urged me to go to her alma mater, Swarthmore, which had offered a full scholarship. UCLA, where George Abell was the undergraduate advisor, was definitely a better choice for a potential astronomer.

## 5. Conclusion

If there is a moral to these folk tales, it has three parts:

- Observers be careful, the theorists are watching you;
- Theorists take care, the observers are watching you; and
- Both observers and theorists be warned the historians are watching you.

### Acknowledgement

I am grateful to the organizers for the opportunity to put my thoughts on these subjects together for the IAU Symposium 314, to Ms. Alison Lara for her, as always, expert, transformation of a typed original into a keyboarded, formatted compuscript, and to all the colleagues living and dead from whom I have ruthlessly plagiarized this information. To the best of my knowledge, everything here is at least broadly true (as Stephen Hawking is supposed to have said about the film *Theory of Everything*) but it could be better documented and perhaps some day will be.

## References

Bailes, M., Lyne, A. G., & Shemar, S. L. 1991, *Nature*, 352, 311

Blaauw, A. 1952, *BAN*, 11, 405

Black, D. C. 1995, *ARA&A*, 33, 359

, Borucki, W. J. 2015. Rep. Prog. Physics (in press)

Borucki, W. J. & Summers, A. L. 1984, *Icarus*, 58, 121

Campbell, B., Walker, G. A. H., & Yang, S. 1988, *ApJ*, 331, 902

Charbonneau, D., Brown, T. M., Dunham, E. W., *et al.* 2004, in *The Search for Other Worlds*, S. S. Holt & D. Deming, eds., *AIP Conf. Proc.* 713, 151

Eggen, O. J. 1957, The Observatory, 77, 229

Eggen, O. J. 1965, in *Galactic Structure*, A. Blaauw & M. Schmidt, eds., 111

Eggen, O. J. 1970, *PASP*, 82, 99

Eggen, O. J. 1993, *ARA&A*, 31, 1

Eggen, O. J. 1998, *AJ*, 115, 2397

Jacobs, W. S. 1855, *Plurality of Worlds*, and *MNRAS*, 15, 228

Jones, H. R. A. 2004, in *The Search for Other Worlds*, S. S. Holt & D. Deming, eds., *AIP Conf. Proc.* 713, 17

Kapteyn, J. 1905. *Rep. Br. Assoc.*, p. 257

Latham, D. W., Stefanik, R. P., Mazeh, T., Mayor, M., & Burki, G. 1989, *Nature*, 339, 38

Marcy, G. W. & Butler, R. P. 1996, *ApJL*, 464, L147

Mayor, M. & Queloz, D. 1995, *Nature*, 378, 355

Navarro, J. F., Helmi, A., & Freeman, K. C. 2004, *ApJL*, 601, L43

Oort, J. H. 1926, The Observatory, 49, 302

Oort, J. H. 1932, *BAN*, 6, 249

Proctor, R. 1869, *Proc. Roy. Soc.* 18, 169

Proctor, R. 1870, *Phil. Trans. Roy.* Soc.

Russell, H. N., Dugan, R. S., & Stewart, R. M. 1927, *JRASC*, 21, 119

Schwarzschild, K. 1907, 1908. Goettingen Nachrichten

Schwarzschild, K. & Hertzspring, E. 1939. *Astron. Nach.* 196, 9

See, T. J. J. 1896, *AJ*, 16, 17

See, T. J. J. 1899, *AJ*, 20, 33

Torres, G., *et al.* 2004, in *The Search for Other Worlds*, S. S. Holt & D. Deming, eds., *AIP Conf. Proc.* 713, 165

Trimble, V. 1997, in *Star Formation Near and Far*, S.S. Holt & L.G. Mundy, eds., *AIP Conf. Ser.* 393, 15

Trimble, V. 2004, in *The Search for Other Worlds*, S. S. Holt & D. Deming, eds., *AIP Conf. Proc.* 713, 3

Trimble, V., *et al.* 2001, *PASP*, 113, 131

van de Kamp, P. 1986, *Space Science Rev.*, 43, 211

Wolszczan, A. & Frail, D. A. 1992, *Nature*, 355, 145

Zuckerman, B. & Song, I. 2004, *ARA&A*, 42, 685

Zuckerman, B. 2015, in 19th European Workshop on White Dwarfs, P. Dufour *et al.*, eds., *ASP Conf. Ser.* 493, 291

*Young Stars & Planets Near the Sun*
*Proceedings IAU Symposium No. 314, 2015*
*J. H. Kastner, B. Stelzer, & S. A. Metchev, eds.*

© International Astronomical Union 2016
doi:10.1017/S1743921315005955

# Star Formation in the Local Milky Way

## Charles J. Lada[1]

[1]Smithsonian Astrophysical Observatory
email: clada@cfa.harvard.edu

**Abstract.** Studies of molecular clouds and young stars near the sun have provided invaluable insights into the process of star formation. Indeed, much of our physical understanding of this topic has been derived from such studies. Perhaps the two most fundamental problems confronting star formation research today are: 1) determining the origin of stellar mass and 2) deciphering the nature of the physical processes that control the star formation rate in molecular gas. As I will briefly outline here, observations and studies of local star forming regions are making particularly significant contributions toward the solution of both these important problems.

**Keywords.** stars: formation

## 1. Introduction

In this contribution I will describe what I consider to be two of the most fundamental problems that confront star formation research and how observations of the local Milky Way are providing critical insights into their resolution. These two basic questions are: 1) what is the origin of stellar mass? and 2) what are the physical processes that control the rate at which the gaseous interstellar medium (ISM) transforms a significant part of itself into stellar form?

Stars are the fundamental objects of the astronomical universe. They convert hydrogen, the primary product of the big bang, into the heavy elements of the periodic table. Through stellar evolution they control the evolution of all stellar systems from stellar clusters to the largest galaxies. They provide the sites for planetary systems and the energy necessary for the development and existence of life. Consequently, knowledge of their origins is both of compelling intrinsic interest as well as essential for understanding the closely related problems of planet formation and galaxy formation and evolution.

## 2. A Predictive Theory of Star Formation

In my view the ultimate goal of star formation research is the development of a predictive theory of stellar origins. By that, I mean the construction of a theory that, given a limited set of initial conditions, is able to predict the basic physical properties of stars, the rate at which these stars form, and how this rate varies in time. Probably the most fundamental physical properties of stars that should be predicted by any respectable theory of stellar origins are their compositions, luminosities, temperatures, sizes, and masses. The first of these properties is trivially predicted since we have *a priori* empirical knowledge of the composition of the ISM from which the stars themselves form. Additionally, the beautiful and powerful theory of stellar structure and evolution, developed in the twentieth century, can already predict the luminosities, temperatures, and sizes of stars, once their initial masses and compositions are specified. This leaves stellar mass as the one fundamental stellar property for which we have yet no physical explanation and which a star formation theory first and foremost must explain. Star formation rates are known to vary considerably within and between galaxies (e.g., Kennicutt 1998) and between local molecular clouds (e.g., Lada *et al.* 2010). However, little is understood about the physical processes that control the star formation rate (SFR) in interstellar gas. These processes and how they vary in space and time are what control the evolution of all stellar systems from star clusters to galaxies. In doing so they drive cosmic evolution itself. A general theory of star formation must be able to predict the SFR and its environmental and temporal variations in molecular clouds and galaxies.

## 3. The Origin of Stellar Mass

### 3.1. *The Stellar Initial Mass Function*

According to the theory of stellar structure and evolution, once formed the subsequent life history of a star is entirely predetermined by the only two parameters: the star's initial mass and, to a lesser extent, its chemical composition. The frequency distribution of stellar masses at birth, or the initial mass function (IMF) of stars, thus plays a pivotal role in the evolution of all stellar systems. In the absence of a general theory that can predict stellar masses and the IMF, these must be first determined empirically. The first attempt to determine the IMF was by Salpeter (1955) more than half a century ago. He derived the IMF from the luminosity function of local field stars (i.e., stars within $\sim 500$ pc of the sun) by converting stellar luminosities to masses using empirical mass-luminosity relations. He corrected for losses in the stellar number counts due to stellar evolution by assuming a constant SFR over the age of the Galaxy. Salpeter demonstrated that between 1 and 10 $M_\odot$ the IMF had the form of a power-law with an index of -1.3 (when adopting the conventional logarithmic mass binning for this function).

Numerous subsequent determinations of the local field star IMF increased the range of distances and masses over which the IMF could be determined, with masses now ranging from OB stars to the hydrogen burning limit and somewhat below (e.g., Scalo 1978, Kroupa 2002, Chabrier 2003). In addition, infrared studies of local, very young star clusters provided independent determinations of the IMF over an even larger dynamic range in mass, from the deuterium burning limit to OB stars (e.g., Muench *et al.* 2002). These studies have found that the shape of the IMF is lognormal-like and exhibits a broad peak between 0.1 and 0.5 $M_\odot$ suggesting a characteristic mass associated with star formation of about 0.25 $M_\odot$. Less than $\sim 25$ % of the objects formed in the Trapezium star forming event were found to have substellar masses (Muench *et al.* 2002). Moreover, the fact that the form of the field star IMF so closely matches that of an embedded cluster such as the Trapezium in Orion (see Figure 1) is quite profound, since field stars were formed over billions of years of Galactic history and over a large volume of space (perhaps as much as a kpc in extent) while an embedded cluster like the Trapezium formed its stars in only a few million years in a volume perhaps ten million times smaller. This indicates that the functional form of the IMF in the disk of the Milky Way is very likely universal in both space and time.

Another stellar property that should be met by a predictive theory of star formation is that of stellar multiplicity. It is well established that many stars in the Milky Way are contained in multiple systems which are mostly binaries. In recent years it has been shown that stellar multiplicity is a strong function of spectral type and stellar mass. The *single* star fraction ranges from only about 1-20% for OB stars to as much as 70-80 % for M stars and objects near the hydrogen burning limit. Since about 75% of the stars that make up the standard IMF are M stars, it is apparent that *the typical outcome of the star formation process is a single M dwarf star* (Lada 2006).

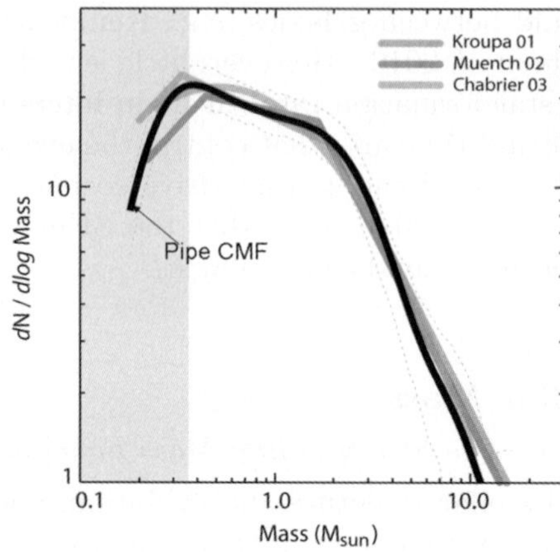



**Figure 1.** Comparison between the CMF of the Pipe molecular cloud and the IMFs derived for field stars and the young Trapezium cluster in Orion. The IMFs have all been scaled (i.e., shifted) higher in mass by a factor of 3 to match the Pipe CMF. The close similarity in shape between the CMF and IMF upon shifting the latter function suggests that the IMF derives directly from the CMF with a star formation efficiency of approximately 30%.

### 3.2. *The Dense Core Mass Function*

Observations of the cold ISM at infrared and millimeter wavelengths indicate that stars form from dense molecular filaments, clumps and cores in cold molecular clouds (e.g., Lada 1992, Beichman *et al.* 1986, Yun & Clemens 1990, Lada *et al.* 2010). The physical conditions in this dense molecular gas can be thought of as the initial conditions for a predictive theory of star formation. These conditions would include the masses, sizes, temperatures, densities, pressures, kinematics, and compositions of dense cores, particularly starless ones. Studies of nearby ( < 150 pc) local clouds have provided a wealth of information concerning these cores and the process of star formation within them. For the purposes of this paper I will now only describe what we know of dense core masses, their mass spectrum and its relation to the stellar IMF. Investigations of the (dense) core mass function, or CMF, have a much more recent history than studies of the IMF. Only in the last decade or so have improved observational capabilities (e.g., wide-field, near-infrared extinction mapping, millimeter-wave detector arrays, space-based far-infrared and sub-millimeter observations, increased computational power, etc.) enabled measurements of dense core masses over a sufficiently large dynamic range to make useful comparisons to the IMF. Earlier observations of nearby clouds suggested that the CMF could be described as a power-law with general similarity to the IMF and with a hint at a flattening toward lower masses (e.g., $\rho$ Ophiuchi: Motte *et al.* 1998). Later, more sensitive observations were able to measure a clear departure from a power-law form at low core masses as well the presence of a broad peak in the CMF (e.g., Pipe Nebula: Alves *et al.* 2007; Aquila Rift: André *et al.* 2010). These observations showed that the shapes of the CMF and IMF were quite similar, as can be observed in Figure 1. However the CMF was found to peak at higher masses ($\sim$ 1–2 $M_\odot$) than the IMF. *These facts suggest that the IMF directly derives from the CMF with an efficiency of $\sim$ 25-30%* (Alves *et al.* 2007). Individually, dense cores appear to be the direct progenitors of stars.

If the above considerations are correct, then the question of the origin of stellar masses and the IMF becomes the question of the origin of the CMF and dense core masses. The origin of core masses is tied to the process of molecular cloud fragmentation and thus to the evolution of cloud structure. The characteristic mass of the CMF likely represents a

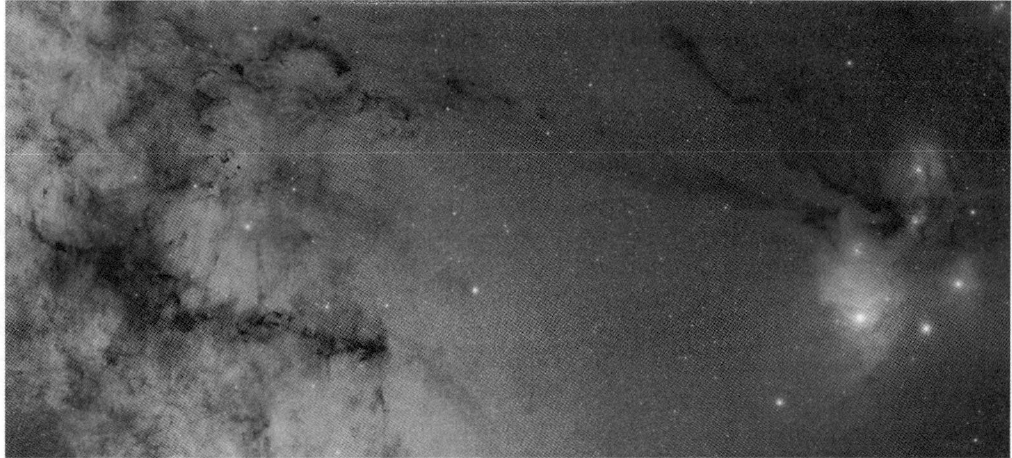

**Figure 2.** The Pipe (left) and Ophiuchus (right) dark nebulae. These nearest examples of massive, cold molecular clouds are at the same distance and are roughly similar in mass and extent, yet they display dramatically different levels of star formation activity. **Credit**: ESO/S. Guisard

characteristic mass scale for cloud fragmentation. In the nearby Pipe Nebula, Lada *et al.* (2008) noted that the characteristic mass of the CMF was approximately the critical Bonnor-Ebert mass for core population in that cloud. This suggests that, at least in the Pipe Nebula, the CMF originated from a Jeans-like process of thermal fragmentation in a pressurized medium. Whether or not such an explanation can account for the core formation process in other clouds and environments remains to be seen. However, it is interesting to consider that if cores at the peak of the CMF have masses comparable to the critically stable, Bonnor-Ebert mass, they will predominately form single stars once they become unstable and collapse. Moreover, with typical star formation efficiencies of 20-30%, such cores will produce stars whose final masses will have values near the observed peak of the IMF.

## 4. The Star Formation Rate

Figure 2 is a stunning image and is one of my favorites. It contains two of the most intriguing local molecular clouds, the Pipe Nebula and the Ophiuchi cloud. Located at nearly the same distance (D $\approx$ 125 pc), in nearly the same volume of Galactic space, these clouds are of very similar mass ($\sim 10^4$ $M_\odot$) and size (L $\sim$ 15-20 pc). Yet, as the image clearly shows, they differ substantially in their level of star formation activity. The Ophiuchi cloud is ablaze in star formation while the Pipe appears cold, dark and quiescent. How is it that two clouds so similar in mass, size, and age and in such close proximity, can be so different in their star formation? Interestingly, the ages of the stellar populations in the two clouds are essentially the same (2-3 Myr; Covey *et al.* 2010). Given the universal nature of the functional form of the IMF, the difference in stellar content and appearance of the two clouds must therefore be due to drastically differing rates of star formation within them.

Traditionally, researchers who study star formation in the Milky Way have not devoted much attention to the question of understanding the SFR in molecular gas. But for extragalactic astronomy, the SFR plays a critical role. For example, it is the primary tracer or metric used to describe the evolution of galaxies over cosmic time (e.g, Madau *et al.* 1998) and to investigate the connection between the level of star formation and the gas content of galaxies (Kennicutt 1998). It has become increasingly apparent to me that the SFR also plays a critical role in GMCs and knowledge of the factors that control the SFR in local molecular clouds is necessary for the development of a comprehensive and predictive theory of star formation. Moreover, understanding the underlying physics that controls the SFR in local molecular gas is likely a key stepping stone to understanding the nature and evolution of galaxies as well.

To address this issue we will take another look at the stellar IMF. Over the past 50 years research on the IMF has almost exclusively dealt with measuring its functional form, its extent in mass, and the degree to which these quantities vary in space and time. However, it is also instructive to consider what information can be gleaned from the amplitude or normalization of the IMF. As an illustrative example, let us consider the log-normal form of the IMF for a single stellar population, like a young cluster. We can write the IMF as

$$\xi(logm_*) = C_0 \exp[-(\log(m_*/m_c))^2/2\sigma^2]. \tag{4.1}$$

Here $m_*$ is the stellar mass, $m_c$ is the characteristic mass, and $\sigma$ is the width of the log normal function. The normalization coefficient is given by $C_0 = N_*(2\pi)^{-1/2}\sigma^{-1}$, where $N_*$ is the total number of stars formed in the cluster (or stellar population) being considered. This coefficient and thus the IMF are directly connected to the SFR since $N_* = b_*\Delta\tau_{sf}$ and SFR $= b_* < m_* >$, where $b_*$ is the average birthrate (yr$^{-1}$) of stars in the cluster, $\Delta\tau_{sf}$ the duration of star formation in the cluster (yr) and $< m_* >$ the average mass ($M_\odot$) of a star in the cluster. The central point here, however, is that in local regions of star formation $N_*$ and $\Delta\tau_{sf}$ are observable quantities and *we can directly measure the SFR in local clouds*. This is fundamentally different than the situation in more distant regions and in galaxies where the SFR has to be determined indirectly, usually requiring the aid of population synthesis models (e.g., Kennicutt 1998).

We can now quantify the absolute and relative SFRs in the clouds of Figure 2. There are over 300 young stellar objects (YSOs) within the Ophiuchi cloud (e.g., Wilking *et al.* 2008), while only 21 are found in the Pipe (Forbrich *et al.* 2009; Forbrich *et al.* 2010). Since the ages of the stellar populations in the two clouds are essentially the same their star formation rates must differ by a factor of $\approx$ 15. These two objects are not the only clouds to display such large variations in SFRs. Based on complete and accurate measurements of the masses and stellar contents of a nearly complete sample of molecular clouds within 0.5 kpc of the sun, Lada *et al.* (2010) demonstrated that the specific star formation rates (i.e., the SFR per unit cloud mass or sSFR) of local clouds vary by more than an order of magnitude. This variation is independent of cloud mass over a range of two orders of magnitude in cloud mass, a result hinted at in early CO observations of more distant and massive Milky Way clouds (Mooney & Solomon 1988).

That the SFRs of local GMCs can be so accurately measured and are found to vary so significantly is an exciting development for star formation research. This is because we can measure the basic physical properties of the local clouds in exquisite detail. Thus astronomers are in an excellent position to determine those physical properties that set the SFR and decipher the processes that control the SFR in molecular gas.

Earlier observations of the Orion B GMC have long suggested that star formation occurs almost exclusively in extended regions of dense gas (e.g., Lada 1992). In their study of the SFRs in the local cloud sample, Lada *et al.* (2010) discovered a relatively tight scaling relation between the mass of high opacity gas and the global SFR. This scaling relation is shown in Figure 3. This scaling is a linear power-law relation, that is, SFR $\propto$ M$_{dense}$. This has been interpreted to indicate that the SFR in a cloud is directly

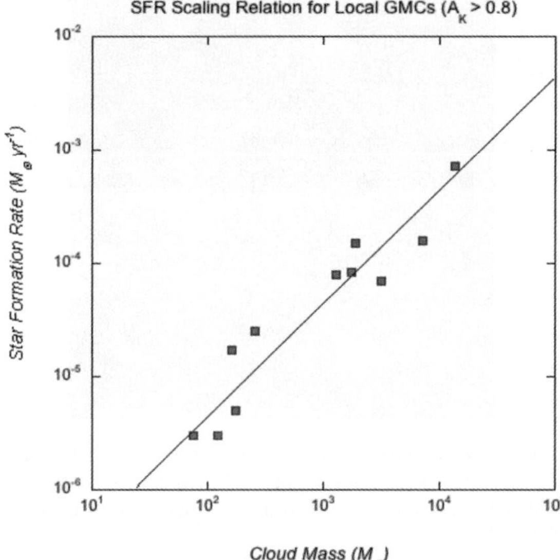



**Figure 3.** The scaling relation between SFR and high extinction (i.e., A$_K$ $\geqslant$ 0.8 magnitudes), presumably dense, gas mass for local GMCs. Adapted from Lada *et al.* (2010). See text.

controlled by the amount of high extinction (and presumably dense) material contained within it. Indeed, recent observational studies of the Orion A GMC (Lombardi *et al.* 2014) and the *Spitzer* C2D local sample of dark clouds (Evans *et al.* 2014; Heiderman & Evans 2015) have shown that roughly 90% of the protostellar objects in these clouds are found projected on high opacity (i.e., A$_K$ $\geqslant$ 0.8 magnitudes) material, providing further evidence of the tight relation between extended dense material and star formation.

The linear scaling between dense gas mass and the SFR in local clouds is reminiscent of a similar global relation found for galaxies by Gao and Solomon (2004) who showed that FIR luminosities of galaxies (including disk galaxies and nuclear starbursts) were linearly correlated with the luminosities of HCN molecular-line emission. The FIR luminosity is a proxy for the SFR and HCN emission is a tracer of the dense ($n_{H_2} > 10^4$ cm$^{-3}$) component of molecular gas. Subsequent observations by Wu *et al.* (2005) comparing FIR and HCN luminosities of massive GMCs in the Milky Way also showed a linear correlation between the two quantities and this relation was found to extrapolate smoothly to that found by Gao & Solomon (2004) for galaxies, spanning a range in scale of over nine orders of magnitude. Moreover, Lada *et al.* (2012) found that the local GMCs also fit on this relation (after application of the appropriate calibrations for the conversion of FIR and HCN luminosities to SFRs and gas masses, respectively).

The scaling relationship between the SFR and dense molecular gas mass found in local GMCs also appears to characterize star formation in galaxies. This similarity suggests that we are observing a similar physical process in star forming environments across all spatial scales. The linear scaling between SFR and dense gas mass suggests that the rate of star formation is directly controlled by the amount of dense gas that can be assembled in any star forming region. Thus, the physical process that controls the assembly of dense material in a cloud and dictates the dense gas fraction likely also controls the cloud's SFR. As mentioned earlier, the evolution of the internal structure of this dense material is also likely responsible for generating the CMF and thus ultimately the stellar IMF.

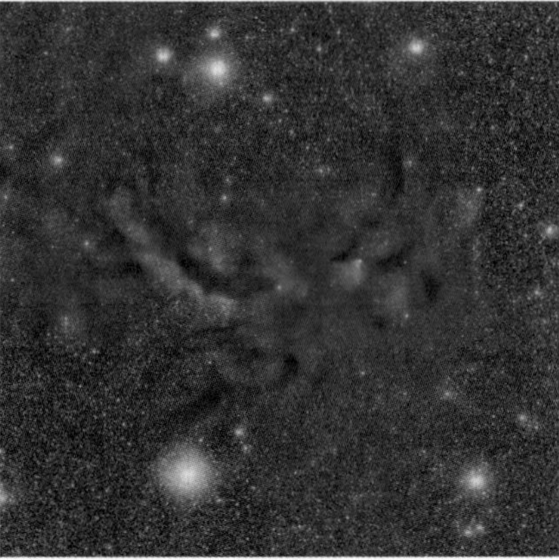

**Figure 4.** Optical Image of B 59, the most massive dense core in the Pipe Nebula. With 12 YSOs buried in this core, it is the most active site of star formation in the Pipe Nebula and an excellent candidate for producing a future moving group of young stars in the vicinity of the sun. (Image courtesy of J. Alves)

## 5. Concluding Remarks

In the previous paragraphs of this contribution I have presented a synopsis of two of the most fundamental problems confronting modern star formation research and have reviewed how studies of the local region of the Milky Way have produced critically important information that has led to some progress toward understanding these issues. I would be remiss, however, not to include some discussion of gains achieved in our understanding of the star formation process that pertain more directly to the subjects of this conference, young stars and in particular small moving groups of young stars near the sun. I will conclude this contribution with the following brief discussion conerning the origin of these small, local stellar moving groups.

Stars form in dense gas with a star formation efficiency (SFE $= \frac{M_*}{M_*+M_{gas}}$) of about 30% in dense cores, and about 10% in the more extended dense gas containing dense cores. Since only about 1-10% of the mass of GMCs is in the form of dense gas, the global SFEs of GMCs can range between 0.1 - 2% but are typically observed to be on the order of 1-2 %. For the local sample of Lada *et al.* (2010), <SFE> $= 1\% \pm 0.8\%$. Stars and stellar groups form in bound regions of GMCs where the majority of the binding mass is gaseous. With such low efficiencies, cloud dispersal results in the production of unbound stellar groups (Lada 1987). This can be seen from simple application of the virial theorem, that is, $M_{tot}v^2 = GM_{tot}^2/R$, where $M_{tot} = M_* + M_{gas}$. Initially, the stars have the virial velocities of the system, that is, $v_* = v = (GM_{tot}/R)^{1/2}$, and if the gas is quickly removed from the system the escape velocity for the stars is $v_{esc} = (GM_*/R)^{1/2}$. So the system cannot remain bound unless $v_* < v_{esc}$ which in turn requires SFE $> 50\%$. This is in essence why GMCs generally spawn unbound expanding OB associations. The stellar expansion velocities resulting from this process are $\approx v_*$. Bound or loosely bound stellar clusters can emerge from the more massive dense cloud cores containing embedded clusters or even small stellar groups where the SFE can reach 20-30% within the volume in which the stellar groups have formed and where the gas removal is more adiabatic (e.g., Lada *et al.* 1984).

Do we know of any dense cores in nearby clouds that are producing small clusters that could be the progenitors of local moving groups of young stars? I would like to propose here one possible candidate source. It is known as Barnard 59 and is the most massive core in the Pipe Nebula. An optical image of this core is shown in Figure 4. Its mass is about 20 $M_\odot$ (Román-Zúñiga *et al.* 2009) and it contains a small cluster of 12 YSOs (Brooke *et al.* 2007) with a total stellar mass of approximately 6-8 $M_\odot$. The core SFE = 23-28%. Since there are no massive stars forming in this core, its disruption (possibly by the generation of outflows from the embedded stars) is likely not to be very violent. These conditions are ideal for the production of a small, loosely bound group of stars when the cluster emerges following the anticipated gradual and adiabatic-like dissipation of the cloud core.

## References


Alves, J., Lombardi, M., & Lada, C. J. 2007, *A&A*, 462, L17

André, P., Men'shchikov, A., Bontemps, S., *et al.* 2010, *A&A*, 518, L102

Beichman, C. A., Myers, P. C., Emerson, J. P., *et al.* 1986, *ApJ*, 307, 337

Brooke, T. Y., Huard, T. L., Bourke, T. L., *et al.* 2007, *ApJ*, 655, 364

Chabrier, G. 2003, *PASP*, 115, 763

Covey, K. R., Lada, C. J., Román-Zúñiga, C., *et al.* 2010, *ApJ*, 722, 971

Evans, N. J., II, Heiderman, A., & Vutisalchavakul, N. 2014, *ApJ*, 782, 114

Forbrich, J., Lada, C. J., Muench, A. A., Alves, J., & Lombardi, M. 2009, *ApJ*, 704, 292

Forbrich, J., Posselt, B., Covey, K. R., & Lada, C. J. 2010, *ApJ*, 719, 691

Gao, Y. & Solomon, P. M. 2004, *ApJ*, 606, 271

Heiderman, A. & Evans, N. J., II 2015, *ApJ*, 806, 231

Kennicutt, R. C., Jr. 1998, *ARA&A*, 36, 189.

Kroupa, P. 2002, Science, 295, 82

Lada, C. J., Margulis, M., & Dearborn, D. 1984, *ApJ*, 285, 141

Lada, C. J. 1987, Star Forming Regions, 115, 1

Lada, C. J. 2006, *ApJ*, 640, L63

Lada, C. J., Muench, A. A., Rathborne, J., Alves, J. F., & Lombardi, M. 2008, *ApJ*, 672, 410

Lada, C. J., Lombardi, M., & Alves, J. F. 2010, *ApJ*, 724, 687

Lada, C. J., Forbrich, J., Lombardi, M., & Alves, J. F. 2012, *ApJ*, 745, 190

Lada, E. A. 1992, *ApJ*, 393, L25

Lombardi, M., Bouy, H., Alves, J., & Lada, C. J. 2014, *A&A*, 566, A45

Madau, P., Pozzetti, L., & Dickinson, M. 1998, *ApJ*, 498, 106

Mooney, T. J. & Solomon, P. M. 1988, *ApJ*, 334, L51

Motte, F., Andre, P., & Neri, R. 1998, *A&A*, 336, 150

Muench, A. A., Lada, E. A., Lada, C. J., & Alves, J. 2002, *ApJ*, 573, 366

Román-Zúñiga, C. G., Lada, C. J., & Alves, J. F. 2009, *ApJ*, 704, 183

Salpeter, E. E. 1955, *ApJ*, 121, 161

Wilking, B. A., Gagné, M., & Allen, L. E. 2008, Handbook of Star Forming Regions, Volume II, 351

Wu, J., Evans, N. J., Gao, Y., Solomon, P. M., Shirle, Y. L., & Vanden Bout, P. A. 2005, *ApJ*, 635, L173

Yun, J. L. & Clemens, D. P. 1990, *ApJ*, 365, L73

*Young Stars & Planets Near the Sun*
*Proceedings IAU Symposium No. 314, 2015*
*J. H. Kastner, B. Stelzer, & S. A. Metchev, eds.*

© International Astronomical Union 2016
doi:10.1017/S1743921315006547

# A Brief History of the Study of Nearby Young Moving Groups and Their Members

## Joel H. Kastner

Laboratory for Multiwavelength Astrophysics, Center for Imaging Science, and School of Physics and Astronomy, Rochester Institute of Technology;
email: jhk@cis.rit.edu

**Abstract.** Beginning with the enigmatic (and now emblematic) TW Hya, the scutiny of individual stars and star-disk systems has both motivated and benefitted from the identification of nearby young moving groups (NYMGs). I briefly outline the emergence of this relatively new subfield of astronomy over the past two decades, and offer a few examples illustrating how the study of NYMGs and their members enables unique investigations of pre-main sequence stellar evolution, evolved protoplanetary disks, and young exoplanets.

**Keywords.** protoplanetary disks, stars: pre-main sequence, stars: individual (TW Hya, V4046 Sgr, HR 4796, $\beta$ Pic)

---

## 1. TW Hya: a Classical T Tauri Star without a Birthplace

The identification and investigation of *nearby young moving groups* (NYMGs) — i.e., loose associations of stars that lie within $\sim$100 pc of Earth and have ages up to $\sim$100 Myr — arguably began with the identification of the late-type, emission-line field star TW Hydrae as a "T Tauri star far from any dark cloud" (Rucinski & Krautter 1983). This comprehensive study by Rucinski & Krautter establishing the classical T Tauri nature of TW Hya was motivated by the work of Henize (1976) and Herbig (1978), and led directly to the identification of additional examples of "isolated T Tauri stars" (de la Reza *et al.* 1989; Gregorio-Hetem *et al.* 1992). These studies yielded the identification of a literal handful of late-type field stars with T Tauri-like characteristics — in particular, deep Li absorption and apparent mid-infrared excesses† — in the general vicinity of TW Hya.

In their landmark paper, de la Reza *et al.* (1989) speculated on the potential nature of these lonely T Tauri stars, suggesting they were "kinematically related" and likely formed *in situ*. Furthermore, on the basis of these stars' high galactic latitudes, de la Reza *et al.* suggested (without assigning a specific distance range) that TW Hya and its apparent cohort may be relatively close to Earth. The last assertion seemed particularly intriguing, given the potential for "close-up" studies of planetary systems in formation that might be afforded by these and other "isolated" T Tauri stars. Following up on this hunch, Zuckerman *et al.* (1995) established that the thermal infrared excess from dust around TW Hya (and a few other "isolated," young, IR-excess stars) was accompanied by the presence of residual orbiting molecular (CO) gas, suggestive of the recent cessation of an epoch of planet formation. But the distance and age of TW Hya remained very poorly constrained.

---

† It is fascinating to note, in retrospect, that two of the five original members of the TW Hya Association were only observed in H$\alpha$ and Li, and hence identified as young, by de la Reza *et al.* (1989) and Gregorio-Hetem *et al.* (1992) thanks to chance alignments with IR-bright background galaxies.

With the release of the ROSAT (X-ray) All-Sky Survey (RASS), we could address these unknowns. The RASS X-ray data firmly established the youth of TW Hya and its (four) nearby siblings, yet also suggested the five were older than typical of T Tauri stars in dark clouds; we guesstimated (sic) the ages of the five stars as $\sim$20 Myr (Kastner *et al.* 1997). With this age estimate, we could then place firmer constraints on the stars' distances, and determined that they were a mere $\sim$40–60 pc from Earth. This range was promptly confirmed by the newly released Hipparcos parallax distances to TW Hya ($D = 54$ pc) and one of its four siblings, HD 98800 ($D = 45$ pc). In light of their similar ages, distances, and kinematics, we took the (perhaps unwise) leap of christening this little ragtag group of young stars the TW Hya Association (TWA), and declared the TWA "the nearest region of recent star formation" (Kastner *et al.* 1997).

Fortunately for us, many subsequent studies — beginning with Webb *et al.* (1999), and extending to the work of Murphy *et al.* presented at this meeting — have dramatically expanded the TWA's known membership, have confirmed its mean distance as $\sim$50 pc, and have honed estimates of its age, which is now generally accepted as $\sim$8 Myr†. Meanwhile, TW Hya itself has gone on to become the Crab Nebula‡ of late-stage pre-main sequence accretion and protoplanetary disk studies, with nearly 1000 `simbad` references (95% of them since 1997) as of the writing of this review.

## 2. The (young association) link between the TWA and $\beta$ Pic

Not long after the naming of the TWA, Jura *et al.* (1998) noted that the A stars HR 4796A, $\beta$ Pic, and 49 Cet — all massive debris disk hosts — were underluminous for their colors relative to the field A star population, suggesting all three are young. The youth of the HR 4796AB system had already been established by Stauffer *et al.* (1995) on the basis of the position of the M-type star HR 4796B *above* the main sequence. Based on the (relative) sky proximity of HR 4796AB to the original five stars of the TWA, Jura *et al.* (1998) further speculated that HR 4796A may also be a TWA member (an assertion subsequently confirmed by Webb *et al.* 1999).

A strong implication of the Jura *et al.* (1998) analysis was that $\beta$ Pic and the stars of the TWA, though not kinematically or spatially associated, were similarly young. At the time, this was somewhat of a revelation, given that some previous estimates had put the age of $\beta$ Pic at a few 100 Myr (Barrado y Navascués *et al.* 1999 and refs. therein). The youth of $\beta$ Pic was soon firmly established by Barrado y Navascués *et al.* (1999), who honed lists of stars potentially comoving with $\beta$ Pic and identified just three M star "survivors," all of age $\sim$20 Myr. This work constituted the first identification of the $\beta$ Pic Moving Group ($\beta$PMG) — its original membership, like that of the original TWA, could be counted on one hand. Not long thereafter, the wider availability and improved application of space velocity data would begin to dramatically expand the known membership of the $\beta$PMG (Zuckerman *et al.* 2001).

At the risk of stating the obvious, the general approach pioneered by Zuckerman (e.g., Zuckerman & Webb 2000; Zuckerman *et al.* 2001), Song (e.g., Song *et al.* 2002), Torres (Torres *et al.* 2006), and a few other investigators — i.e., selecting candidate young stars on the basis of proper motions and/or X-ray fluxes, and following up spectroscopically to confirm signatures of youth and obtain $UVW$ velocities — has been refined and improved

---

† This (traceback-based) age estimate seems relatively robust (Ducourant *et al.* 2014). However, given the many important caveats regarding ages of NYMGs that have been raised at this meeting (see, e.g., reviews by Mamajek and Bell), the age of the TWA certainly bears revisiting via other methods.

‡ With apologies, and thanks, to David Wilner.

tremendously over the past $\sim$15 years. In particular, statistically rigorous treatments of stellar kinematics (e.g., Malo *et al.* 2013) have replaced the original, painstaking, "by-hand" searches of the Hipparcos and Tycho catalogs, and the Galex UV sky survey data have replaced the RASS as a superior means to select active (hence, possibly young) late-type stars in the field (Shkolnik *et al.* 2011; Rodriguez *et al.* 2013). Thus, the techniques first brought to bear to identify and confirm members of the TWA and $\beta$PMG have now been successfully applied to identify (and/or expand the memberships of) another dozen or so established and candidate NYMGs (see, e.g., Zuckerman *et al.* 2011 and review by Mamajek in these Proceedings).

## 3. The impact of nearby young moving group studies

I summarize here three recent examples of studies involving $\beta$PMG members that highlight the utility of studies of NYMGs and their members for purposes of improving our understanding of pre-main sequence (pre-MS) stellar evolution, evolved protoplanetary disks, and young exoplanets.

### 3.1. *Understanding early stellar evolution*

The comprehensive study of the $\beta$PMG by Mamajek & Bell (2014) well illustrates how a NYMG can serve as a testbed both for the astrophysics of pre-main sequence stars and techniques for age-dating stars and stellar associations. Mamajek & Bell compiled and critically compared a dozen $\beta$PMG age determinations involving roughly a half-dozen different techniques, including placement on pre-MS star isochrones, kinematic traceback, Li depletion, and variants thereof. They conclude that the age of the $\beta$PMG should be revised upwards, from the widely quoted $\sim$12 Myr (Zuckerman *et al.* 2001; Torres *et al.* 2006) to $\sim$23 Myr (which is, ironically, closer to the original estimate by Barrado y Navascués *et al.* 1999). However, this refinement in the age of the $\beta$PMG is perhaps less interesting than the conclusion by Mamajek & Bell (2014) that, at least in application to the $\beta$PMG stars, the Li depletion boundary and isochronal age estimation techniques are superior to kinematic methods of age determination. Given the interesting, fundamental problems in stellar astrophysics that are inherent in the former two methods (as highlighted in talks by Bell, Song, Somers and others at this meeting), it is clear that the tests of Li depletion boundary and isochronal age determinations offered by studies of NYMGs will become increasingly important, moving forward.

### 3.2. *Investigating (evolved) protoplanetary disks*

Though it remains understudied, especially relative to TW Hya, the $\beta$PMG member system V4046 Sgr is nearly as close ($D \sim 73$ pc) and, arguably, at least as interesting†. This hierarchical multiple system features a close, near-equal-components, solar-mass binary that is orbited by, and actively accreting from, a molecule-rich circumbinary disk. The disk is surprisingly extensive and massive (outer CO radius $\sim$350 AU and estimated $H_2$ mass $\sim$0.09 $M_\odot$, Rosenfeld *et al.* 2013; Fig. 3.2, left panel) given the advanced age of V4046 Sgr and, hence, potentially affords the opportunity to study late stages in the process of planet formation around a close binary. Recently, Rapson *et al.* (2015) used the new Gemini Planet Imager (GPI) to obtain near-infared coronagraphic/polarimetric (scattered-starlight) images of the innermost disk regions (Fig. 3.2, right panel). Thanks to the combination of GPI's exquisite performance and the proximity of V4046 Sgr, these

† For a brief overview of V4046 Sgr, see the May 2013 Star Formation Newsletter: http://www.ifa.hawaii.edu/~reipurth/newsletter/newsletter245.pdf.

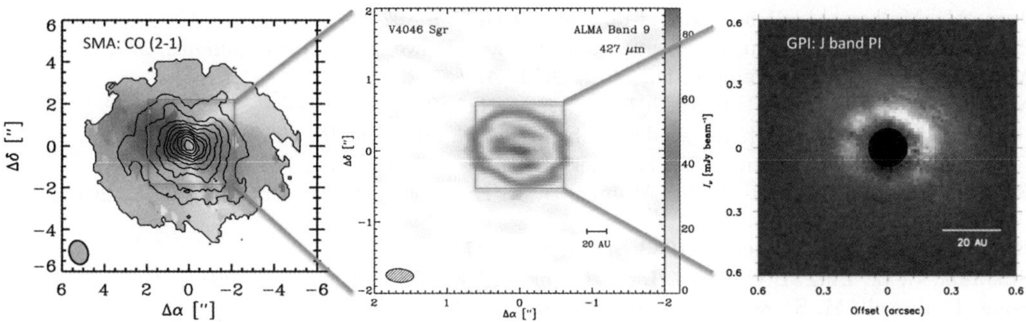


**Figure 1.** The V4046 Sgr disk as imaged in CO line emission by the Submillimeter Array (left; from Rosenfeld *et al.* 2013), in 427 $\mu$m continuum emission by ALMA during Early Science (Cycle 0) operations (center; from Andrews *et al.*, in prep.), and in polarized intensity at 1.25 $\mu$m by the Gemini Planet Imager in its coronagraphic/polarimetric mode (right; from Rapson *et al.* 2015).

images probe the structure of the dust disk to within $\sim$7 AU of the central binary, at a jaw-dropping $\sim$3 AU resolution. The GPI imaging reveals concentric rings of submicron-sized dust grains that are contained within the ring of submm-sized grains responsible for thermal emission in submm continuum imaging. The structures seen in these (GPI and ALMA) images of V4046 Sgr (Fig. 3.2) are strikingly similar to those seen in synthetic images derived from simulations describing newborn planets clearing gaps within a protoplanetary disk (see, e.g., Fig. 7 in Dong *et al.* 2015).

### 3.3. *Discovering and characterizing young exoplanets*

Because young gas giant planets are self-luminous in the infrared, NYMG members offer the best targets for extreme adaptive optics imaging of giant exoplanets. Indeed, direct imaging of nearby, young stars represents the *only* means that will be available in the near future to detect exoplanets in wide ($>$5 AU) orbits around their host stars (see, e.g., review by Chauvin in these Proceedings). This potential for NYMG members — especially those hosting dusty debris disks — to yield direct-imaging detections of massive exoplanets is, of course, amply demonstrated by the detection and subsequent intensive study of $\beta$ Pic b (e.g., Bonnefoy et al. 2014). With GPI and SPHERE now having demonstrated their capabilities, we appear to be on the verge of a trove of such direct-imaging discoveries of planets orbiting NYMG member stars.

### Acknowledgements

JHK's research on young stars near Earth is supported by NSF grant AST 1108950 and NASA Astrophysics Data Analysis Program grant NNX12AH37G to RIT.

### References

Barrado y Navascués, D., Stauffer, J. R., Song, I., & Caillault, J.-P. 1999, *ApJL*, 520, L123
Bonnefoy, M., Marleau, G.-D., Galicher, R., *et al.* 2014, *A&A*, 567, L9
de la Reza, R., Torres, C. A. O., Quast, G., Castilho, B. V., & Vieira, G. L. 1989, *ApJL*, 343, L61
Dong, R., Zhu, Z., & Whitney, B. 2015, *ApJ*, 809, 93
Ducourant, C., Teixeira, R., Galli, P. A. B., *et al.* 2014, *A&A*, 563, A121
Gregorio-Hetem, J., Lepine, J. R. D., Quast, G. R., Torres, C. A. O., & de La Reza, R. 1992, *AJ*, 103, 549
Henize, K. G. 1976, *ApJS*, 30, 491
Herbig, G. H. 1978, in *Problems of Physics and Evolution of the Universe*, 171

Jura, M., Malkan, M., White, R., *et al.* 1998, *ApJ*, 505, 897

Kastner, J. H., Zuckerman, B., Weintraub, D. A., & Forveille, T. 1997, *Science*, 277, 67

Malo, L., Doyon, R., Lafrenière, D., *et al.* 2013, *ApJ*, 762, 88

Mamajek, E. E. & Bell, C. P. M. 2014, *MNRAS*, 445, 2169

Rapson, V. A., Kastner, J. H., Andrews, S. M., *et al.* 2015, *ApJL*, 803, L10

Rodriguez, D. R., Zuckerman, B., Kastner, J. H., *et al.* 2013, *ApJ*, 774, 101

Rosenfeld, K. A., Andrews, S. M., Wilner, D. J., Kastner, J. H., & McClure, M. K. 2013, *ApJ*,
    775, 136

Rucinski, S. M. & Krautter, J. 1983, *A&A*, 121, 217

Shkolnik, E. L., Liu, M. C., Reid, I. N., Dupuy, T., & Weinberger, A. J. 2011, *ApJ*, 727, 6

Song, I., Bessell, M. S., & Zuckerman, B. 2002, *A&A*, 385, 862

Stauffer, J. R., Hartmann, L. W., & Barrado y Navascues, D. 1995, *ApJ*, 454, 910

Torres, C. A. O., Quast, G. R., da Silva, L., *et al.* 2006, *A&A*, 460, 695

Webb, R. A., Zuckerman, B., Platais, I., *et al.* 1999, *ApJL*, 512, L63

Zuckerman, B., Forveille, T., & Kastner, J. H. 1995, *Nature*, 373, 494

Zuckerman, B. & Webb, R. A. 2000, *ApJ*, 535, 959

Zuckerman, B., Song, I., Bessell, M. S., & Webb, R. A. 2001, *ApJL*, 562, L87

Zuckerman, B., Rhee, J. H., Song, I., & Bessell, M. S. 2011, *ApJ*, 732, 61

*Young Stars & Planets Near the Sun*
Proceedings IAU Symposium No. 314, 2015
J. H. Kastner, B. Stelzer, & S. A. Metchev, eds.

© International Astronomical Union 2016
doi:10.1017/S1743921315006250

# A Pre-Gaia Census of Nearby Stellar Groups

## Eric E. Mamajek

Department of Physics & Astronomy, University of Rochester, Rochester, NY, 14627-0171
email: emamajek@pas.rochester.edu

**Abstract.** The nearest, youngest groups of stars to the Sun provide important samples of age-dated stars for studying circumstellar disk evolution, imaged exoplanets, and brown dwarfs. I briefly comment on the status of the known stellar groups within 100 pc: $\beta$ Pic, AB Dor, UMa, Car-Near, Tuc-Hor and $\beta$ Tuc nucleus, Hyades, Col, TW Hya, Car, Coma Ber, 32 Ori, $\eta$ Cha, and $\chi^1$ For. I also discuss some poorly characterized groups and "non-groups." Grades for 2015 of *Pass*, *Satisfactory*, or *Fail* are assigned to the groups for the purposes of age-dating stars and brown dwarfs. I speculate that Tuc-Hor could have provided a supernova $\sim$60 pc away $\sim$2.2 Myr ago which showered the Earth with traces of $^{60}$Fe-bearing dust.

**Keywords.** open clusters and associations: general — solar neighborhood — stars: distances

## 1. Introduction

The stellar groups within 100 pc offer unique samples to investigate the results of the star and planet formation process at ages $\sim$10$^7$-10$^9$ yr. Members of these groups have provided some of the first and best examples of imaged dusty debris disks, imaged extrasolar planets, and young substellar objects. Given the page limits, I will just summarize some under-appreciated and new aspects of these groups (and apologize for the lack of figures). Recent discussions on these groups can be found in Zuckerman & Song (2004), Torres *et al.* (2008), Riedel *et al.* (2014), Malo *et al.* (2014), and Gagne *et al.* (2014). **Table 1** is a compilation of the stellar clusters and associations within 100 pc (with best estimates of velocities and ages). For those interested in adopting ages to stars based on their membership to one of these kinematic groups (and assuming the star exhibits some secondary indicators hinting at coevality with the group), I assign grades of *Pass* (§2), *Satisfactory* (§3), or *Fail* (§4). Table 1 only contains the *Pass* and *Satisfactory* groups.

## 2. Physical Groups (Grade: Pass)

The **Ursa Major, Hyades, Coma Ber**, and $\eta$ **Cha** groups are clearly real *clusters*. The young ($\sim$10 Myr) $\eta$ Cha cluster's density is roughly $\sim$30 M$_\odot$ pc$^{-3}$ - the densest of any cluster within 100 pc - while the older clusters ($\sim$0.5 Gyr) are lower density ($\sim$0.3-3 M$_\odot$ pc$^{-3}$), but all exceed the local disk density ($\sim$0.1 M$_\odot$ pc$^{-3}$). The $\sim$10 Myr-old **TW Hya** *association* is a well-characterized group of $\sim$3 dozen stars (Kastner *et al.* 1997, Mamajek 2005, Weinberger et al. 2013, Ducourant *et al.* 2014), and will not be discussed further. The $\beta$ **Pic** *association* was recently reviewed by Mamajek & Bell (2014, and references therein). Ages of <20 Myr can be discounted as the 6 A-type members plus the F0 member 51 Eri are all on the ZAMS. For the "classic" age of 12 Myr adopted for $\beta$ Pic for most of the past decade, *all* isochrones would predict that late A- and early F-type members should be very much pre-MS (alas, they are not). MS turn-on ages and Li depletion boundary ages appear to be in agreement with a mean age of $23 \pm 3$ Myr.

Table 1 lists the mean distance to the **AB Dor** *nucleus* based on revised Hipparcos parallaxes (van Leeuwen 2007) for nuclear members listed by Zuckerman et al. (2004).

**Table 1.** Catalog of Stellar Groups Within 100 pc

| Group ... | Dist pc | Ref. ... | $U$ km/s | $V$ km/s | $W$ km/s | $\sigma_U, \sigma_V, \sigma_W$ km/s | $\sigma_v$ km/s | Ref. ... | Age Myr | Ref. ... |
|---|---|---|---|---|---|---|---|---|---|---|
| $\beta$ Pic | $\sim$15$^a$ | 1 | -10.9 | -16.0 | -9.2 | 0.3, 0.3, 0.3 | 1.5 | 2 | $23 \pm 3$ | 2 |
| AB Dor | $20.1 \pm 1.6$ | 3 | -7.6 | -27.3 | -14.9 | 0.4, 1.1, 0.3 | 1.0 | 3 | $150^{+50}_{-30}$ | 4 |
| UMa | $25.2 \pm 0.3$ | 5 | 14.6 | 1.8 | -8.6 | 0.4, 0.7, 1.0 | 1.4 | 5 | $530 \pm 40$ | 6 |
| Car-Near | $33 \pm 1$ | 5 | -24.8 | -18.2 | -2.3 | 0.7, 0.7, 0.4 | 1.3 | 5 | $\sim$200 | 7 |
| $\beta$ Tuc | $43 \pm 1$ | 5 | -9.6 | -21.6 | -0.7 | 1.0, 1.3, 0.6 | 1.1 | 5 | $45 \pm 4$ | 4 |
| Tuc-Hor | $\sim$48 | 9 | -10.6 | -21.0 | -2.1 | 0.2, 0.2, 0.2 | 1.1 | 8 | $45 \pm 4$ | 4 |
| Hyades | $46.5 \pm 0.5$ | 10 | -42.3 | -19.1 | -1.5 | 0.1, 0.1, 0.2 | 0.3 | 11 | $750 \pm 150$ | 6 |
| Columba | $\sim$50 | 1 | -12.2 | -21.3 | -5.6 | 1.1, 1.2, 0.9 | ... | ... | $42 \pm 5$ | 4 |
| TW Hya | $53 \pm 2$ | 12 | -11.2 | -18.2 | -5.1 | 0.4, 0.4, 0.4 | 0.8 | 12 | $10 \pm 3$ | 4 |
| Carina | $\sim$65 | 1 | -10.5 | -22.4 | -5.8 | 1.0, 0.6, 0.1 | ... | ... | $45 \pm 10$ | 4 |
| Coma Ber | $87 \pm 1$ | 10 | -2.4 | -5.5 | -0.6 | 0.1, 0.1, 0.1 | 0.4 | 13 | $560 \pm 90$ | 14 |
| 32 Ori | $92 \pm 2$ | 4 | -11.8 | -18.5 | -8.9 | 0.4, 0.4, 0.3 | $\sim$1 | 5 | $22 \pm 4$ | 4 |
| $\eta$ Cha | $94 \pm 1$ | 15 | -10.2 | -20.7 | -11.2 | 0.2, 0.1, 0.1 | 1.5 | 15 | $11 \pm 3$ | 4 |
| $\chi^1$ For | $99 \pm 6$ | 5 | -13.1 | -22.1 | -3.7 | 0.4, 0.5, 1.1 | ... | 5 | $\sim$50? | 5 |

**Notes:** Velocities are quoted on the standard Galactic coordinate system where $U$ is towards the Galactic center, $V$ is towards Galactic rotation ($\ell = 90°$), and $W$ is towards the north Galactic pole (e.g. Johnson & Soderblom 1987). $\sigma_U$, $\sigma_V$, $\sigma_W$ are uncertainties in the mean velocities, not velocity dispersion ($\sigma_v$ is an estimate of the intrinsic 1D velocity dispersion). **References and Notes:** 1) Does not have well-defined concentration. Distance is to centroid estimated by Malo *et al.* (2014). 2) Mamajek & Bell (2014). 3) Barenfeld *et al.* (2013). 4) Bell, Mamajek, & Naylor (2015), see also Bell (this volume). 5) This work or Mamajek (unpublished). 6) Brandt & Huang (2015). Brandt & Huang (2015) have recently revised the Hyades age upward, however the -1$\sigma$ uncertainty quoted encapsulates recent younger ($\sim$650 Myr) estimates e.g. de Bruijne et al. (2001). 7) Zuckerman *et al.* (2006). 8) Kraus *et al.* (2014). 9) mean kinematic distance to 120 Tuc-Hor members from Kraus et al. (2014) calculated using UCAC4 proper motions and space motion from Kraus. 10) van Leeuwen (2009). 11) de Bruijne *et al.*(2001). 12) Mean distance and velocity using astrometry from Ducourant *et al.* (2014), but omitting interlopers TWA 14, 15, 19, & 22. Velocity agrees well with Weinberger et al. (2013). Velocity dispersion from Ducourant *et al.* (2014) and Mamajek (2005). 13) Calculated using astrometry from van Leeuwen (2009) and radial velocity from Mermilliod et al. (2009), and velocity dispersion from Mermilliod *et al.* (2009). 14) Silaj & Landstreet (2014). 15) Murphy *et al.* (2013).

Omitting HIP 26369 due to large parallax uncertainty, the mean distance to the rest of the nuclear members is $20.1^{+1.7}_{-1.4}$ pc, with intrinsic 1$\sigma$ dispersion of $\pm4$ pc. Barenfeld *et al.* (2013) showed that the group must be >120 Myr old, and the new analysis by Bell *et al.* (2015) estimates an isochronal age of roughly $150^{+50}_{-30}$ Myr. *There is no astrophysical support for ages as young as $\sim$50-100 Myr that are commonly cited.* While the nuclear stars show remarkably coherent motions (1D dispersion $\sim$1 km s$^{-1}$), and are obviously concentrated, the status of the rest of the AB Dor membership (i.e. the *"stream"*) is unfortunately murkier (grade: S/F?). A spectroscopic analysis by Barenfeld et al. (2013) showed that only roughly half of purported AB Dor members sampled outside the nucleus were consistent with being co-chemical (i.e. possibly co-natal), hence the AB Dor group may share motions with other young stars of similar ages from different birthsites.

Zuckerman *et al.* (2006) discovered a nearby $\sim$200 Myr-old group dubbed **Car-Near**. The mean distance to the Zuckerman nuclear members using revised Hipparcos astrometry is $32.7 \pm 1.2$ pc, with median RV = $+17.5 \pm 0.8$ km s$^{-1}$. The intrinsic velocity dispersion is only $1.3 \pm 0.5$ km s$^{-1}$. The group is puny ($\sim$8 M$_\odot$), but its inferred density is just below the local disk density. No objects are hotter than F1, and a thorough search of the Hipparcos catalog finds no plausible B/A members.

Kraus *et al.* (2014) have shown the **Tuc-Hor** group to be a much larger entity ($>10^2$ stars) than previously appreciated. Projections of the total stellar population based on the members found so far range from $\sim$200-400 (A. Kraus, priv. comm., Gagne et al. 2014, and calculation by author). The distribution of stars in Tuc-Hor appears somewhat filamentary and sheet-like over tens of parsecs - even at age $\sim$40 Myr. While the kinematic data are consistent with a velocity dispersion of only $\sim$1 km s$^{-1}$, the small extent of the group in $Z$ suggests a dispersion of $<$0.2 km s$^{-1}$, analogous to that seen for small scale structures in the Taurus clouds. Combining the membership lists of Malo *et al.* (2014) and Kraus *et al.* (2014), one starts to see substructure in Tuc-Hor (see also Fig. 4 of Zuckerman et al. 2001). Tuc-Hor is draped across the southern sky, with several ill-defined clumps which may constitute subgroups: 1) a small clump of members in Pavo associated with the massive star Peacock ($\alpha$ Pav; B2.5; $\sim$6 M$_\odot$; $\alpha, \delta \simeq 306°$, -57°; $\sim$55 pc); 2) another small clump in Indus associated with HR 8352 (HIP 108195; F1; $\alpha, \delta \simeq 329°$, -62°; $\sim$46 pc); 3) another small clump associated with DS Tuc (HIP 116748; G5; $\alpha, \delta \simeq 355°$, -69°; $\sim$46 pc); 4) the original $\beta$ **Tuc** nucleus (discovered by Zuckerman & Webb 2000, listed separately in Table 1; $\alpha, \delta \simeq 8°$, -63°; $\sim$43 pc); 5) to the heart of the **Horologium** subgroup centered on Achernar (B3; $\alpha, \delta \simeq 24°$, -57°; $\sim$43 pc)†; 6) a small clump centered on $\eta$ Hor (A6; $\alpha, \delta \simeq 39°$, -52°.5; $\sim$46 pc); and 7) a small, diffuse clump near $\epsilon$ Hyi (B9; $\alpha, \delta \simeq 40°$, -70°; $\sim$47 pc). Tuc-Hor does not appear so much as one large group as an ensemble of evaporating subgroups, with many of the low-mass members in the immediate vicinity of more massive stars (indeed, likely within the tidal radii, constituting unstable "trapezia"). Precise astrometry and further characterization of the membership of the Tuc-Hor complex with Gaia should yield a remarkable picture of a dynamical "missing link" between star-forming regions and the field population.

Given the number of 2-8 M$_\odot$ Tuc-Hor members known ($\sim$7), a Salpeter IMF would predict $\sim$1 star with mass $>$8 M$_\odot$. Tuc-Hor may have eked out forming a star hotter than B2 which would have undergone supernova in the recent past. *It is possible that Tuc-Hor, and not Sco-Cen, was responsible for the supernova which produced the 2.2 Myr-old* $^{60}$*Fe signal in sea floor ferromanganese crusts, and contributed to sweeping out the Local Bubble* (e.g. Fry *et al.* 2015). Tucana was only slightly further away ($\sim$60 pc) from the Sun 2.2 Myr ago, well positioned to provide a $\sim$8 M$_\odot$ supernova close enough to Earth for its $^{60}$Fe-enriched dust to pollute the Earth.


## 3. Likely Physical Groups; More Work Needed (Grade: Satisfactory)

The **Carina** and **Columba** groups are reported to be of similar ages and kinematics to Tuc-Hor (Torres et al. 2008). Bell *et al.* (2015) estimate new consistent isochronal ages of $45^{+11}_{-7}$ and $42^{+6}_{-4}$ Myr for Car and Col, respectively. I have tentatively included these in Table 1, however they clearly require further study.

**32 Ori:** This is a new group from Mamajek (2007) that will be discussed further in Bell et al. (2015) and Shvonski *et al.* (in prep.). Bell *et al.* was able to find a consistent age for a preliminary membership list ‡, of $22 \pm 4$ Myr.

† The $\sim$8.7 M$_\odot$ Achernar binary has *7(!)* K7-M5 Tuc-Hor members from Kraus *et al.* (2014) projected within its estimated tidal radius 2.8 pc $\simeq$ $\sim$3°.7: 2MASS J01344601-5707564, J01375879-5645447, J01504543-5716488, J01380311-5904042, J01505688-5844032, J01521830-5950168, J01275875-6032243. They display a clear Li-depletion boundary between $\sim$M4-M4.5 which could be used to independently age-date Achernar.

‡ 32 Ori, HD 35656, HD 35714, HD 36338, HD 35499, HD 35695, HD 245059, 2MASS J05200029+0613036, 2MASS J05203182+0616115, 2MASS J05234246+0651581, V1874 Ori, 2MASS J05253253+0625336, 2MASS J05194398+0535021. Group is centered near $\alpha, \delta \simeq 82°$, +6°, with proper motion $\mu_\alpha, \mu_\delta \simeq +9$, -35 mas yr$^{-1}$ and mean radial velocity $+18.5 \pm 0.4$ km s$^{-1}$.

The **Alessi 13** ($\chi^1$ **For**) cluster (Dias et al. 2002) appears to consists of at least a dozen stars with masses in the range 0.9-2.4 $M_\odot$ (total mass $\sim$18 $M_\odot$) - anchored by the A1 stars $\chi^1$ For and $\chi^3$ For A ($\alpha, \delta \simeq 51°$, -36°; $\mu_\alpha$, $\mu_\delta \simeq$ +37, -4 mas yr$^{-1}$). Within a volume of $\sim$100 pc$^3$, these stars have a density of $\sim$0.2 $M_\odot$ pc$^{-3}$ - roughly double the local disk density. Kharchenko *et al.* (2013) estimated an isochronal age of $\sim$525 Myr, but the MS turn-off is ill-defined. However, the X-ray emitting stars¶ *all* appear to have saturated emission ($\log(L_X/L_{bol} \simeq$ -3.4), more suggestive of an age more like $\sim10^{7.5}$ yr. Indeed, Alessi 13's velocity and position are suggestive that it may be part of the Tuc-Hor/Col/Car complex, which itself may be related to the Cas-Tau association.

## 4. Unphysical Groups or Streams (Grade: Fail)

While there are convincing cases of $\sim$40 Myr-old stars in **Argus** membership lists, Bell *et al.* (2015) were unable to assign an ambiguous isochronal age to the group. The scatter in HR diagram positions is highly suggestive that Argus is either seriously contaminated by interlopers, and/or may not constitute a coeval group (i.e. is a *stream*).

Zuckerman *et al.* (2013) proposed a new group of $\sim$14 systems dubbed **Oct-Near** with age $\sim$30-100 Myr. The group has a wide range in velocity, most importantly V and W ($>$5 km s$^{-1}$). The members do not appear to clump on the sky. The uniqueness of the velocities of these young stars is pointed out by Zuckerman *et al.*, however it probably warrants *stream* status at this point, and further investigation is needed.

**Her-Lyr** was defined by Gaidos (1998) and Fuhrmann (2004), with more recent adjustments by Lopez-Santiago *et al.* (2006) and Eisenbeiss *et al.* (2013). Fuhrmann has demonstrated that there is a clump in velocity space of young-ish stars within 25 pc, however given the large velocity dispersion (variously quoted at $\sim$3-4 km/s), the stars in most of these lists would not stay within each other's vicinity for very long. Lopez-Santiago *et al.* (2006) said their Her-Lyr members were *"chosen by their kinematics assuming a total dispersion of $\pm$6 km s$^{-1}$ in U and V, respectively... [t]he value of the dispersion has been chosen equal to that of the $\sim$200 Myr old Castor MG... coeval with the Hercules-Lyra Association. No restriction in the W component has been imposed in this first selection."* This is a recipe for selecting stars with a wide range of ages and birthsites (and whose sample mean age is likely to be of limited utility). There is remarkably little continuity in the published Her-Lyr membership lists. Only three stars have withstood the scrutiny of Fuhrmann, Lopez-Santiago *et al.*, and Eisenbeiss *et al.* as Her-Lyr "members": HD 10008, HD 166, and HD 206860, with the latter two being the sole surviving members from the original Gaidos study! Eisenbeiss *et al.* (2013) does show that the gyrochronology ages for these three stars are somewhat clustered (231, 315, 296 Myr, respectively). Extrapolation of the mass function for Her-Lyr by Eisenbeiss et al. (2013) predicts a population of $\sim$25 $\sim$200-300 Myr Her-Lyr M dwarf members within 25 pc. The spectroscopic and kinematic survey of local ($<$25 pc) X-ray-bright M dwarfs by Shkolnik *et al.* (2012) was designed to discover just such young M dwarfs. Their survey yielded only a *single* candidate Her-Lyr M dwarf. Thus far, Her-Lyr may constitute a *stream*, but I concur with Brandt *et al.* (2014) that membership to Her-Lyr should not be used for age-dating stars.

Mamajek *et al.* (2013) and Zuckerman et al. (2013) independently argued that the **Castor Moving Group** is unphysical. Purported Castor "members" have a wide velocity dispersion ($\sim$3-6 km/s). The velocities of the key members (e.g. Fomalhaut, Vega, etc.) are sufficiently well-constrained that one can confidently conclude that they were

¶ TYC 7027-715-1, TYC 7027-852-1, TYC 7026-185-1.

not near each other as recently as 10 Myr ago, let alone hundreds of Myr ago. The group should probably be considered the *Castor Stream*. The **IC 2391 Supercluster** and **Local Association** are streams that suffer from the same problems as Her-Lyr and Castor. Membership to these groups and adoption of group ages is unhelpful for age-dating stars.

The nearest Cepheid is the famous F8 supergiant **Polaris**, which is at least a triple system along with two F dwarfs. Turner (2004) proposed that Polaris belonged to a larger group†, and I've included it here as there are a range of distance estimates between ∼90-130 pc (Turner 2009). However, the recent trig parallax from van Leeuwen (2007) places it at ∼129 pc. Turner (2009) presented a CMD for Polaris's neighbors, but no kinematic analysis. If one adopts the Polaris space motion of Wielen et al. (2000) $(U, V, W) = (-14.2 \pm 1.2, -28.0 \pm 0.8, -5.4 \pm 1.0)$ km s$^{-1}$, and convert it to a convergent point solution of $\alpha, \delta = 107°, -31°, S_{tot} = 31.9$ km s$^{-1}$, one finds that Turner's stars are clearly not sharing motion (peculiar velocities of typically ∼5-10 km/s; too large to stay near Polaris over tens of Myr.). I have seen no evidence for a "Polaris cluster".

Chereul *et al.* (1999) reported the discovery of three new "loose clusters" in a Hipparcos study of density-volume inhomogeneities among nearby ($d < 125$ pc) A-F type stars. The closest of these ($d \simeq 89$ pc), was dubbed "Pegasus 2" by Chereul *et al.*, but appears as "**Chereul 3**" in WEBDA‡ and Dias et al. (2002). WEBDA currently lists Chereul 3 as the 3rd closest *open cluster*, while Dias *et al.* (2002) call it a *moving group*, and exclude it from their open cluster catalog. Despite their proximity, the Chereul cluster candidates have gone largely unstudied since their discovery. The memberships for these groups were not listed in Chereul et al. (1999), however they were kindly provided to the author by E. Chereul (priv. comm.). While Fig. 10 of Chereul *et al.* 1999 shows 8 members, the list from Chereul (priv. comm.) contained 6 Hipparcos systems¶. The revised Hipparcos parallaxes for these 6 stars indicate that they are at a range of distances (roughly ±20 pc rms about a mean distance of 95 pc). Unfortunately, their radial velocities in the compiled catalog of Gontcharov *et al.* (2006) have a large scatter (range: -38 to +16 km/s). Attempts to determine a robust convergent point solution for this sample which results in reasonable agreement between observed and predicted radial velocities, and trig and kinematic parallaxes, were unsuccessful. The HR diagram positions of the Chereul 3 stars are consistent with A/F-type main sequence stars drawn from a wide range of ages (0.4-2.5 Gyr). HIP 104338 and 104616 (54' separation) could comprise a wide physical pair (similar RVs and astrometry), but there is little to suggest any relation between the other purported members. I conclude that Chereul 3 is likely to be unphysical.

**Chereul 2:** Chereul *et al.* (1999) also reported a ∼800 Myr-old group dubbed "Pegasus 1", but listed as "Chereul 2" by WEBDA and Dias *et al.* (2002). E. Chereul (priv. comm.) listed 14 candidate members‖, for which I estimate mean distance 91 pc and E(B-V) = 0.04. However the parallaxes are consistent with intrinsic spread ±15 pc and the radial velocities range from -28.6 to +14.4 km s$^{-1}$, with no obviously clumping. The HR diagram positions are consistent with isochronal ages between the ZAMS and ∼$10^{9.4}$ yr, hence both the color-mag and kinematic data are consistent with Chereul 2 being unphysical.

Latyshev (1977) reported a candidate nearby open cluster in UMa (dubbed **Latyshev**

---

† Including HD 7283, 40335, 45919, 52908, 103435, 108862, & 118285.

‡ http://www.univie.ac.at/webda/

¶ Chereul 3 members: HIP 103652, 104338, 105902, 106488, 106783, 107120. One of these is resolved into two components by Hipparcos (HIP 103652) and one has an obvious wide separation companion (HIP 106781 is 39" from HIP 106783). Two A/F-type stars missing from Chereul's list which appear in Fig. 10 of Chereul *et al.* 1999 are HIP 104430 and 104616.

‖ Chereul 2 members: HIP 102299, 103261, 103652, 103813, 104771, 104884, 105478, 105608, 106362, 107585, 108389, 108439, 108441, 109349, & 110465.

**2** by Archinal & Hynes 2003) comprised of 7 A-type stars of similar proper motions and magnitudes††. With modern astrometry, it is clear that these stars have a wide range of space velocities (8 to 25 $\mathrm{km\,s^{-1}}$), and no consistent convergent point solution can be found which explains the observed range of radial velocities (-19 to -3 $\mathrm{km\,s^{-1}}$). The mean distance to the Hipparcos entries is 110 pc, but the kinematics indicate it is unphysical.

## References


Archinal, B. A. & Hynes, S. J. 2003, Star clusters; Richmond, VA: Willmann-Bell

Barenfeld, S. A., Bubar, E. J., Mamajek, E. E., & Young, P. A. 2013, *ApJ*, 766, 6

Bell, C. P. M., Mamajek, E. E., & Naylor, T. 2015, submitted to *MNRAS*

Brandt, T. D. & Huang, C. X. 2015, *ApJ*, 807, 58

Brandt, T. D., *et al.* 2014, *ApJ*, 786, 1

de Bruijne, J. H. J., Hoogerwerf, R., & de Zeeuw, P. T. 2001, *A&A*, 367, 111

Chereul, E., Crézé, M., & Bienaymé, O. 1999, *A&A*, 135, 5

Dias, W. S., Alessi, B. S., Moitinho, A., & Lépine, J. R. D. 2002, *A&A*, 389, 871

Ducourant, C., Teixeira, R., Galli, P. A. B., *et al.* 2014, *A&A*, 563, A121

Eisenbeiss, T., Ammler-von Eiff, M., Roell, T., *et al.* 2013, *A&A*, 556, A53

Esplin, T. L., Luhman, K. L., & Mamajek, E. E. 2014, *ApJ*, 784, 126

Fry, B. J., Fields, B. D., & Ellis, J. R. 2015, *ApJ*, 800, 71

Fuhrmann, K. 2004, Astronomische Nachrichten, 325, 3

Gagné, J., *et al.* 2014, *ApJ*, 783, 121

Gaidos, E. J. 1998, *PASP*, 110, 1259

Gontcharov, G. A. 2006, Astronomy Letters, 32, 759

Johnson, D. R. H. & Soderblom, D. R. 1987, *AJ*, 93, 864

Kastner, J. H., Zuckerman, B., Weintraub, D. A., & Forveille, T. 1997, *Science*, 277, 67

Kharchenko, N. V., et al. 2013, *A&A*, 558, A53

Kraus, A. L., Shkolnik, E. L., Allers, K. N., & Liu, M. C. 2014, *AJ*, 147, 146

Latyshev, I. N. 1977, Astronomicheskij Tsirkulyar, 969, 7

Lejeune, T. & Schaerer, D. 2001, *A&A*, 366, 538

López-Santiago, J., M. J. *et al.* 2006, *ApJ*, 643, 1160

Malo, L., Artigau, É., Doyon, R., *et al.* 2014, *ApJ*, 788, 81

Mamajek, E. E. 2005, *ApJ*, 634, 1385

Mamajek, E. E. 2007, *IAU Symposium*, 237, 442

Mamajek, E. E., Kenworthy, M. A., Hinz, P. M., & Meyer, M. R. 2010, *AJ*, 139, 919

Mamajek, E. E., Lawson, W. A., & Feigelson, E. D. 1999, *ApJL*, 516, L77

Murphy, S. J., Lawson, W. A., & Bessell, M. S. 2013, *MNRAS*, 435, 1325

Pinsonneault, M. H., Terndrup, D. M., Hanson, R. B., & Stauffer, J. R. 2004, *ApJ*, 600, 946

Riedel, A. R., Finch, C. T., Henry, T. J., *et al.* 2014, *AJ*, 147, 85

Shkolnik, E. L., Anglada-Escudé, G., Liu, M. C., *et al.* 2012, *ApJ*, 758, 56

Silaj, J. & Landstreet, J. D. 2014, *A&A*, 566, A132

Torres, C. A. O., Handbook of Star Forming Regions, Volume II, 757

Turner, D. G. 2004, *BAAS*, 36, 744

Turner, D. G. 2009, *AIPC*, 1170, 59

van Leeuwen, F. 2009, *A&A*, 497, 209

Weinberger, A. J., Anglada-Escudé, G., & Boss, A. P. 2013, *ApJ*, 762, 118

Wielen, R., Jahreiß, H., Dettbarn, C., Lenhardt, H., & Schwan, H. 2000, *A&A*, 360, 399

Zuckerman, B., Bessell, M. S., Song, I., & Kim, S. 2006, *ApJL*, 649, L115

Zuckerman, B. & Song, I. 2004, *ARA&A*, 42, 685

Zuckerman, B., Vican, L., Song, I., & Schneider, A. 2013, *ApJ*, 778, 5


†† SAO 28803, 28866, 28885, 28928, 28843, 28868, and 28891. SAO 28868 appears to be a typo in Latyshev (1977), as the star's properties clearly correspond to SAO 28862 instead.

*Young Stars & Planets Near the Sun*
*Proceedings IAU Symposium No. 314, 2015*
*J. H. Kastner, B. Stelzer, & S. A. Metchev, eds.*

© International Astronomical Union 2016
doi:10.1017/S1743921315006377

# Nearby Young Moving Groups: Statistical Methods and Challenges for Assigning Membership

L. Malo[1,2], J. Gagne[2], R. Doyon[2], D. Lafreniere[2], E. Artigau[2] and L. Albert[2]

[1] Canada-France-Hawaii Telescope, 65-1238 Mamalahoa Hwy, Kamuela Hawaii, 96743, USA.
email: `malo@cfht.hawaii.edu`
[2] Institut de Recherche sur les Exoplanetes (iREx), Universite de Montreal, Departement de Physique, C.P. 6128 Succ. Centre-ville, Montreal, QC H3C 3J7, Canada.

**Abstract.** Young associations, being sparsely populated and relatively close to the Sun, their members are found all over the sky. In the Solar Neighborhood, young moving groups are found within 100 pc with ages ranging from 5 to 120 Myr. While known members of these groups were identified mostly through the Hipparcos data, only the most massive members have been fully characterized so far, and defined the core members. In the last decades, several new candidate members have been identified, using different approaches. Based on the global properties of the core members (kinematics and over luminosity), those methods used several criteria to establish the membership, from qualitative manner to quantitive methods using reduced chi-squared or membership probability. A full confirmation of the membership for those numerous candidates requires radial velocity and parallax measurements to confirm their kinematics, age-dating indicator measurement to assess their youth and multiplicity follow-up to rule out binary objects. In this proceeding, we summarize a general recipe to assign membership, describe the numerous challenges for assigning membership, and end with a discussion on the appropriateness and reliability of the BANYAN I and II tools to assess membership.

**Keywords.** methods: statistical,(Galaxy:) solar neighborhood, stars: low-mass, stars: pre-main sequence, techniques: radial velocities.

---

## 1. Introduction

Nearby young comoving groups (NYMG) are sparse, gravitationally unbound stellar associations comprising a few dozens of stars scattered within $\sim 100$ pc of the Sun with ages ranging from 5 to a few hundred Myr. Comoving group members are characterized by a common position and space motion within the Galaxy. As a result of a projection effect, they display an organized motion in the sky moving toward a convergent point and this can be used to discriminate genuine members from field stars. Since 1958, several methods have been developed to identify new associations and/or new members of a given association (Eggen 1958, Montes *et al.* 2001, Zuckerman & Song 2004, Torres *et al.* 2006). Moreover, Hipparcos mission combined with multiple radial velocity studies led to the identification of the most massive members of the NYMGs, but relatively shallow completeness magnitude prevented the detection of low-mass members. Assuming that NYMG stars follow a typical Initial Mass Function, the vast majority of low-mass members are expected to have gone unnoticed. This led to a number of efforts to extend measurements of proper motion, radial velocity and parallax to the low-mass regime (e.g., Shkolnik *et al.* 2012). To complete the census of NYMG members, kinematics alone is not enough to confirm membership, independent youth indicators are also mandatory to confirm the age of these stars.

## 2. General recipe to assign membership

Identifying new members of NYMGs requires to be careful on the selection of the core members, as well as the determination of their global properties and finally the method used to quantify the membership. The following section summarizes the key elements for each of the essential steps.

### 2.1. *Reference sample*

For the sole purpose of establishing the core membership of the association, one needs to identify strong members (bona fide). Stars should be considered bona fide members of young kinematic groups only when a good measurement of trigonometric distance, proper motion and radial velocity are available, which means that Galactic space velocity and Galactic positions are accurate. Moreover, bona fide members need to display youth indicators which depend on the mass and age, and finally an appropriate location in the HR diagram. Several studies have established lists of bona fide members, such as Zuckerman & Song (2004), Torres *et al.* (2008), Fernandez *et al.* (2008).

### 2.2. *Common properties*

Because of the coeval formation of a given association, all members share, to within a few km s$^{-1}$, a common space velocity within the Galaxy. The galactic space motion of a star, $UVW$, is determined from its sky position ($\alpha$, $\delta$), radial velocity (RV), proper motion, and parallax, using the Johnson & Soderblom (1987) relations.

By virtue of their youth and their coeval formation, the members of young associations have had little time to disperse within the Galaxy. As a result, the positions $XYZ$ of the members of a young association are relatively well confined within the Galaxy. The galactic position of a star, $XYZ$, is determined from its sky position ($\alpha$, $\delta$) and trigonometric distance.

Color-Magnitude Diagrams (CMD) have been a crucial tool for identifying young stars, since they are over-luminous compared to an old population. The color indices used are strongly mass dependant. To discern between young and old low-mass stars wide color indices (e.g. $V - K$, $I_c - J$) are good to provide protection against measurement errors and time variability (Zuckerman & Song 2004). Figure 1 shows CMD for low-mass stars using $M_J$ vs $I_c - J$. Past studies have shown that $B - V$ can be useful for more massive stars (Zuckerman & Song 2004) and color indices in the near and mid-infrared should be used for brown dwarfs and planetary-mass objects (Gagne *et al.* 2014).

### 2.3. *Metric*

A key element in order to quantitatively assess star membership in a given NYMG is confirmed if candidate's $UVWXYZ$ match those of a given NYMG. When radial velocity and parallax are available, one can minimize a reduced $\chi^2$ statistic with several degrees of freedom to quantify the match. This method has been used by Shkolnik *et al.* (2012), Faherty *et al.* (2012), Riedel *et al.* (2014) using the full 6 dimensions ($UVWXYZ$) and applied to 14 NYMGs and young associations.

When radial velocity and parallax are not available, one has to predict the values, which a star should have if it was an actual member of a given NYMG. To predict the radial velocity value, Zuckerman & Song (2004) have developed the good $UVW$ box method which determines the radial velocity required combined with the distance and

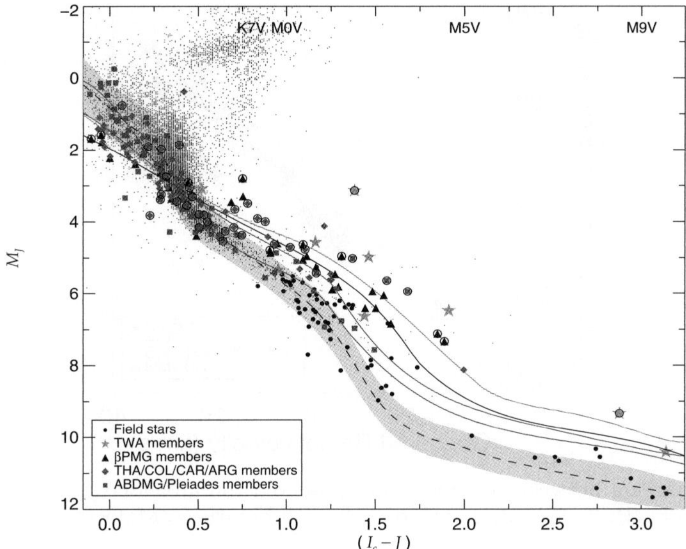

**Figure 1.** Color-magnitude diagram ($M_J$ vs $I_c - J$) for members of TWA (green stars), $\beta$PMG (black triangles), THA, COL, CAR, and ARG (red diamonds), and ABDMG and Pleiades (blue squares). Fields stars are represented by dots and filled black circles, and binary stars are those with black circles superposed on their own symbol (Malo *et al.* 2013).

sky position, to match a given NYMG's galactic space velocity. Moreover, Rodriguez *et al.* (2013) have developed GALNYSS method which includes the convergent point method to derive the radial velocity and kinematic distance for a given NYMG.

To predict the distance, Torres *et al.* (2006) have developed the SACY project, based on the minimization of a merit function to infer the distance to the star. This merit function requires to match the $UVWXYZ$ and the expected brightness as a function of color index.

A more robust method, combining all previous works, has been developed by Malo *et al.* (2013) and Gagné *et al.* (2014), by building a kinematic model of a given NYMG to predict the amplitude of the proper motion and radial velocity which a star should have if it were a member of the NYMG. The kinematic model is combined with a Bayesian analysis to predict the most likely distance to the star. Therefore, the Bayesian Analysis for Nearby Young AssociatioNs (BANYAN) tool derives membership probabilities by combining Bayesian inference to empirical models for seven observables characterizing the kinematics ($UVWXYZ$) and photometry of known members. Such models were built for the seven closest and youngest well-defined moving groups : TW Hydrae, $\beta$ Pictoris, Tucana-Horologium, Columba, Carina, Argus and AB Doradus. Also, models were built for a field sample of stars allowing for each candidate member, kinematic and photometric comparisons to an old field population. In addition to the membership probability, statistical predictions for radial velocity and distance, BANYAN includes a binary hypothesis, which should be interpreted as an over-luminosity compared to other young stars, which could be due to the presence of a binary companion, chromospheric activity, or peculiar colors of the star.

More recently, Gagné *et al.* (2014) developed BANYAN II, which includes a more sophisticated kinematic model (ellipsoids with arbitrary rotation) and uses NIR color-magnitude diagrams to specifically target the identification of brown dwarfs and planetary-mass objects in these moving groups. Moreover, expected field and moving

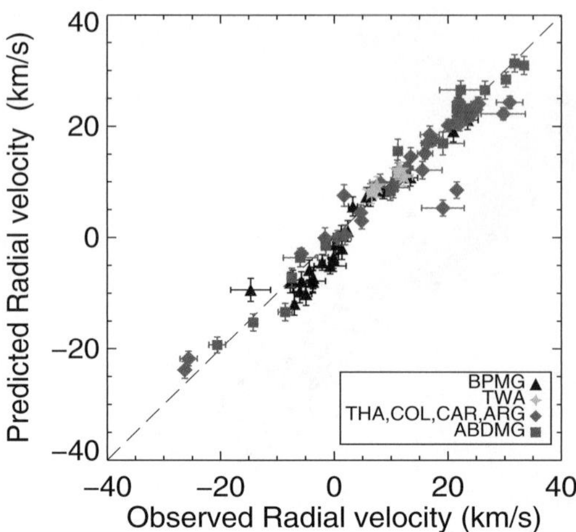

**Figure 2.** Comparison between predicted and observed radial velocities for the 111 candidate members (Pv > 90% and excluding known binaries) (Malo *et al.* (2014)).

group populations are taken into account as prior probabilities, and an extensive false-positive and hit rate analysis is performed based on a Besançon Galactic model (Robin *et al.* in prep).

A web-based tool of the Bayesian techniques are available at :

www.astro.umontreal.ca/~malo and www.astro.umontreal.ca/~gagne .

### 2.4. *Membership confirmation*

We used the BANYAN I and BANYAN II tools to newly identify 200 K5–M5, 175 M4–L7 candidate members to NYMGs, and 39 previously known brown dwarfs with signs of youth (Gagne *et al.* 2014, Gagne *et al.* in press). Although the analysis unveiled new highly probable candidate members to NYMGs, a word of caution is necessary before assigning a firm membership to these objects. These candidates will be attributed the status of bona fide members only after their radial velocity and parallax are measured and shown to be consistent with the radial velocity and statistical distance predicted by our analysis. For this reason, Malo *et al.* (2014) presented RV measurements for 219 K7–M5 candidate members, using high resolution spectroscopy. This follow-up study shows that the predicted radial velocity agrees with the measured value for 130 stars. Figure 2 shows the comparison between predicted and observed values.

Furthermore, a collaboration with A. Riedel (Riedel *et al.* 2014) and the CTIOPI group, as well as recent parallax measurements in the literature (Dupuy & Liu 2012; Shkolnik *et al.* 2012; Liu *et al.* 2013; Dittmann *et al.* 2014) confirmed statistical distance predictions from BANYAN for 30 candidate members.

In order to confirm the youth of the candidates, signs of youth depending on the spectral type should be measured. For solar type stars, coronal activity indicators such as X-ray and UV emission are often used (Soderblom 2010). For low-mass dwarfs, chromospheric and coronal activity indicators, lithium abundance and rotation are common youth indicators. In the case of brown dwarfs, spectroscopic indicators of low-gravity are preferred, such as the equivalent width of Na I, KI, the depth of VO band-heads, as well as the shape of the H-band continuum.

Last, because the interpretation of the observed luminosity is different in the case of an unresolved multiple system, RV monitoring and high contrast imaging should be pursued to identify binary systems within the proposed bona fide members, before claiming final candidate membership.

## 3. Challenges

Assigning candidate membership in NYMGs is a complex puzzle, which requires huge amount of tools and knowledge. There are at least three challenges which we faced when identifying and confirming candidate membership. First, quantitative assessment requires a precise determination of the global properties for NYMG members, which are strongly dependant on the method used and the sample of bona fide members. As shown in Gagne *et al.* (2014), parametrization of $UVWXYZ$ using ellipsoids can better reproduce the shape of the clusters.

A second challenge when assigning membership is the missing radial velocity and parallax information, for the vast majority of the candidates. Given that one needs to develop such tools to predict radial velocity and parallax values which a star should have if the star was a member of a given NYMG. Furthermore, these values have to be measured with a high precision and should validate the predictions.

Finally, independently of the method used to assess membership, a contamination rate simulation should be performed and the contamination rate should be taken into account, when claiming the final candidate membership. To minimize the contamination rate, analysis methods should use star's kinematics ($UVWXYZ$) complemented with the photometric properties (over-luminosity) and analysis methods should include not only NYMGs but also an old field population for comparison. BANYAN tools have performed such contamination analysis showing false alarm rate between 5–15% depending on the type of stars and NYMGs considered (Malo *et al.* 2013, Gagne *et al.* 2014)

## 4. Future prospects

The main limitation of the current BANYAN tool is that it does not include young association members located beyond 100 pc such as Eta Cha (Zuckerman & Song 2004), the Scorpius-Centaurus complex (de Zeeuw *et al.* 1999) and the Pleiades (Jeffries 1995). The main reason for this is that the construction of reliable spatial and kinematic models for these associations requires precise measurements of RV and distance for a significant set of credible members. Those RV and parallax measurements are still not available. However, Rizzuto *et al.* (2011), Rizzuto *et al.* (2012) have been focusing on combining bayesian analysis and global properties of Scorpius-Centaurus complex members, resulting in strong candidate members, for which predicted values will be available soon. The *Gaia* mission will provide such an opportunity by providing measurements of trigonometric distance for objects with $V \leqslant 20$. However, RV measurements from *Gaia* mission will be limited to a precision of a few $\mathrm{kms}^{-1}$ for faint low-mass stars and brown dwarfs. This brings out the important role that ground-based RV surveys reaching precisions of less than 1 $\mathrm{kms}^{-1}$ will play in characterizing the kinematics of red objects. With accurate RV and parallax measurements for all stars in hand, the determination of the $UVW$, $XYZ$ for young group members beyond 100 pc will become possible. Moreover, the discovery of new NYMGs within 100 pc will be now possible, helping with a better understanding of the solar neighborhood formation. Current and next generation instruments such as GRACES (Chene *et al.* 2014) at Gemini-North, SPIRou at CFHT (Delfosse *et al.* 2013) and CARMENES (Quirrenbach *et al.* 2014) are opening the door to precise radial velocity measurements leading to more accurate galactic space velocities, when combined with *Gaia* parallaxes, which are required to identify new NYMGs in the Solar Neighborhood.

## Acknowledgement

The authors would like to thank the organizers of the Young Stars Near the Sun meeting, for providing the opportunity to have this full day discussion about how to identify and confirm star's membership; and the participants for engaging in a productive discussion. We thank our collaborators: Jackie Faherty, Adric Riedel, Kelle Cruz and Gregory Feiden. This work was supported through grants from the Canada-France-Hawaii Telescope, the Fond de Recherche Quebecois – Nature et Technologie and the Natural Science and Engineering Research Council of Canada.

## References

Chene, A-N., Padzer, J., Barrick, G., *et al.* 2014, in SPIE, 9151, id. 915147

de Zeeuw, P. T., Hoogerwerf, R., de Bruijne, J. H. J., Brown, A. G. A., & Blaauw, A. 1999, *AJ*, 117, 354

Delfosse, X., Donati, J.-F., Kouach, D., *et al.* 2013, in SF2A-2013: Proceedings of the Annual meeting of the French Society of Astronomy and Astrophysics, ed. L. Cambresy, F. Martins, E. Nuss, & A. Palacios, 497

Dittmann, J. A., Irwin, J. M., Charbonneau, D., & Berta-Thompson, Z. K. 2014, *AJ*, 784, 156

Dupuy, T. J. & Liu, M. C. 2012, *ApJS*, 201, 19

Eggen, O. J. 1958, *MNRAS*, 118, 65

Faherty, J. K, Burgasser, A. J, Walter, F. M, *et al.* 2012, *ApJ*, 752, 56

Fernandez, D., Figueras, F., & Torra, J. 2008, *A&A*, 480, 735

Gagné, J., Lafrenière, D., Doyon, R., Malo, L., & Artigau, É. 2014, *ApJ*, 783, 121

Jeffries, R. D. 1995, *MNRAS*, 273, 559

Johnson, D. R. H. & Soderblom, D. R. 1987, *AJ*, 93, 864

Liu, M. C., Dupuy, T. J., & Allers, K. N. 2013, *AN*, 334, 1

Malo, L., Doyon, R., Lafrenière, D., *et al.* 2013, *ApJ*, 762, 88

Malo, L., Artigau, É., Doyon, R., *et al.* 2014, *ApJ*, 788, 81

Montes, D., Lopez-Santiago, J., Galvez, M. C., *et al.* 2001, *MNRAS*, 328, 45

Quirrenbach, A., Amado, P. J., Caballero, J. A., *et al.* 2014, in SPIE, 9147, id 91471F

Riedel, A. R., Finch, C. T., Henry, T. J., *et al.* 2014, *AJ*, 147, 85

Rizzuto, A. C., Ireland, M. J., & Robertson, J. G. 2011, *MNRAS*, 416, 3108

Rizzuto, A. C., Ireland, M. J., & Zucker, D. B. 2012, *MNRAS*, 421, 97

Robin, A. C., Marshall, D. J., Schultheis, M., & Reylé, C. 2012, *A&A*, 538, A106

Rodriguez, D. R., Zuckerman, B., Kastner, J. H., *et al.* 2013, *AJ*, 774, 101

Shkolnik, E. L., Anglada-Escude, G., Liu, M. C., *et al.* 2012, *AJ*, 758, 56

Soderblom, David. R. 2010, *ARA&A*, 48, 581

Torres, C. A. O., Quast, G. R., da Silva, L., *et al.* 2006, *A&A*, 460, 695

Torres, C. A. O., Quast, G. R., da Melo, C. H. F., *et al.* 2008, in Handbook of Star forming Regions, Vol. II: The Southern Sky, ed. B. Reipurth (ASP Monograph Publ., Vol. 5; San Francisco, CA: ASP), 757

Zuckerman, B. & Song, I. 2004, *ARA&A*, 42, 685

*Young Stars & Planets Near the Sun*
*Proceedings IAU Symposium No. 314, 2015*
*J. H. Kastner, B. Stelzer, & S. A. Metchev, eds.*

© International Astronomical Union 2016
doi:10.1017/S1743921315006419

# LACEwING: Lessons from a New Moving Group Code

## A. R. Riedel[1]

[1] The College of Staten Island
email: **adric.riedel@csi.cuny.edu**
CUNY-The College of Staten Island
Hunter College
The American Museum of Natural History

**Abstract.** With the recent accelerating rate of discoveries in the field of nearby young stars, the ability to identify new nearby young stars is as important as ever, and membership identification codes will continue to perform a vital role in scientific research. In the process of creating a new moving group membership identification code - LocAting Constituent mEmbers In Nearby Groups (LACEwING): we have compiled a few pointers relevant to astronomers trying to use codes like LACEwING to locate young stars.

**Keywords.** stars:low-mass, stars:pre-main-sequence, galaxy:open clusters and associations

## 1. Introduction

One of the most important methods used in characterizing young stars, and the only method that can establish membership in specific nearby young moving groups, is their kinematics. Kinematics exploits the property that young stars (at ages less than a billion years) are still tracing the space motion of the gas from which they formed. Several membership identification codes are available publically: BANYAN (Malo *et al.* 2013), BANYAN II (Gagne *et al.* 2014); a convergence code (Rodriguez *et al.* 2013); and (soon), LACEwING (Riedel *et al.* in prep)†.

These codes use the six basic kinematic elements (RA, DEC, $\pi$, $\mu_{RA}$, $\mu_{DEC}$, and RV) to predict probabilities of membership in nearby young moving groups, and can handle incomplete data. It is, however, important to know how to use and interpret the results of LACEwING (and codes like it) to obtain the most accurate results.

## 2. Overview of LACEwING

The LACEwING code computes membership probabilities in 10 moving groups ($\epsilon$ Chameleon, TW Hydra, $\beta$ Pic, Octans, Tucana-Horologium, Columba, Argus, AB Doradus, Hercules-Lyra, and Ursa Major) and 4 open clusters ($\eta$ Chameleon, The Pleiades, Coma Berenices, and The Hyades). The groups are represented as freely oriented triaxial ellipsoids, which are computed from new lists of bona-fide (high confidence) members taken from existing lists (e.g. Malo *et al.* 2013), and filtered using epicyclic tracebacks to limit membership to only the stars that were plausibly within the group's boundaries at the time of formation.

The LACEwING code matches stars to groups by computing up to four metrics of membership, matching the proper motion, distance, radial velocity, and spatial location of the star to those estimated for a group member at the same RA and DEC. Those

† See also https://github.com/ariedel/lacewing

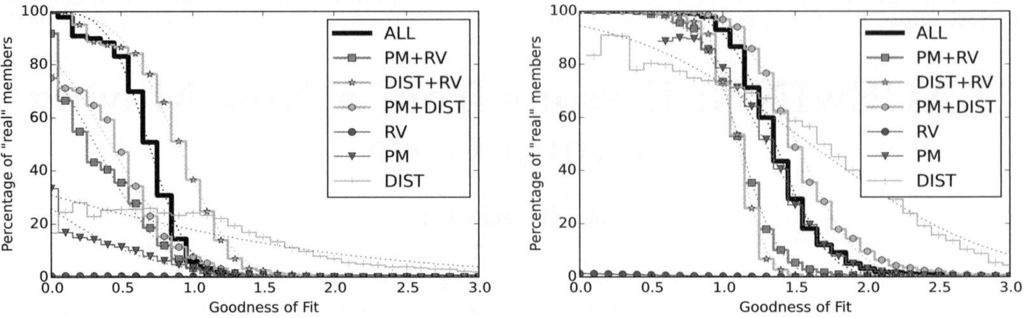

**Figure 1.** Histograms of matches to the simulated $\beta$ Pictoris Moving Group (left) and Coma Berenices open cluster (right), showing the percentage of stars in each bin that are genuine members of the group/cluster. Curves fit to these histograms (dotted lines) are used to translate the LACEwING code's goodness-of-fit parameters to actual membership probabilities.

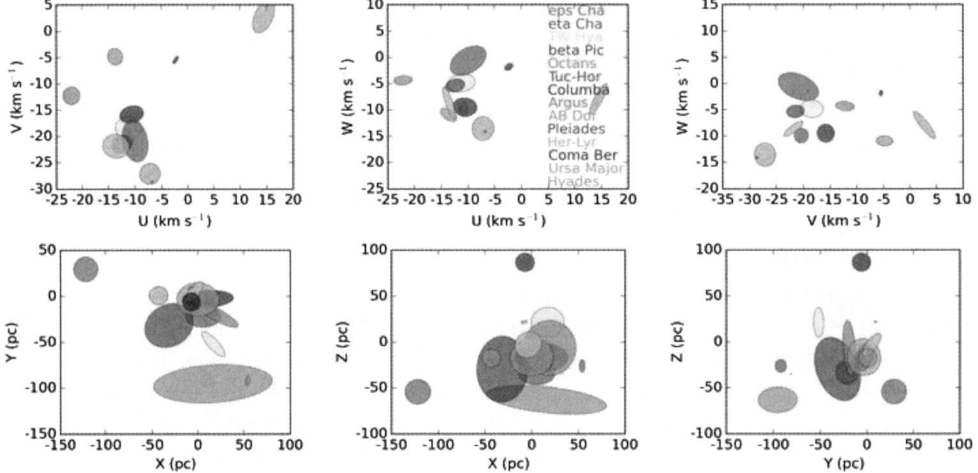



**Figure 2.** 2D projections of the freely oriented triaxial ellipsoids that constitute LACEwING's kinematic models (in $UVW$ motion and $XYZ$ spatial position) of the 14 nearby young moving groups and open clusters. As shown in these figures, some groups have overlapping $UVW$ space velocities or $XYZ$ space positions. This can make it difficult to determine to which group individual stars belong.

metrics are combined into a single numerical goodness-of-fit value. The goodness-of-fit values are transformed into membership probabilities by creating a large simulation (using 32 million stars; Riedel *et al.* submitted) of the Solar Neighborhood, which is then binned by goodness-of-fit value relative to a group. This is shown for $\beta$ Pic and Coma Ber in Fig. 1, where all seven different possible combinations of data (RA and DEC are assumed to be known) are given as different colors. Curves are fit to those distributions to provide membership probabilities.

## 3. Lessons from LACEwING

### 3.1. Groups Are Not Equally Easy to Identify

Some groups (like the Coma Berenices open cluster) have very distinctive $UVW$ space velocities and space positions (Fig. 2), which translate into uniquely identifying proper motion and radial velocity estimates for any given star.

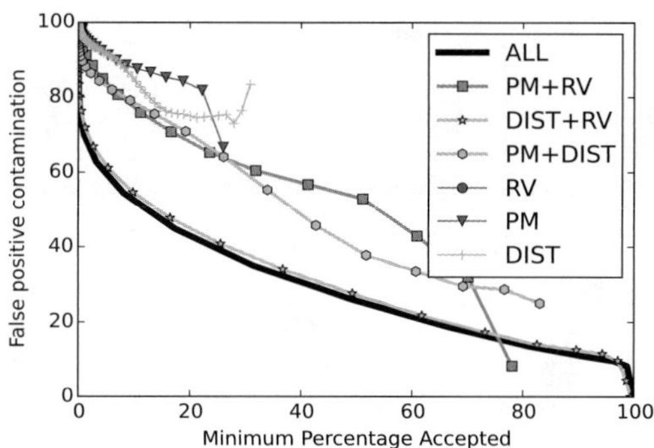

**Figure 3.** In this figure we see the cost of lowering the membership probability threshold: the false positive contamination of the sample increases. The best way to reduce the false positive contamination is to obtain more of the 6 kinematic elements for the sample of stars, but even in this case, additional methods (spectroscopic confirmation, for example) will be needed to weed out the sample.

As an example, the large simulation of stars in Fig. 1 shows us that for the case in which all kinematic elements are available (RA, DEC, plus proper motion, distance, and RV), a goodness-of-fit value of 0.75 relative to $\beta$ Pic yields a roughly 50% probability that the object is a member of $\beta$ Pic. The stars matched to Coma Ber show that, for the equivalent case using all available data, a goodness-of-fit of 0.75 yields a nearly 100% probability that the object is a member of Coma Ber.

## 3.2. Additional Kinematic Data Vastly Improves Membership Probabilities

Fig. 1 shows the seven different data availability scenarios LACEwING expects: RA and DEC plus all combinations of proper motion, radial velocity, and distance. For $\beta$ Pic, the proper motion method (red dashed line) demonstrates that even in the best possible case of a goodness-of-fit $= 0$, the probability of membership is only 35%, i.e. 65% of those perfect matches to $\beta$ Pic are actually (simulated) members of other moving groups and the generic field population. Adding a radial velocity (pink triangles) increases the best-case probability to 80%/20% contamination.

Another good reason to obtain more kinematic information is to reduce the false positive contamination (Fig. 3) of a survey. With only proper motion and a fixed cutoff of 20% probability, over 80% of the objects in a survey will be contaminants (to be weeded out by some other technique, such as spectroscopy). Having complete data for the star of interest reduces the contamination rate to just above 40%.

## 3.3. Not All Young Stars Belong to a Known Moving Group

Virtually all lithium is primordial and easily fused by objects over roughly 60 Jupiter masses. Therefore, a low mass object with detectable lithium must either be a brown dwarf, or a young star. Riedel *et al.* (submitted) assembled a sample of roughly 500 lithium-detected nearby stars and used those stars as a test sample for LACEwING, reasoning that these should be the best nearby young moving group members. However, nearly half the sample did not match any of the known nearby young moving groups LACEwING tests for. The same lithium-detected sample was run through other moving group codes—BANYAN I and II, and the Convergence code—with similar results. Some lithium-detected stars were not sorted into moving groups (Fig. 4).

There are three potential reasons for this: 1.) There are additional nearby young stellar groups not currently known or accounted for; 2.) there is a young field population not part of any previously identified groups; 3.) our current understanding of the kinematics

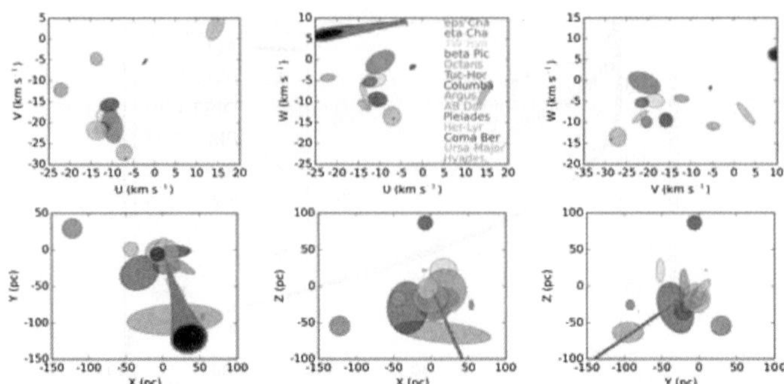


**Figure 4.** 2MASS 0447−7412 has a significant lithium equivalent width, 50±20 mÅ(Murphy & Lawson 2015), a proper motion, a radial velocity, and no distance. If we generate a range of reasonable distances for the star, it could spatially be within the Octans (cyan) moving group, but it cannot possibly share the space velocity of Octans, and cannot be a member (as previously noted by Murphy)

of moving groups is incomplete. The fact that multiple moving group codes (each with a different algorithm and kinematic model of the nearby young moving groups) could not identify large numbers of stars as members of nearby young moving groups suggests that the explanation probably involves some combination of the first two points. Ultimately, the message is clear: young stars do not need to be members of a known nearby young moving group. "None of the above" can be a correct answer.

## 4. Conclusions

As we enter the third decade of nearby young star research, there is much more to be done from the standpoint of membership statistics. Moving group identification codes such as LACEwING will continue to provide essential tools in studies of stellar and planetary formation and evolution, but care must still be taken in interpreting their results.

## References

Gagné, J., Lafrenière, D., Doyon, R., Malo, L., & Artigau, È 2014, *ApJ*, 783, 121
Malo, L., Doyon, R., & Lafrenière, D.; *et al.* 2013, *ApJ*, 762, 88
Murphy, S. J. & Lawson, W. A. 2015, *MNRAS*, 447, 1267
Riedel, A.R., Blunt, S.C., Rice, E.L. *et al.* submitted, *ApJ*
Rodriguez, D. R., Zuckerman, B., Kastern, J. H., *et al.* 2013, *ApJ*, 774, 101

## Discussion

SHKOLNIK: What about chemical abundance tagging?

RIEDEL: LACEwING does only one thing: it uses kinematic and spatial data to evaluate potential memberships in moving groups. It should be used alongside spectroscopic methods to confirm that the star(s) in question is/are actually young, and have appropriate ages for the group LACEwING suggests is the most probable host.

ZUCKERMAN: Why isn't Carina-Near in the code?

RIEDEL: There wasn't enough in the literature about it. The same goes for Cas-Tau, 32 Ori, and Alessi 13. But they can be added fairly easily: new ellipsoids, new simulation.

*Young Stars & Planets Near the Sun*
*Proceedings IAU Symposium No. 314, 2015*
*J. H. Kastner, B. Stelzer, & S. A. Metchev, eds.*

© International Astronomical Union 2016
doi:10.1017/S1743921315006559

# Chemical tagging of FGK stars:
# Testing the Membership of Young Stellar Kinematics Groups

D. Montes[1], H.M. Tabernero[1,2], and J.I. González Hernández[1,3,4]

[1] Dpto. Astrofísica, Facultad de CC. Físicas, Universidad Complutense de Madrid, E-28040 Madrid, Spain
email: dmontes@ucm.es
[2] Universidad de Alicante, DFISTS, E-03080 Alicante, Spain
[3] Instituto de Astrofísica de Canarias, E-38205 La Laguna, Tenerife, Spain
[4] Universidad de La Laguna, Dept. Astrofísica, E-38206 La Laguna, Tenerife, Spain

**Abstract.** In this contribution talk we summarize the results of our ongoing project of detailed analysis of the chemical content (*chemical tagging*) as a promising powerful method to provide clear constraints on the membership of FGK kinematic candidates to stellar kinematic groups of different ages that can be used as an alternative or complementary to the methods that use kinematics, photometry or age indicators. This membership information is very important to better understand the star formation history in the solar neighborhood discerning between field-like stars (associated with dynamical resonances (bar) or spiral structure) and real physical structures of coeval stars with a common origin (debris of star-forming aggregates in the disk). We have already applied the chemical tagging method to constrain the membership of FGK candidate stars to the Hyades supercluster and the Ursa Major moving group and in this contribution we present the preliminary results of our study of the Castor moving group.

**Keywords.** Galaxy: open clusters and associations, Stars: fundamental parameters, Stars: abundances, Stars: kinematics and dynamics, Stars: late-type

## 1. Introduction

Stellar kinematic groups (SKGs) –superclusters (SCs) and moving groups (MGs)– are kinematic coherent groups of stars (Eggen 1994) that might share a common origin. Among them, the youngest SKGs are: the Hyades SC (600 Myr), the Ursa Major MG (Sirius SC, 300 Myr), the Local Association or Pleiades MG (20 to 150 Myr), the IC 2391 SC (35-55 Myr), the Castor MG (200 Myr), and Hercules-Lyra (see Montes *et al.*, 2001; López-Santiago *et al.* 2006; Klutsch *et al.* 2014 and references therein). More recently other very young SKGs (TW Hya, $\beta$ Pic, Tuc-Hor, AB Dor, Columba and Carina) have been identified with kinematics close to the Local Association and others close to other MGs like Argus to IC 2391 and Octans, Octans-Near to Castor (see Zuckerman & Song 2004; Torres *et al.* 2008; Montes 2010, 2015 and references therein). Even new associations are identified, such as the ASYA, All Sky Young Association (Torres, Quast, & Montes 2015). In order to constrain the membership of kinematic candidate stars to these SKGs additional information is needed. Recently, several methods have been developed to take into account kinematics and photometry simultaneously: SACY survey (Torres *et al.* 2008), BANYAN survey (Malo *et al.* 2013), LACEwING (Riedel 2015). Using different age indicators such as the lithium line at 6707.8 Å, the chromospheric and coronal activity level, and gyrochronology, it is possible to quantify the contamination by younger or older field stars (López-Santiago *et al.* 2010; Maldonado *et al.* 2010). However, the detailed

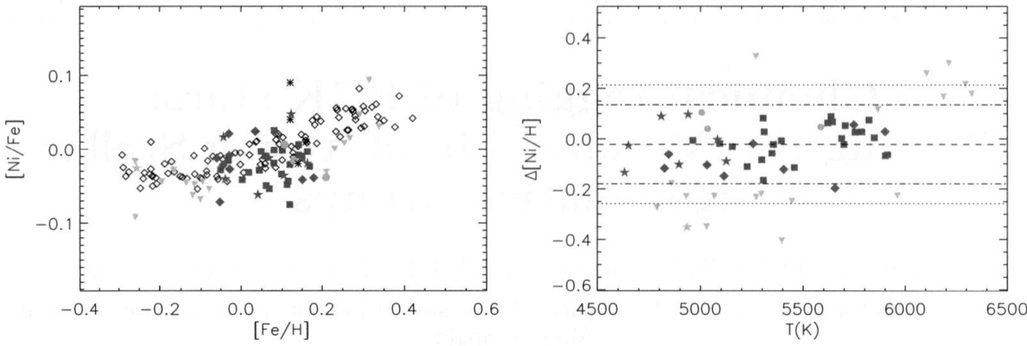

**Figure 1.** *Left panel*: [Ni/Fe] vs. [Fe/H] for the Hyades SC sample, open diamonds represent the thin disc data (González Hernández *et al.* 2010). *Right panel*: $\Delta$[Ni/H] (differential abundances for Ni) vs. $T_{\rm eff}$ for the Hyades SC sample. Dashed-dotted lines represent 1-rms over and below the median for our sample, whereas dotted lines represent the 1.5-rms level. Dashed lines represent the mean differential abundance. In both panels, red diamonds are our stars compatible to within 1-rms with the Fe abundance but not for all elements, blue squares and blue starred symbols are the candidates selected as compatible with Hyades SC membership. Green downward-pointing triangles show stars incompatible with Hyades SC membership. For more details see Tabernero *et al.* (2012).

analysis of the chemical content (*chemical tagging*) is another powerful method that provides clear constraints on the membership to these structures (Freeman & Bland-Hawthorn 2002; Tabernero *et al.* (2012) and references therein). Studies of open clusters such as the Hyades and Collinder 261 have found high levels of chemical homogeneity, showing that chemical information is preserved within the stars and that the possible effects of any external sources of pollution are negligible.

## 2. Observations and spectroscopic analysis

Spectroscopic observations were obtained at the 1.2 m Mercator Telescope at the *Observatorio del Roque de los Muchachos* (La Palma, Spain) during several observing runs. Fundamental stellar parameters ($T_{\rm eff}$, $\log g$, $\xi$, and [Fe/H]) are determined using an automatic code (StePar) which takes into account the sensitivity of iron $EW$s measured in the spectra, see Tabernero *et al.* (2012). Chemical abundances were calculated using the equivalent width ($EW$) method in a fully differential way by comparing them with those derived for a solar spectrum. A total of 20 elements were analyzed: Fe, the $\alpha$-elements (Mg, Si, Ca, and Ti), the Fe-peak elements (Cr, Mn, Co, and Ni), the odd-Z elements (Na, Al, Sc, and V) and the s-process elements (Cu, Zn, Y, Zr, Ba, Ce and Nd). We determined differential abundances $\Delta$[X/H] (*chemical tagging*) by comparing our measured abundances with those of a reference star known to be a member of the SKG under study on a line-by-line basis.

## 3. Hyades supercluster

We have applied the chemical tagging method to constrain the membership of 61 FGK candidate stars to the Hyades SC. Fig. 1 shows the results for [Ni/Fe] vs. [Fe/H] and $\Delta$[Ni/H] (differential abundances for Ni) vs. $T_{\rm eff}$. For more details see Tabernero *et al.* (2012). We found that 28 of the 61 stars analyzed have homogeneous abundances for all the elements we considered (a 46% of the studied stars) and are compatible with the

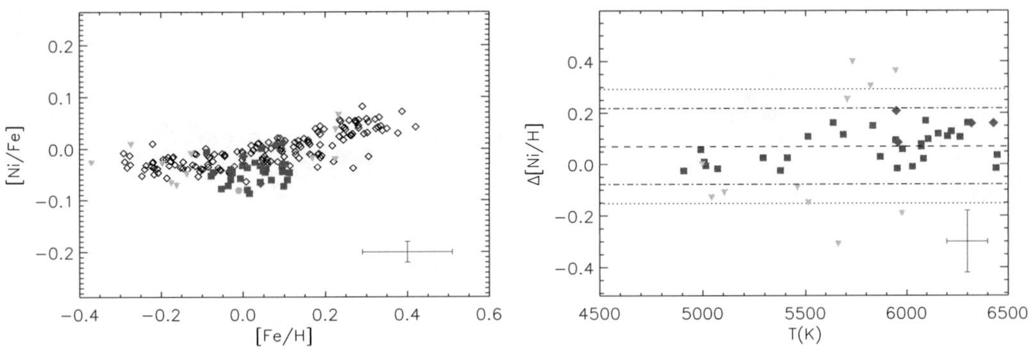

**Figure 2.** As Fig. 1 for the Ursa Major MG sample, see Tabernero *et al.* (2015).

Hyades isochrone, as expected if they have evaporated from the Hyades cluster. The large membership percentage that we find in this work (46 %) compared with those of other authors demonstrates the importance of the sample selection and a detailed chemical analysis.

## 4. Ursa Major moving group

We have applied the chemical tagging method to constrain the membership of 44 FGK candidate stars to the Ursa Major MG. Fig. 2 shows the results for [Ni/Fe] and $\Delta$[Ni/H] for the Ursa Major MG sample as in Fig.1. For more details see Tabernero *et al.* (2015). Our chemical tagging analysis indicates that the Ursa Major MG is less affected by field star contamination than other moving groups (such as the Hyades SC). We find a roughly solar iron composition [Fe/H]=0.03 ± 0.07 dex for the finally selected stars, whereas the [X/Fe] ratios are roughly sub-solar except for super-solar Barium abundance. We conclude that 29 out of 44 (i.e. 66 %) candidate stars share a similar chemical composition. In addition, we find that the abundance pattern of the Ursa Major MG is different from that of the Hyades SC.

## 5. Castor moving group

We present here the preliminary results of our chemical tagging study (Tabernero *et al.* 2015, in prep) of 19 FGK candidate stars to the Castor MG following the same method used for the Hyades SC and the Ursa Major MG (see Fig. 3). This analysis will contribute to clarify the controversial nature of this MG (see Mamajek *et al.* 2013 and Mamajek 2015) and the possible relation with the Octans and Octans-Near associations.

## 6. Conclusions

We have computed in an homogeneous way the fundamental stellar parameters and differential abundances of FGK candidate stars to different SKGs that allowed us to apply the *chemical tagging* method and constrain their membership and quantify the contamination by younger or older field stars. We have already applied this method to the Hyades SC, the Ursa Major and Castor MGs, and we plan to apply it to additional spectroscopic observations of other younger SKGs that will contribute to a better understanding of the star formation history in the solar neighborhood discerning between field population stars and real physical structures of coeval stars with a common origin. The

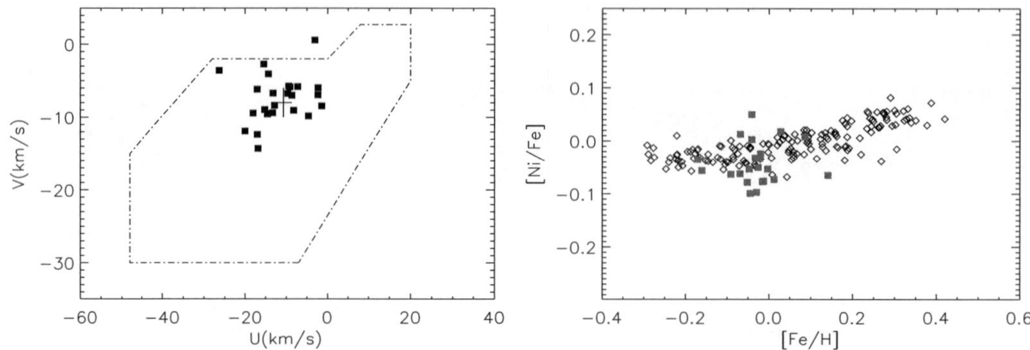

**Figure 3.** *Left panel*: *UV* galactic velocity components for the Castor MG sample. *Right panel*: [Ni/Fe] vs. [Fe/H] for the Castor MG sample, open diamonds as Fig. 1 and Fig. 2 and red squares represent our sample. For more details see Tabernero *et al.* (2015, in prep).

*chemical tagging* is therefore a powerful method to confirm known SKGs or identify new ones using large high resolution spectroscopic surveys like the Gaia-ESO survey, GES (Gilmore *et al.* 2012) and GALAH (De Silva *et al.*, 2015). See also promising approaches (Mitschang *et al.* 2013) and limitations (Blanco-Cuaresma *et al.* 2015) of the method.

## Acknowledgements

This work was supported by the Univ. Complutense de Madrid (UCM) and the Spanish Ministry of Economy and Competitiveness (MINECO) under grants AYA2011-30147-C03-02, AYA2011-29060 and Severo Ochoa Program SEV-2011-0187.

## References

Blanco-Cuaresma, S., Soubiran, C., Heiter, U., *et al.* 2015, *A&A*, 577, A47
De Silva, G. M., Freeman, K. C., Bland-Hawthorn J., *et al.* 2015, *MNRAS*, 449, 2604
Eggen, O. J. 1994, *Galactic and Solar System Optical Astrometry*, 191
Freeman, K. & Bland-Hawthorn, J. 2002, *ARA&A*, 40, 487
Gilmore, G., Randich, S., Asplund, M., *et al.* 2012, *The Messenger*, 147, 25
González Hernández, J. I., Israelian, G., Santos, N. C. *et al.* 2010, *ApJ*, 720, 1592
Klutsch, A., Freire Ferrero, R., Guillout, P., *et al.* 2014, *A&A*, 567, A52
López-Santiago, J., Montes, D., Crespo-Chacón, I., *et al.* 2006, *ApJ*, 643, 1160
López-Santiago, J., Montes, D., Gálvez-Ortiz, M. C. *et al.* 2010 *A&A*, 514, A97
Maldonado J., Martínez-Arnáiz R., Eiroa C., *et al.* 2010 *A&A*, 521, A12
Malo, L., Doyon, R., Lafrenière, D., *et al.* 2013, *ApJ*, 762, 88
Mamajek, E. E., Bartlett, J. L., Seifahrt, A., *et al.* 2013, *AJ*, 146, 154
Mamajek, E. E. 2015, *IAUS 314*, This volume, arXiv:1507.06697
Mitschang, A. W., De Silva, G., Sharma, S., & Zucker, D. B. 2013, *MNRAS*, 428, 2321
Montes, D., López-Santiago, J., Gálvez, M. C., *et al.* 2001, *MNRAS*, 328, 45
Montes, D. 2010, *IAU Special Session SpS7, Highlights of Astronomy*, 15, E1
Montes, D. 2015, *Highlights of Spanish Astrophysics VIII*, p. 606
Riedel, A. R. 2015, *IAUS 314*, This volume, arXiv:1506.06685
Tabernero, H. M., Montes, D., & González Hernández, J. I. 2012, *A&A*, 547, A13
Tabernero, H. M., Montes, D., González Hernández, *et al.* 2015, *A&A*, in press, arXiv:1409.2348
Torres, C. A. O., Quast, G. R., Melo, C. H. F., *et al.* 2008, *H. Star Form. Regions*, Vol. II, 757
Torres, C. A. O., Quast, G. R., & Montes, D. 2015, *IAUS 314*, This volume
Zuckerman, B. & Song, I. 2004, *ARA&A*, 42, 685

*Young Stars & Planets Near the Sun*
*Proceedings IAU Symposium No. 314, 2015*
*J. H. Kastner, B. Stelzer, & S. A. Metchev, eds.*

© International Astronomical Union 2016
doi:10.1017/S1743921315006213

# The Isochronal Age Scale of Young Moving Groups in the Solar Neighbourhood

**Cameron P. M. Bell[1], Eric E. Mamajek[1] and Tim Naylor[2]**

[1]Department of Physics & Astronomy, University of Rochester, Rochester, NY 14627, USA
email: `cbell@pas.rochester.edu`
[2]School of Physics and Astronomy, University of Exeter, Exeter EX4 4QL, UK

**Abstract.** We present a self-consistent, absolute isochronal age scale for young ($\lesssim 200\,\mathrm{Myr}$), nearby ($\lesssim 100\,\mathrm{pc}$) moving groups, which is consistent with recent lithium depletion boundary ages for both the $\beta$ Pic and Tucana-Horologium moving groups. This age scale was derived using a set of semi-empirical pre-main-sequence model isochrones that incorporate an empirical colour-$T_{\mathrm{eff}}$ relation and bolometric corrections based on the observed colours of Pleiades members, with theoretical corrections for the dependence on $\log g$. Absolute ages for young, nearby groups are vital as these regions play a crucial role in our understanding of the early evolution of low- and intermediate-mass stars, as well as providing ideal targets for direct imaging and other measurements of dusty debris discs, substellar objects and, of course, extrasolar planets.

**Keywords.** stars: evolution – stars: formation – stars: pre-main sequence – stars: fundamental parameters – techniques: photometric – solar neighbourhood – open clusters and associations: general – Hertzsprung-Russell and C-M diagrams

## 1. Introduction

Over the past couple of decades several hundred low- and intermediate-mass stars have been identified within $\simeq 100\,\mathrm{pc}$ of the Sun as part of a concerted effort to identify and characterise the young solar neighbourhood. The distribution of these stars on the sky is not uniform, instead they comprise dispersed, predominantly unbound associations in which the members share a common space motion. Although the age ordering of these young associations is well-constrained i.e. the AB Dor moving group is older than the Tucana-Horologium moving group (Tuc-Hor), which in turn is older than the $\beta$ Pic moving group (BPMG), the absolute ages of these groups are still under-constrained.

Theoretical model isochrones are (arguably) the most commonly used method of age-dating young (presumably coeval) stellar populations. Although the use of maximum-likelihood fitting techniques to subjectively fit model isochrones has been used for more distant Galactic clusters (e.g. NGC 2547 in Naylor & Jeffries 2006), such methods have not yet been applied to the young groups within $\simeq 100\,\mathrm{pc}$. Additionally, for these young groups there has been a distinct lack of homogeneity when it comes to the fitting of model isochrones, particularly in terms of the models adopted and the photometric bandpasses used to construct the colour-magnitude diagrams (CMDs). Furthermore, recent lithium depletion boundary (LDB) ages, which are advocated in the recent review of young stellar ages by Soderblom *et al.* (2014) to provide our best chance of establishing a *reliable* and *robust* age scale in the range $20 - 200\,\mathrm{Myr}$, have been demonstrated to be systematically older when compared to isochronal ages (see e.g. Kraus *et al.* 2014).

In this contribution we provide a self-consistent, absolute isochronal age scale for young ($\lesssim 200\,\mathrm{Myr}$) moving groups within $100\,\mathrm{pc}$ of the Sun. This age scale is based on homogeneous fitting of photometric data in the $M_V, V - J$ CMD using a maximum-likelihood fitting statistic in conjunction with semi-empirical model isochrones. We find that the isochronal ages for both the BPMG and Tuc-Hor are consistent with recent LDB ages, thereby instilling confidence in our isochronal age scale.

## 2. Sample of young, nearby moving groups

In this contribution we focus on the young, nearby moving groups within 100 pc; specifically the AB Dor moving group, Argus association, BPMG, Carina association, Columba association, $\eta$ Cha cluster, Tuc-Hor, TW Hya association (TWA), and 32 Ori group (see also the recent review by Torres $et\ al.$ 2008). For each group, we have assembled a list of members and candidate members (hereafter simply referred to together as 'members') from the literature. For the inclusion of candidate members, we require that they have a membership probability of $\geq$ 90% as calculated using the Bayesian Analysis for Nearby Young AssociatioNs (BANYAN; see e.g. Malo $et\ al.$ 2013). Furthermore, we require that any candidate members have a measured radial velocity and/or trigonometric parallax which are/is consistent with membership for a given young group.

Our sample of members (not accounting for unresolved multiples) includes 89 members of the AB Dor moving group, 27 members of Argus, 97 members of the BPMG, 12 members of Carina, 50 members of Columba, 18 members of $\eta$ Cha, 189 members of Tuc-Hor, 30 members of TWA, and 14 members of 32 Ori.

Whilst all of our members have counterparts in the Two-Micron All-Sky Survey Point Source Catalog (2MASS PSC; Cutri $et\ al.$ 2003), near-IR CMDs are not ideal for age determination primarily because for $2500 \lesssim T_{\rm eff} \lesssim 4000\,{\rm K}$ the loci of young clusters becomes vertical ($J - K_{\rm s} \simeq 0.9\,{\rm mag}$), and hence degenerate with age. Therefore we supplement our near-IR photometry with $V$-band data and derive ages for our sample of groups in the $M_V, V - J$ CMD so as to minimise the effects of circumstellar material on the $K_{\rm s}$ magnitudes for the youngest groups (especially $\eta$ Cha and TWA).

For assigning distances to each star in our list of members we prefer to adopt trigonometric parallax measurements, however when this is not an option (or where the parallax estimates appear to be in error or have unusually large uncertainties) we either adopt a kinematic distance from the literature or derive one using the 'moving cluster' method (see e.g. Mamajek 2005).

## 3. Semi-empirical model isochrones

There are large discrepancies between pre-main-sequence (pre-MS) tracks and observational data (such as dynamical masses, age trends, etc.; see e.g. Hillenbrand & White 2004; Feiden this volume). In Bell $et\ al.$ (2012) we discussed the reasons why the Pleiades represents an ideal benchmark cluster for the comparison of the models to the data, however chief among these is the fact that both the distance and age have been determined independently of the use of model isochrones. Fig. 1, which shows the $V, V - I_{\rm c}$ CMD of the Pleiades with several commonly adopted sets of model isochrones overlaid, effectively demonstrates the aforementioned discrepancies. At $T_{\rm eff} \lesssim 4300\,{\rm K}$, the models overestimate the flux in the optical by up to a factor of two. Hence, if we are to derive consistent ages for young stellar populations using pre-MS model isochrones we must first perform some form of empirical correction to the models for $T_{\rm eff}$ lower than $\simeq 4300\,{\rm K}$.

In Bell $et\ al.$ (2013, 2014) we introduced a method of creating semi-empirical pre-MS model isochrones using the observed colours of young stars in the Pleiades, in addition to incorporating theoretical corrections for the dependence on the surface gravity ($\log g$). Briefly, we used binary systems with well-constrained dynamical masses to demonstrate

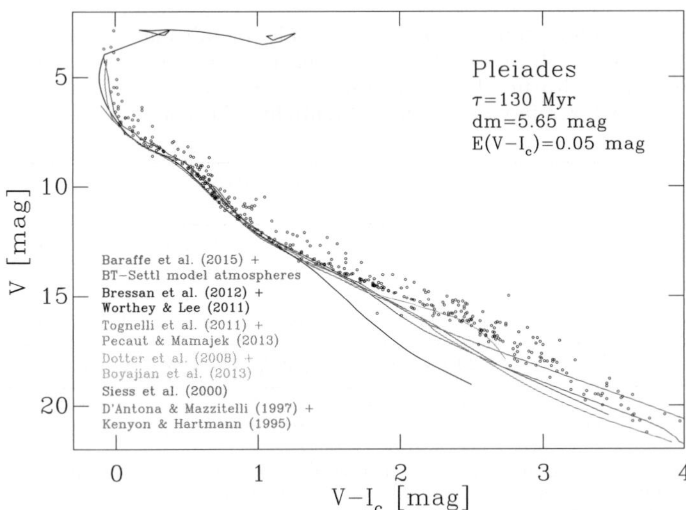


**Figure 1.** The $V, V - I_c$ CMD of the Pleiades with several sets of model isochrones overlaid. The upper reference refers to the interior models, whereas the lower reference corresponds to the colour-$T_{\rm eff}$ relation and bolometric corrections adopted to transform the model into CMD space.

that the model predictions for the $K_{\rm s}$-band are essentially correct. This then allows us to use the $K_{\rm s}$-band magnitude along our fiducial locus as a $T_{\rm eff}$ indicator, but also ensures that our semi-empirical model isochrones have a reliable mass scale tied to that of the fiducial cluster. On the assumption that the $K_{\rm s}$-band magnitudes predicted by the models are correct, we can then use CMDs with the $K_{\rm s}$-band on both axes (e.g. $K_{\rm s}, V - K_{\rm s}$) to calculate any discrepancy between the fiducial locus and the model isochrone, which in this case would correspond to the requisite empirical correction in the $V$-band necessary to fit the Pleiades at a given age and distance. This empirical correction is then applied to the theoretical bolometric correction grid (calculated from atmospheric models and a function of $T_{\rm eff}$ and $\log g$) at the appropriate $T_{\rm eff}$ irrespective of its $\log g$. This process is then repeated for other photometric bandpasses to create sets of semi-empirical bolometric corrections. The semi-empirical pre-MS model isochrones used in this contribution are based on existing stellar interior models coupled with these newly derived semi-empirical bolometric corrections. In this contribution we adopt four sets of interior models, namely those of Dotter *et al.* (2008), Tognelli *et al.* (2011), Bressan *et al.* (2012), and Baraffe *et al.* (2015; hereafter referred to as Dartmouth, Pisa, PARSEC, and BHAC15 respectively)†.

## 4. Isochronal age scale for young, nearby groups

We use the $\tau^2$ fitting statistic of Naylor & Jeffries (2006) and Naylor (2009) to derive ages from the $M_V, V - J$ CMDs of our sample of young groups. The $\tau^2$ fitting statistic is particularly well-suited to this task because it allows for the effects of binarity, yields reliable uncertainties on the derived parameters and provides a goodness-of-fit test. We have made some modifications to the statistic since those of Naylor (2009), however for the purposes of this contribution we refrain from detailing these and instead refer the reader to Bell *et al.* (2015).

† A subset of these models is available via the Cluster Collaboration isochrone server
`http://www.astro.ex.ac.uk/people/timn/isochrones/`

**Table 1.** Ages for the young groups in our sample. The penultimate row lists our final adopted age for each group for which the associated uncertainties represent the statistical and systematic uncertainties added in quadrature. The final row shows literature LDB ages for the BPMG and Tuc-Hor (see Section 5 for references) which highlights the consistency between the two age diagnostics.

| Model | Group age (Myr) | | | | | | | | |
|---|---|---|---|---|---|---|---|---|---|
| | AB Dor | Argus[1] | BPMG | Carina | Columba | $\eta$ Cha | Tuc-Hor | TWA | 32 Ori |
| BHAC15 | $145^{+889}_{-5}$ | $69^{+19}_{-8}$ | $25 \pm 1$ | $46^{+16}_{-3}$ | $44^{+8}_{-3}$ | $12 \pm 1$ | $46^{+1}_{-2}$ | $10 \pm 1$ | $23^{+3}_{-2}$ |
| Dartmouth | $135^{+15}_{-9}$ | $60^{+64}_{-3}$ | $23 \pm 1$ | $49^{+13}_{-5}$ | $43^{+8}_{-3}$ | $10 \pm 1$ | $48^{+1}_{-2}$ | $7^{+2}_{-1}$ | $20^{+5}_{-1}$ |
| PARSEC | $151^{+24}_{-18}$ | $60^{+268}$ | $25^{+1}_{-2}$ | $45^{+7}_{-12}$ | $43^{+3}_{-4}$ | $14 \pm 1$ | $46 \pm 1$ | $13 \pm 1$ | $25^{+2}_{-5}$ |
| Pisa | $166^{+74}_{-23}$ | $55^{+235}_{-5}$ | $20^{+1}_{-2}$ | $41^{+8}_{-6}$ | $38 \pm 3$ | $8 \pm 1$ | $40^{+1}_{-2}$ | $9 \pm 1$ | $18 \pm 1$ |
| **Adopted** | $\mathbf{149^{+51}_{-19}}$ | – | $\mathbf{24 \pm 3}$ | $\mathbf{45^{+11}_{-7}}$ | $\mathbf{42^{+6}_{-4}}$ | $\mathbf{11 \pm 3}$ | $\mathbf{45 \pm 4}$ | $\mathbf{10 \pm 3}$ | $\mathbf{22^{+4}_{-3}}$ |
| LDB age | – | – | $24 \pm 5$ | – | – | – | $40 \pm 3$ | – | – |

*Notes:*
[1] No final adopted age is given as it is unclear whether the stars in our list of members represent a single population of coeval stars (see Section 5). Note also that the PARSEC models only provide an upper limit on the age of the association and therefore we do not provide a lower age uncertainty in the table.

In essence the $\tau^2$ fitting statistic can be viewed as a generalisation of the $\chi^2$ statistic to two dimensions on the basis that both the model isochrone and photometric data are two-dimensional distributions; specifically uncertainties in colour and magnitude for the data and the widening of the model isochrone due to the effects of binarity. As in Naylor & Jeffries (2006), we define the $\tau^2$ statistic as

$$\tau^2 = -2 \sum_{i=1,\,N} \ln \iint U_i(c - c_i, m - m_i)\, \rho(c, m)\, \mathrm{d}c\,\mathrm{d}m, \qquad (4.1)$$

where $U_i$ represents the two-dimensional uncertainty function and $\rho$ the probability distribution of the expected model in CMD space. We create our model distributions using a Monte Carlo method to simulate $10^6$ stars over a given mass range and populate these using a broken power law mass function. For the inclusion of binaries we assume a uniform fraction of 50%. Note that the best-fit age is remarkably insensitive to the adopted binary fraction. Our full grid of models range from $\log(\text{age}) = 6.0 - 10.0$ in steps of $\Delta\log(\text{age}) = 0.01\,\text{dex}$. For each model/data combination, we then calculate the resultant $\tau^2$ until we find the model which minimises the above function. This is equivalent to maximising the 'overlap' between the data points and the model. Fig. 2 illustrates an example of the best-fitting $M_V, V - J$ CMDs of Tuc-Hor, and Table 1 shows the best-fit ages for each young group in our sample.

As part of our modification to the statistic, we now assign prior membership probabilities to each individual star in a given catalogue. Consequently, our best-fitting model returns posterior membership probabilities. For a given group in our sample, we expect a negligible age spread (or equivalently luminosity spread), and hence we can then use these posteriors to identify stars which appear to be non-members (based solely on CMD position) in an effort to further refine the membership lists of these young group.

## 5. Discussion

To assess the reliability of our self-consistent, absolute isochronal age scale, we must first compare our derived ages to what are considered well-constrained ages for the same groups. Model-independent methods (such as kinematic 'traceback' or 'expansion' ages) should offer such a comparison, however studies have demonstrated that the ages inferred

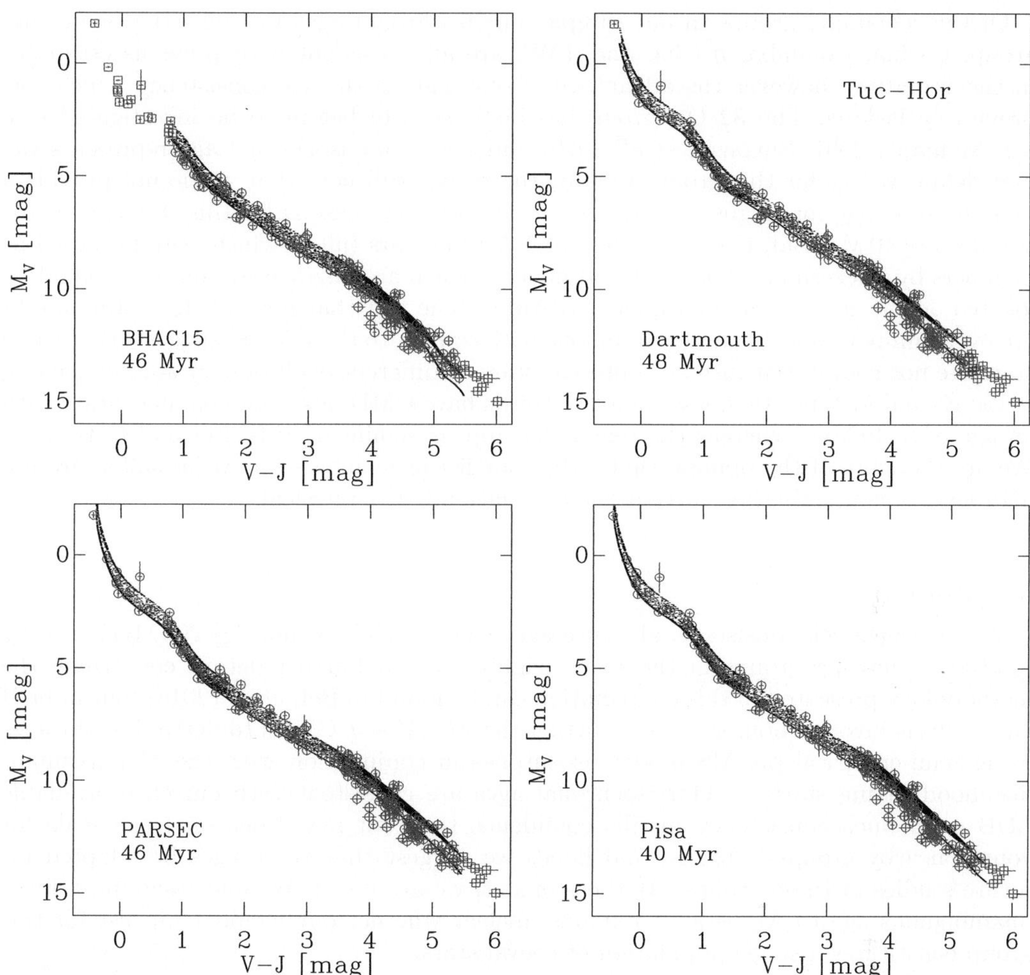

**Figure 2.** Best-fitting $M_V$, $V - J$ CMDs of Tuc-Hor. The red circles represent fitted data, whereas the blue squares denote objects which are removed prior to fitting as they lie outside the area of CMD space covered by the grid of models. **Top left:** BHAC15. **Top right:** Dartmouth. **Bottom left:** PARSEC. **Bottom right:** Pisa.

from kinematic information alone are simply too unreliable and even unreproducible when similar analyses have been performed using improved astrometric data (see e.g. Mamajek & Bell 2014). LDB ages have been argued to be both *accurate* and *precise* to just a few Myr (see the discussion in Soderblom *et al.* 2014), and have recently been calculated for both the BPMG and Tuc-Hor‡. The LDB in the BPMG has been identified by both Binks & Jeffries (2014) and Malo *et al.* (2014b), who derive an age consistent with $24 \pm 5$ Myr. Note that our age for the BPMG is in excellent agreement with the $22 \pm 3$ Myr derived by Mamajek & Bell (2014) which was based on an analysis of the A-, F- and G-type members of the group. In addition, Kraus *et al.* (2014) identified the LDB in Tuc-Hor and derived an age commensurate with $40 \pm 3$ Myr. Our isochronal ages for both of these groups are consistent with the LDB ages.

‡ Recent analyses (see e.g. Somers this volume) have demonstrated that LDB ages may be underestimated by $\simeq 10 - 20\%$ as a result of neglecting the effects of starspots in the evolution of pre-MS stars.

Of the remaining groups in our sample, our isochronal ages for the AB Dor moving group, Carina, Columba, $\eta$ Cha, and TWA are all consistent with previous estimates in the literature, however the advantage of our scale is the self-consistency which was previously lacking. The 32 Ori group has only recently begun to be investigated (see e.g. Mamajek 2007; Shvonski *et al.* 2010) and hence our isochronal age represents the first definitive age for this group. Finally, the reader will note that we do not provide a final adopted age for Argus, and there are two primary reasons for this. First, if Argus is indeed $\simeq 40\,$Myr-old, then the group of 5 A-type stars (all of which were proposed as members by Zuckerman *et al.* 2011 and none of which are unresolved binaries) should be located on the zero-age main-sequence (ZAMS). The fact that none of these stars are (4 are over-luminous and 1 is under-luminous with respect to the ZAMS) suggests that these stars are not coeval, but instead represent stars at different evolutionary stages. Second, of the K- and M-type stars, less than two thirds have CMD positions commensurate with an age of $\simeq 40\,$Myr, whereas the remainder appear significantly fainter and thus older. We are therefore of the opinion that either our list of members for Argus suffers from a high level of contamination or that the association is not physical.


## 6. Summary

We present a self-consistent, absolute age scale for eight young ($\lesssim 200\,$Myr), nearby ($\lesssim 100\,$pc) moving groups in the solar neighbourhood. Further details concerning the methodology presented in this contribution can be found in Bell *et al.* (2015), but in brief our analysis involves homogeneously fitting the $M_V, V - J$ CMDs to derive best-fit ages using semi-empirical pre-MS models isochrones in conjunction with the $\tau^2$ maximum-likelihood fitting statistic. Our isochronal ages are consistent with currently available LDB ages. Such consistency instills confidence that our new isochronal age scale for young, nearby groups is robust and hence we suggest that these ages be adopted for future studies of these groups. At the moment, we are uncomfortable assigning a final, unambiguous age to Argus as it remains unclear whether our membership list for this group constitutes a single population of coeval stars.


## References

Baraffe, I., Homeier, D., Allard, F., & Chabrier, G. 2015, *A&A*, 577, 42
Bell, C. P. M., Naylor, T., Mayne, N. J., *et al.* 2012, *MNRAS*, 424, 3178
Bell, C. P. M., Naylor, T., Mayne, N. J., *et al.* 2013, *MNRAS*, 434, 806
Bell, C. P. M., Rees, J. M., Naylor, T., *et al.* 2014, *MNRAS*, 445, 3496
Bell, C. P. M., Mamajek, E. E., & Naylor, T. 2015, submitted to *MNRAS*
Binks, A. S. & Jeffries, R. D. 2014, *MNRAS*, 438, L11
Bressan, A., Marigo, P., Girardi, L., *et al.* 2012, *MNRAS*, 427, 127
Cutri, R. M., Skrutskie, M. F., van Dyk, S., *et al.* 2003, *VizieR Online Data Catalog: 2MASS All-Sky Catalog of Point Sources*, 2246, 0
Dotter, A., Chaboyer, B., Jevremović, D., *et al.* 2008, *ApJS*, 178, 89
Hillenbrand, L. A. & White, R. J. 2004, *ApJ*, 604, 741
Kraus, A. L., Shkolnik, E. L., Allers, K. N., & Liu, M. C. 2014, *AJ*, 147, 146
Malo, L., Doyon, R., Lafrenière, D., *et al.* 2013, *ApJ*, 762, 88
Malo, L., Artigau, É., Doyon, R., *et al.* 2014a, *ApJ*, 788, 81
Malo, L., Doyon, R., Feiden, G. A., *et al.* 2014b, *ApJ*, 792, 37
Mamajek, E. E. 2005, *ApJ*, 634, 1385
Mamajek, E. E. 2007, *IAU Symposium 237, Triggered Star Formation in a Turbulent ISM*, 442
Mamajek, E. E. & Bell, C. P. M. 2014, *MNRAS*, 445, 2169

Naylor, T. & Jeffries, R. D. 2006, *MNRAS*, 373, 1251

Naylor, T. 2009, *MNRAS*, 399, 432

Shvonski, A. J., Mamajek, E. E., Meyer, M. R., & Kim, J. S. 2010, *Bulletin of the American Astronomical Society*, *Vol. 42*, #428.22

Soderblom, D. R., Hillenbrand, L. A., Jeffries, R. D., *et al.* 2014, *Protostars and Planets VI*, University of Arizona Press, 219

Tognelli, E., Prada Moroni, P. G., & Degl'Innocenti, S. 2011, *A&A*, 533, 109

Torres, C. A. O., Quast, G. R., Melo, C. H. F., & Sterzik, M. F. 2008, *Handbook of Star Forming Regions, Volume II: The Southern Sky*, 757

Zuckerman, B., Rhee, J. H., Song, I., & Bessell, M. S. 2011, *ApJ*, 732, 61

## Discussion

G. CHABRIER: More of a (nasty) comment than a question, sorry. I'm not sure I believe your age scale given the methodology of creating the semi-empirical model isochrones you have described. First, some of the models you have adopted in your analysis have internal inconsistencies with regard to the interior physics (e.g. the stellar interior/atmospheric boundary condition) and should therefore not be adopted in the first place. Second, you derive semi-empirical $T_{\rm eff}$-bolometric correction relations at the age of the Pleiades and then apply these at younger ages, however there is no reason to believe that such a relation will be valid in this age regime. Third, your models rest almost entirely on a well-constrained age (and distance) for the Pleiades.

C. BELL: I appreciate your concerns, however how are we to determine the ages of young stellar populations if none of the models fit the data? Given this discrepancy the *only* option is to make some sort of empirical correction to the models. I'll answer your points in reverse order. First, the Pleiades does have a well-constrained age of $\simeq 130\,{\rm Myr}$ which has been derived from both the high- and low-mass population using distinct techniques and methods that rely on different aspects of stellar evolution, and so I believe the age we adopt is robust. Second, true we do derive the $T_{\rm eff}$-bolometric correction relation at $130\,{\rm Myr}$, however we attempt to account for the $\log g$ dependence using information from the atmospheric models; specifically how the bolometric corrections vary as a function of $\log g$. Third, again true, the majority of models do suffer from internal inconsistencies, however even if we adopt those with fully consistent treatments (such as the most recent BHAC15 models), there is still a large discrepancy between the models and the data for the Pleiades at low $T_{\rm eff}$. Furthermore, any internal inconsistencies are effectively 'tuned out' as a result of the empirical corrections we derive and apply to the theoretical $T_{\rm eff}$-bolometric correction relation, and so these should not have a significant effect on the derived ages.

B. ZUCKERMAN: I'm surprised at the older ages you derive for Argus. I respect Beto Torres and collaborators enough to find their age of $\sim 40\,{\rm Myr}$ hard to reconcile with those you derive.

C. BELL: I agree, it is perturbing. Interestingly, if we take the Argus members from Torres *et al.* (2008) and run these through the BANYAN analysis tool we find that only 11 of the 29 stars (38%) appear to be high-probability ($\geqslant 90\%$) candidate members. Furthermore, of these, several have calculated kinematic distances which differ from those given in Torres *et al.* by more than $2 - 3\sigma$. Whether this is highlighting an underlying issue with the Torres *et al.* membership list or the BANYAN kinematic distances remains unclear.

J. GAGNÉ: On a related note, are the members you adopt for Argus taken directly from the BANYAN analyses of Malo *et al.*?

C. BELL: Yes, our membership list for Argus is comprised from the studies of Malo *et al.* (2013, 2014a,b). However, we *only include* candidate members if they have a measured diagnostic (e.g. radial velocity or trigonometric parallax) which is consistent with Argus membership and *do not include* the other high-probability candidate members with no such measured diagnostic.

*Young Stars & Planets Near the Sun*
Proceedings IAU Symposium No. 314, 2015
J. H. Kastner, B. Stelzer, & S. A. Metchev, eds.

© International Astronomical Union 2016
doi:10.1017/S1743921315006134

# The BANYAN All-Sky Survey for Brown Dwarf Members of Young Moving Groups

Jonathan Gagné[1], David Lafrenière[1], René Doyon[1],
Jacqueline K. Faherty[2,3,4], Lison Malo[6,1], Kelle L. Cruz[3,5],
Étienne Artigau[1], Adam J. Burgasser[7], Marie-Eve Naud[1],
Sandie Bouchard[1], John E. Gizis[8] and Loïc Albert[1]

[1]Institut de Recherche sur les Exoplanètes (iREx), Université de Montréal, Département de Physique, C.P. 6128 Succ. Centre-ville, Montréal, QC H3C 3J7, Canada.
email: jonathan.gagne@astro.umontreal.ca
[2]Department of Terrestrial Magnetism, Carnegie Institution of Washington, DC 20015, USA
[3]Department of Astrophysics, AMNH, Central Park West at 79th Street, New York, NY 10024
[4]Hubble Fellow
[5]Department of Physics & Astronomy, Hunter College, City University of New York, 695 Park Avenue, NY 10065, USA
[6]Canada-France-Hawaii Telescope, 65-1238 Mamalahoa Hwy, Kamuela, HI 96743, USA
[7]CASS, UCSD, 9500 Gilman Dr., Mail Code 0424, La Jolla, CA 92093, USA
[8]Dept. of Physics and Astronomy, Univ. of Delaware, 104 The Green, Newark, DE 19716, USA

**Abstract.** We describe in this work the *BASS* survey for brown dwarfs in young moving groups of the solar neighborhood, and summarize the results that it generated. These include the discovery of the 2MASS J01033563–5515561 (AB)b and 2MASS J02192210–3925225 B young companions near the deuterium-burning limit as well as 44 new low-mass stars and 69 new brown dwarfs with a spectroscopically confirmed low gravity. Among those, $\sim 20$ have estimated masses within the planetary regime, one is a new L4$\gamma$ bona fide member of AB Doradus, three are TW Hydrae candidates with later spectral types (L1–L4) than all of its previously known members and six are among the first contenders for low-gravity $\geqslant$ L5 $\beta/\gamma$ brown dwarfs, reminiscent of WISEP J004701.06+680352.1, PSO J318.5338–22.8603 and VHS J125601.92–125723.9 b. Finally, we describe a future version of this survey, *BASS-Ultracool*, that will specifically target $\geqslant$ L5 candidate members of young moving groups. First experimentations in designing the survey have already led to the discovery of a new T dwarf bona fide member of AB Doradus, as well as the serendipitous discoveries of an L9 subdwarf and an L5 + T5 brown dwarf binary.

**Keywords.** brown dwarfs — methods: data analysis — proper motions — stars: kinematics and dynamics — stars: low-mass

## 1. Introduction

Young Moving Groups (YMGs) are ideal laboratories to study the low-mass end of the initial mass function and obtain age-calibrated samples of brown dwarfs to understand the evolution of their fundamental and atmospheric properties over time. However, the identification of substellar members of YMGs is challenging because of their intrinsic faintness and sparse distribution on the sky. These YMGs include e.g. the TW Hydrae association (TWA; 5–15 Myr; de la Reza *et al.* 1989; Kastner *et al.* 1997), Tucana-Horologium (THA; 20–40 Myr; Torres *et al.* 2000) and AB Doradus (ABDMG; 110–130 Myr; Zuckerman *et al.* 2004).

Malo *et al.* (2013) developed the Bayesian Analysis for Nearby Young AssociatioNs (BANYAN) tool to identify new low-mass star members of YMGs in the solar neighborhood. The biggest power of this tool is that its use of Bayes' theorem allows it to

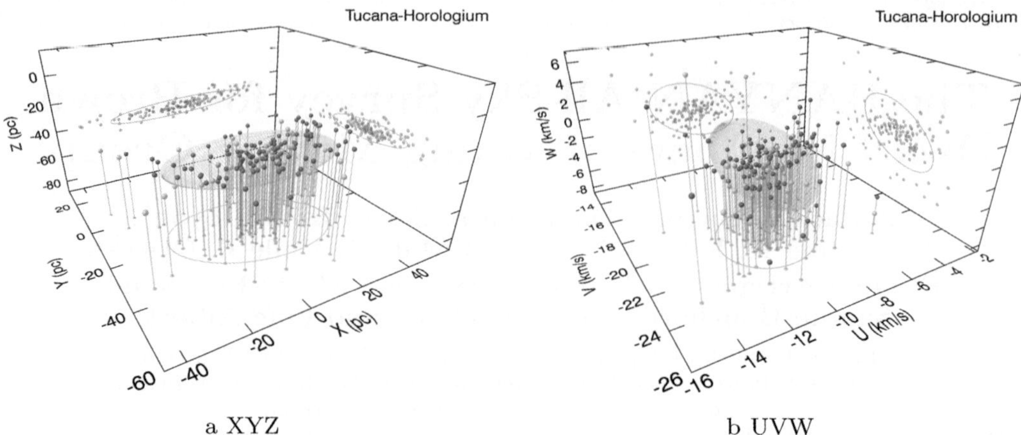



**Figure 1.** Most probable galactic positions XYZ and space velocities UVW based on BANYAN II statistical distances and RVs for all *BASS* candidate members of THA (red points), compared with bona fide members (green points), as well as the spatial and kinematic ellipsoid models used in BANYAN II (orange ellipsoids; see Gagné *et al.* 2014a). All points and models are projected on the three normal planes for a better clarity.

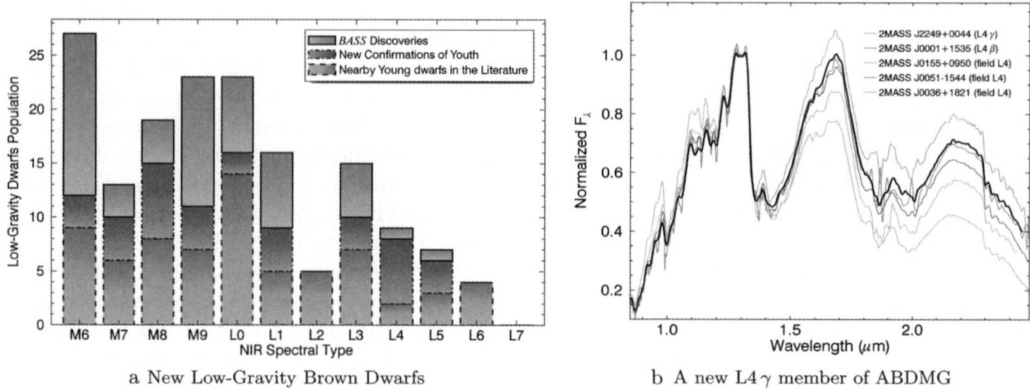



**Figure 2.** Panel a : Near-infrared spectral types for all known low-gravity dwarfs and those presented in this work. The *BASS* survey has contributed significantly in increasing the number of known low-gravity M6–L5 dwarfs. Panel b : Near-infrared spectrum of a L4 γ ABDMG bona fide member, compared with various field and low-gravity L4 brown dwarfs. Its *H*-band continuum has a triangular shape and a red continuum, which are signs of a low gravity.

derive membership probabilities even in scenarios where not all kinematic information is available. In particular, radial velocity and distance measurements are not available for most low-mass stars, which is primarily due to their relative faintness.

Our team built on this tool to create BANYAN II (Gagné *et al.* 2014a), an updated version of the tool that is designed to search for brown dwarfs, contains improved models of the field and of YMGs, and is more efficient such that it can be applied to large data sets. This allowed us to identify 24 new candidate members of YMGs as well as a new M7.5 bona fide member of Tucana-Horologium among known low-gravity brown dwarfs.

## 2. The *BASS* Survey

Equipped with the BANYAN II tool, we have undertaken the *BANYAN All-Sky Survey* (*BASS*; Gagné *et al.* 2015a), a survey for new brown dwarf candidate members of YMGs from a cross-match of the *Two Micron All-Sky Survey* (*2MASS*; Skrutskie *et al.* 2006) and the *AllWISE* catalog based on the *WISE* survey (Wright *et al.* 2010; Kirkpatrick *et al.* 2014) outside of the galactic plane. We used this cross-match to calculate proper motions with a typical precision of $\sim 15\,\mathrm{mas\,yr^{-1}}$ and to built an input list of 98 970 high-proper motion objects with colors that are consistent with $\geqslant$ M5 dwarfs in the solar neighborhood. We then used the BANYAN II tool to identify 228 new late-type candidate members of YMGs, including 79 potential brown dwarfs and 22 potential isolated planetary-mass objects. A cross-match with the literature allowed us to estimate our rate of false positives at around $\sim 13\%$ in this survey. An additional list of 275 candidates (the *Low-Priority BASS*, or *LP-BASS*, sample) was generated with the same method except with more permissive color cuts.

In Figure 1, we display the most probable galactic positions (XYZ) and space velocities (UVW) for all candidate members of THA that we identified in the *BASS* survey. The BANYAN II tool outputs the most likely statistical distances and radial velocities for each target, which allowed us to derive their most probable XYZ and UVW. We note that our candidate members are likely located in the same loci as bona fide members of THA. In the XYZ plane, they also follow the plane-shaped distribution of THA members, a property of THA members that was first observed by Kraus *et al.* (2014).

As part of this survey, we have identified a low-gravity M6 candidate member of THA with a 12–14 $M_{\mathrm{Jup}}$ companion at a separation of 84 AU (Delorme *et al.* 2013), as well as an additional low-gravity M6 $\gamma$ + L4 $\gamma$ system (with a separation of 160 AU) also in THA (Artigau *et al.* 2015). We also identified the latest-type L dwarf candidate member of TWA (Gagné *et al.* 2014b) and a $\sim 10\,M_{\mathrm{Jup}}$ low-gravity mid-L candidate member of Argus (Gagné *et al.* 2014c) in the first stage of the near-infrared spectroscopic follow-up of the *BASS* sample. More recently, we presented the spectroscopic follow-up of a significant portion of *BASS* (Gagné *et al.* 2015c, in press), revealing 44 new low-mass stars and 69 brown dwarfs with spectroscopic confirmations of low gravity (Figure 2a). A new L4 $\gamma$ bona fide member of AB Doradus was also discovered as part of this work (see Figure 2b).

We used these new discoveries and known low-gravity brown dwarfs to create new color-magnitude diagrams (e.g., Figure 3b) and investigate their fundamental properties using the BT-Settl atmosphere models (Allard *et al.* 2013). We identified an unexpectedly large number of 12–14.5 $M_{\mathrm{Jup}}$ spectroscopically confirmed low-gravity candidate members of THA, where our current follow-up is most sensitive, compared with its main-sequence stellar population (Figure 3a). This represents an over-density of a factor $36.4^{+16.6}_{-12.5}$. We find that systematic errors in the atmosphere models are not likely to cause such an over-density, and that the most likely explanation would be a combination of small number statistics and young interlopers from YMGs other than THA. It will be necessary to obtain radial velocity and parallax measurements to assess whether this is the case or if THA really has an up-turn in the low-mass end of its IMF.

We are currently working on an updated version of BANYAN II that will include additional young associations with slightly older ages (up to $\sim 500\,\mathrm{Myr}$) and combine this with a cross-match of *2MASS*, *AllWISE*, *SDSS* and *UKIDSS* to identify new $\geqslant$ L5 young brown dwarfs, up into the T dwarfs regime. Our first tests in constructing this survey have already allowed us to identify a new T dwarf bona fide member in AB Doradus. It

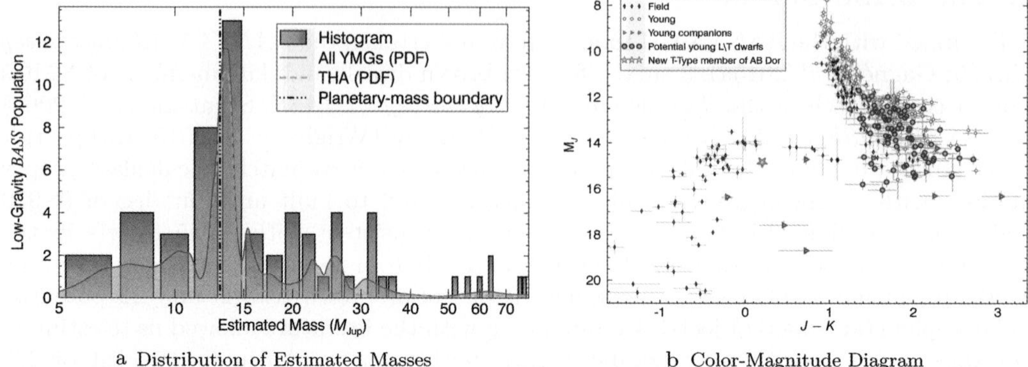

a   Distribution of Estimated Masses      b   Color-Magnitude Diagram

**Figure 3.** Panel a : Estimated masses for low-gravity dwarfs in our sample. The continuous probability density functions (PDFs) provide a histogram-like continuous distribution that include measurement errors and are independent of binning. Panel b : MKO $J - K$ versus $M_J$ color–magnitude diagram for field and low-gravity brown dwarfs. A few known substellar companions, new YMG candidates at the L/T transition from a preliminary version of *BASS-Ultracool* and a new bona fide T dwarf member of ABDMG are also shown.

is shown in a color-magnitude diagram in Figure 3b; this result is presented in Gagné *et al.* (2015b).

## 3. Conclusions

We reviewed in this work some discoveries from the BANYAN All-Sky Survey that aims at the identification of new low-gravity brown dwarf members of moving groups in the solar neighborhood, and outlined future plans to locate low-gravity T dwarfs in the upcoming years. These discoveries will serve as important benchmarks to understand the atmospheric properties of giant, gaseous exoplanets and to characterize the low-mass end of the initial mass function.

## References


Allard, F., Homeier, D., Freytag, B., & Schaffenberger, Rajpurohit, A. S. 2013, Mem. Soc. Astron. Ital., 24, 128

Artigau, É., Gagné, J., Faherty, J., *et al.* 2015, The Astrophysical Journal, 806, 254

Delorme, P., Gagné, J., Girard, J. H., *et al.* 2013, Astronomy & Astrophysics, 553, L5

Gagné, J., Lafrenière, D., Doyon, R., Malo, L., & Artigau, É. 2014a, The Astrophysical Journal, 783, 121

Gagné, J., Faherty, J. K., Cruz, K. L., *et al.* 2014b, The Astrophysical Journal Letters, 785, L14

Gagné, J., Lafrenière, D., Doyon, R., *et al.* 2014c, The Astrophysical Journal Letters, 792, L17

Gagné, J., Lafrenière, D., Doyon, R., Malo, L., & Artigau, É. 2015a, The Astrophysical Journal, 798, 73

Gagné, J., Burgasser, A. J., Faherty, J. K., *et al.* 2015b, The Astrophysical Journal Letters, 808, L20

Gagné, J., Faherty, J. K., Cruz, K. L., *et al.* 2015c, The Astrophysical Journal Supplements, in press. (arXiv:1506.07712)

Kastner, J. H., Zuckerman, B., Weintraub, D. A., & Forveille, T. 1997, Science, 277, 67

Kirkpatrick, J. D., Schneider, A., Fajardo-Acosta, S., *et al.* 2014, The Astrophysical Journal, 783, 122

Kraus, A. L., Ireland, M. J., Cieza, L. A., *et al.* 2014, The Astrophysical Journal, 781, 20

Malo, L., Doyon, R., Lafrenière, D., *et al.* 2013, The Astrophysical Journal, 762, 88 de la Reza, R., Torres, C. A. O., Quast, G., Castilho, B. V., & Vieira, G. L. 1989, Astrophysical Journal, 343, L61

Skrutskie, M. F., Cutri, R. M., Stiening, R., *et al.* 2006, The Astronomical Journal, 131, 1163

Torres, C. A. O., da Silva, L., Quast, G. R., de La Reza, R., & Jilinski, E. 2000, AJ, 120, 1410

Zuckerman, B., Song, I., & Bessell, M. S. 2004, ApJ, 613, L65

## Discussion

KATELYN ALLERS: It seems that you have a lack of low-gravity L2 dwarfs (Figure 2). What explains this ?

AUTHOR: This is partly due to our lack of an $L2\,\beta$ template and the fact that we assign spectral types by comparison to visual templates. We have some low-gravity L2 contenders, but they have a low signal-to-noise ratio, so we refrained from calling them L2 for the moment until we have a good template (i.e. those would be called L1: or L3: right now). This "valley" at L2 becomes much less apparent if we include spectral type uncertainties to generate a continuous histogram (a probability density function) and the difference between the number of L1/L2 or L2/L3 low-gravity objects becomes insignificant at $0.2\sigma$.

MARK PINSONNEAULT: On the over-density of low-mass candidates in Tucana-Horologium; you are exploring a region of the color-magnitude diagram that can be populated by mergers and weird objects that might mimick the properties of the objects you are really searching for.

AUTHOR: I agree that there are such things in that part of the color-magnitude diagram, but I would be really surprised if such a variety of objects could mimick the full spectra of young brown dwarfs at resolutions of $R \sim 100\text{-}6000$. I would believe that for these peculiar contaminants, the strength of absorption lines would be different, that there would be some strong emission lines or that we would see some reddening. Obtaining radial velocities and parallaxes will definitely tell us without a doubt if we are observing such objects, and this is something that I want to do next.

GILLES CHABRIER: I think that those numbers [over-density of low-mass candidates] are not very significant because of small number statistics.

AUTHOR: We used Poisson statistics to estimate the error bars on our measurements and on the over-density factor; it seems that it is a $3\sigma$ result at the moment, however combining this with interlopers from other moving groups could really bring down the significance. For example if only 5/12 of our candidates turn out to be members of other young associations then the over-density would become just a $1\sigma$ result. Our simulations to characterize the false-positive rates in BANYAN II tell us that it's almost impossible that these kinds of candidates are interlopers (all have $> 90\%$ membership in Tucana-Horologium and are confirmed as younger than $\sim 200\,\mathrm{Myr}$), however this assumes that our models of the neighborhood (i.e. the field and its moving groups) are perfect, and this is definitely not the case. We could be missing some groups, and some known moving groups might be shaped differently than what we think, which could thus very well cause a fraction of these 12 objects to be young interlopers.

*Young Stars & Planets Near the Sun*
*Proceedings IAU Symposium No. 314, 2015*
*J. H. Kastner, B. Stelzer, & S. A. Metchev, eds.*

© International Astronomical Union 2016
doi:10.1017/S1743921315006481

# A Pan-STARRS1 Search for Substellar Young Moving Group Members

**Kimberly M. Aller** [1], **Michael C. Liu**[2] **and Eugene A. Magnier**[3]

[1] University of Hawaii, Institute for Astronomy
email: `kaller@ifa.hawaii.edu`
[2] University of Hawaii, Institute for Astronomy
email: `mliu@ifa.hawaii.edu`
[3] University of Hawaii, Institute for Astronomy
email: `eugene@ifa.hawaii.edu`

**Abstract.** Young moving groups (YMGs) are coeval, comoving groups of stars which have migrated from their birthsites after formation. In the substellar regime, YMG members are key benchmarks to empirically define brown dwarf evolution with age and to study the lowest mass end of the initial mass function. We have combined Pan-STARRS1 (PS1) proper motions with optical+IR photometry from PS1, 2MASS and WISE to perform a large-scale ($\approx$30,000 deg$^2$) systematic search for substellar members down to $\approx$10 $M_{Jup}$. We have obtained near-IR spectroscopy of a large sample of ultracool candidate YMG members to assess their youth via gravity-sensitive absorption features. We have identified several new intermediate-gravity candidate members of the AB Dor Moving Group, potentially greatly expanding the substellar membership. These new candidate members bridge the gap between the known low-mass stellar and planetary-mass members and yield valuable insight into the spectral characteristics of young brown dwarfs.

**Keywords.** brown dwarfs, low-mass stars

## 1. Introduction

In the past 15 years, large astrometric surveys such as *Hipparcos* (Perryman *et al.* 1997) have been used to discover young ($\lesssim$100 Myr), coeval stellar associations within 100 pc, which have left their molecular cloud after formation and dispersed into the field (e.g. Zuckerman & Song 2004). These young moving groups (YMGs) provide a valuable evolutionary link between ongoing star formation in molecular clouds ($\sim$1 Myr) and old field stars ($>$1 Gyr). Because YMGs are relatively nearby ($\lesssim$100 pc), they are ideal for characterizing the mass function down to substellar masses. In particular, the substellar YMG members are valuable benchmarks for establishing the relationship between spectral type, age, and luminosity.

Photometry and spectroscopy of young brown dwarfs and directly imaged planets have revealed that they differ from their old field counterparts (e.g. Chauvin *et al.* 2005; Marois *et al.* 2008; Bowler *et al.* 2010, 2013; Patience *et al.* 2010; Barman *et al.* 2011). Young brown dwarfs also have redder NIR colors compared to field brown dwarfs of the same spectral type (e.g. Gizis *et al.* 2012). Recent studies have begun to classify gravity-sensitive spectral features by comparing young and old ultracool dwarfs (e.g. Allers *et al.* 2007; Allers & Liu 2013). A key step towards further characterizing brown dwarf spectral evolution is to determine the substellar spectral sequence in YMGs spanning the ages of $\sim$10–100 Myr. However, until the past few years, very few bona fide brown dwarf YMGs members were known.

## 2. Data and Results

We have used Pan-STARRS1 (PS1) proper motions and photometry from PS1, 2MASS and *WISE* to conduct a large-area search for substellar members of YMGs. PS1 is a unique resource for such work because its large area coverage (30,000 deg$^2$) encompasses the sparsely distributed YMGs; its far-red sensitivity enables detection of $\sim$10$\times$ more objects than 2MASS; and its multi-epoch high quality astrometry can distinguish candidate YMG members from old field stars. We use PS1, 2MASS and *WISE* photometry to construct spectral energy distributions and then estimated spectral types and photometric distances using the template-fitting method of Aller *et al.* (2013). We then used PS1+2MASS astrometry to determine proper motions of our candidate YMG members and followed the methods of Schlieder *et al.* (2012) to screen for candidates with space motions consistent with the known YMGs.

We have obtained follow-up near-IR spectroscopy of several candidates using SpeX (Rayner *et al.* 2003), a medium-resolution near-IR spectrograph (0.8–2.5 $\mu$m) on the 3-meter NASA Infrared Telescope Facility (IRTF). We used the Allers & Liu (2013) classification method to assess the spectral type and youth of our candidates. On this system, spectral types are determined by combining a visual comparison of our objects to the Kirkpatrick *et al.* (2010) spectral standards in the $J$ and $K$ bands with the index-based methods from McLean *et al.* (2003), Slesnick *et al.* (2004), and Allers *et al.* (2007). We discovered two early-L dwarfs ($\approx$30–40 $M_{Jup}$) and two late-M dwarfs ($\approx$60–70 $M_{Jup}$) as candidate substellar members of the AB Dor Moving Group, with intermediate gravities typical of young ultracool dwarfs. We also discovered two other L dwarf candidate AB Dor Moving Group members with spectra that show hints of youth but may be older field objects. Figure 1 shows the NIR spectra of our candidate AB Dor Moving Group members. In Figure 2, we compare one of our AB Dor Moving Group candidates, which we have classified as intermediate gravity (INT-G) on the Allers & Liu (2013) system, with an old field gravity (FLD-G) standard object from Kirkpatrick *et al.* (2010) and a young, very low gravity (VL-G) standard object from Allers & Liu (2013).

Although the known young and dusty field L dwarfs have redder *WISE* colors (W1–W2) and NIR colors (J–K) compared with old field dwarfs of the same spectral type (e.g. Gizis *et al.* 2012), not all of our AB Dor candidates appear to follow this trend. One possible reason that our AB Dor Moving Group candidates members are not significantly redder than their field dwarf counterparts is that they are older ($\sim$125 Myr) than the brown dwarf members of the TW Hydrae or $\beta$ Pic moving groups ($\sim$10–20 Myr), which have gravity classifications of VL-G on the Allers & Liu (2013) system.

## 3. Conclusions

We have combined PS1, 2MASS, and *WISE* photometry with PS1+2MASS proper motions to systematically search for substellar YMG members. We have spectroscopically confirmed the youth of four candidate members of the AB Dor Moving Group and have two other candidate members with tentative spectroscopic signatures of youth. Our AB Dor Moving Group candidate members have NIR colors which are not significantly redder than their field counterparts, suggesting that intermediate-aged (INT-G) brown dwarfs may not be as red as young (VL-G) brown dwarfs of the same spectral type. Our candidate young brown dwarf AB Dor Moving Group members are benchmarks which can be used to characterize the brown dwarf spectral evolution and to study the atmospheres of young exoplanet analogs.

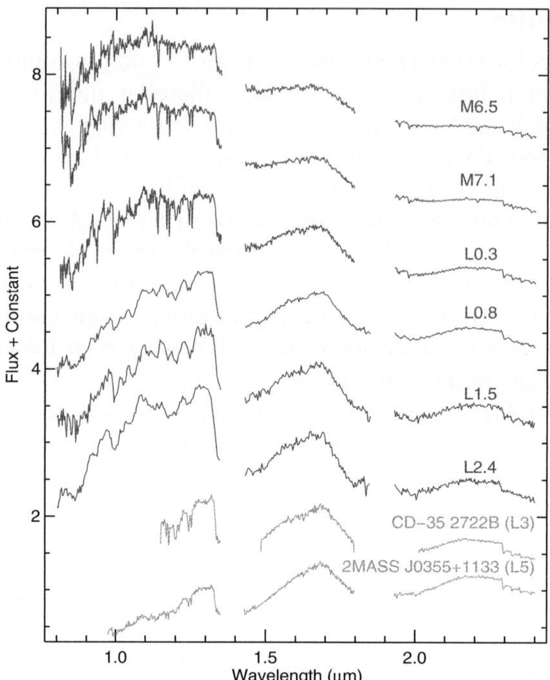

**Figure 1.** NIR spectra of our AB Dor Moving Group candidate members (*red*) compared to two known brown dwarf members (*blue*), CD-35 2722B (Wahhaj *et al.* 2011) and 2MASS J0355+1133 (Liu *et al.* 2013a; Faherty *et al.* 2013). The spectral types of our candidates are determined by combining visual and index-based classifications as in Allers & Liu (2013).

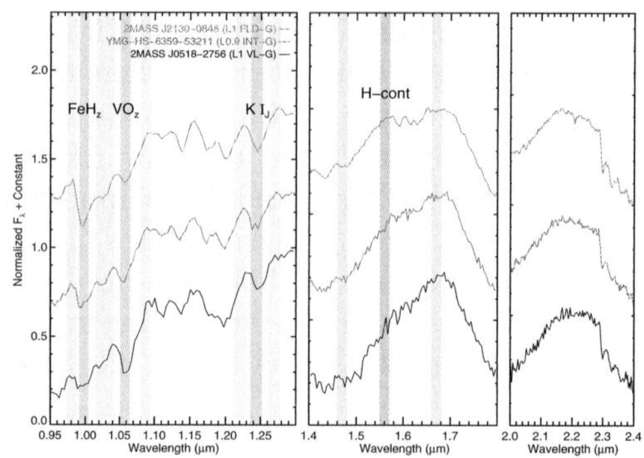

**Figure 2.** NIR low-resolution (R∼130) spectra from IRTF/SpeX of one of our AB Dor Moving Group candidates (INT-G, *red*) compared with a field (FLD-G, *light blue*) and a young (VL-G, *dark blue*) dwarf with a spectral type within half a spectral type from our candidate. We used a field spectral standard from Kirkpatrick *et al.* (2010) and a young standard from Allers & Liu (2013) smoothing all spectra to a common resolution. Wavelength regions used to calculated gravity-sensitive indices from Allers & Liu (2013) are highlighted for FeH$_z$ (*yellow-green*), VO$_z$ (*blue*), KI$_J$ (*purple*), and H-cont (*orange*).

## References


Aller, K. M., *et al.* 2013, *ApJ*, 773, 63
Allers, K. N. & Liu, M. C. 2013, *ApJ*, 772, 79
Allers, K. N., *et al.* 2007, *ApJ*, 657, 511
Barman, T. S., Macintosh, B., Konopacky, Q. M., & Marois, C. 2011, *ApJ*, 733, 65

Bowler, B. P., Liu, M. C., Dupuy, T. J., & Cushing, M. C. 2010, *ApJ*, 723, 850

Bowler, B. P., Liu, M. C., Shkolnik, E. L., & Dupuy, T. J. 2013, *ApJ*, 774, 55

Chauvin, G., Lagrange, A.-M., Dumas, C., Zuckerman, B., Mouillet, D., Song, I., Beuzit, J.-L., & Lowrance, P. 2005, *A&A*, 438, L25

Faherty, J. K., Rice, E. L., Cruz, K. L., Mamajek, E. E., & Núñez, A. 2013, *AJ*, 145, 2

Gizis, J. E., *et al.* 2012, *AJ*, 144, 94

Kirkpatrick, J. D., *et al.* 2010, *ApJS*, 190, 100

Liu, M. C., Dupuy, T. J., & Allers, K. N. 2013a, Astronomische Nachrichten, 334, 85

Marois, C., Macintosh, B., Barman, T., Zuckerman, B., Song, I., Patience, J., Lafrenière, D., & Doyon, R. 2008, *Science*, 322, 1348

McLean, I. S., McGovern, M. R., Burgasser, A. J., Kirkpatrick, J. D., Prato, L., & Kim, S. S. 2003, *ApJ*, 596, 561

Patience, J., King, R. R., de Rosa, R. J., & Marois, C. 2010, *A&A*, 517, A76

Perryman, M. A. C., *et al.* 1997, *A&A*, 323, L49

Rayner, J. T., Toomey, D. W., Onaka, P. M., Denault, A. J., Stahlberger, W. E., Vacca, W. D., Cushing, M. C., & Wang, S. 2003, *PASP*, 115, 362

Schlieder, J. E., Lépine, S., & Simon, M. 2012, *AJ*, 143, 80

Slesnick, C. L., Hillenbrand, L. A., & Carpenter, J. M. 2004, *ApJ*, 610, 1045

Wahhaj, Z., *et al.* 2011, *ApJ*, 729, 139

Zuckerman, B. & Song, I. 2004, *ARA&A*, 42, 685

*Young Stars & Planets Near the Sun*
*Proceedings IAU Symposium No. 314, 2015*
*J. H. Kastner, B. Stelzer, & S. A. Metchev, eds.*

© International Astronomical Union 2016
doi:10.1017/S1743921315006031

# New Low-mass Accretors in the Scorpius-Centaurus OB Association

**Simon J. Murphy**[1,2], **Warrick A. Lawson**[2] **and Joao Bento**[1]

[1]Research School of Astronomy & Astrophysics, Australian National Unversity
email: `simon.murphy@anu.edu.au`
[2]School of Physical, Environmental and Mathematical Sciences,
University of New South Wales Canberra

**Abstract.** We describe the serendipitous discovery of two new lithium-rich M5 members of the Scorpius-Centaurus OB Association (Sco-Cen). Both stars exhibit large 12 and 22 $\mu$m excesses and strong, variable H$\alpha$ emission which we attribute to accretion from circumstellar discs. Such stars are thought to be incredibly rare at the $\sim$16 Myr median age of much of Sco-Cen. The serendipitous discovery of two accreting stars hosting large quantities of circumstellar material may be indicative of a sizeable age spread in Sco-Cen, or further evidence that disc dispersal and planet formation time-scales are longer around lower-mass stars.

**Keywords.** circumstellar matter, open clusters and associations: individual (Sco-Cen), stars: low-mass, stars: pre-main sequence

## 1. Introduction

Nearby stars with ages of <20 Myr are ideal laboratories in which to investigate the end stages of star and planet formation. It is over this time period that gas-rich protoplanetary discs which feed giant planet formation and circumstellar accretion dissipate and are replaced by gas-poor, dusty debris discs (Wyatt 2008; Williams & Cieza 2011).

The Scorpius-Centaurus OB Association (Sco-Cen, Sco OB2; Blaauw 1964; de Zeeuw *et al.* 1999; Preibisch & Mamajek 2008) is the closest site of recent massive star formation and the dominant population of pre-main sequence stars in the solar neighbourhood. Sco-Cen has traditionally been divided into three subgroups, each with subtly distinct mean distances, ages and space motions: Upper Scorpius (145 pc), Upper Centaurus Lupus (UCL, 140 pc) and Lower Centaurus Crux (LCC 120 pc; de Zeeuw *et al.* 1999). The subgroups have median ages of approximately 10 Myr, 16 Myr and 17 Myr, respectively (Mamajek *et al.* 2002; Pecaut *et al.* 2012). While the low-mass population of Upper Scorpius has been well-studied in recent years, little is known of the older and larger UCL and LCC subgroups below 1 $M_\odot$. In this contribution we describe the serendipitous discovery of two rare accreting M5 stars in LCC ( 2MASS J12392312–5702400 ) and UCL ( 2MASS J14224891–3623009 ), found during a search for new members of the Octans Association. Further information on the candidate selection procedures, spectroscopic observations and new members of the $\sim$10 Myr-old TW Hydrae Association also identified during this work can be found in Murphy *et al.* (2015, submitted).

## 2. Membership in Sco-Cen

2M1239–5702 and 2M1422–3623 lie within the classical boundaries of LCC and UCL, respectively, with SPM4 proper motions similar to higher-mass members identified with *Hipparcos* (de Zeeuw *et al.* 1999). These proper motions correspond to kinematic

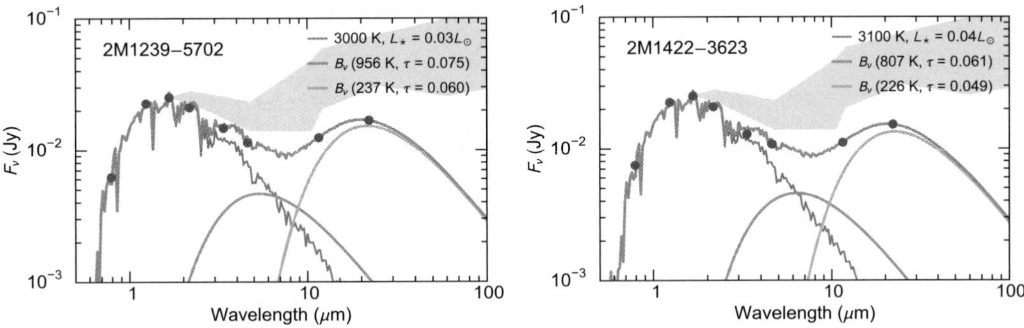



**Figure 1.** SEDs of 2M1239–5702 (left) and 2M1422–3623 (right). Photometry from DENIS, 2MASS and *WISE* is plotted, with a solar-metallicity BT-Settl model fitted to the $\lambda < 3$ $\mu$m data. The *WISE* excesses are fit using blackbodies of the specified temperature and fractional luminosity ($\tau = L_{IR}/L_*$). The shaded region is the interquartile range of Class I K5–M2 sources observed in Taurus from D'Alessio *et al.* (1999), normalised at $H$ (1.6 $\mu$m).

distances of $119^{+13}_{-11}$ pc and $142^{+35}_{-24}$ pc, assuming the mean LCC and UCL space motions of Chen *et al.* (2011), and are in excellent agreement with the mean *Hipparcos* distances of the subgroups. At these distances 2M1239–5702 and 2M1422–3623 are 3 km s$^{-1}$ and 6 km s$^{-1}$, respectively, from the Chen *et al.* mean space motions, with uncertainties of 2–4 km s$^{-1}$ in each velocity component. This is well within the scatter of higher-mass Sco-Cen members with known radial velocities (Murphy *et al.* 2015, submitted). Their colour-magnitude diagram (CMD) positions suggest an age of approximately 10 Myr, similar to TW Hydrae members and consistent with recent age ranges for UCL and LCC (Song *et al.* 2012; Pecaut *et al.* 2012). 2M1422–3623 is 15–20° from the Lupus dark clouds within UCL. Given its large separation from confirmed Lupus sources (Galli *et al.* 2013), we do not believe it is a member of this much younger (age <5 Myr) star-forming region. 2M1239–5702 lies near the centre of LCC, $\sim$1° from the foreground M-giant $\gamma$ Cru.

## 3. Disc and accretion properties

2M1239–5702 and 2M1422–3623 were two of only a handful of spectroscopic candidates in our Octans survey to show excesses in 12 and 22 $\mu$m *Wide-field Infrared Survey Explorer* (*WISE*; Wright *et al.* 2010) photometry. Their remarkably similar spectral energy distributions (SEDs) are plotted in Figure 1. Both are consistent with somewhat evolved, 'homologously-depleted' discs in the process of becoming optically thin, without strong evidence for the inner holes or gaps seen in 'transitional'-type objects (Espaillat *et al.* 2012). Double-temperature blackbody models with a cool $\sim$230 K outer component and a warmer 800–900 K inner component provide a good fit to the *WISE* photometry in both cases. Assuming $L_* = 0.03\,L_\odot$ from integration of the underlying photospheric flux (Figure 1) and large grains, these temperatures correspond to disc radii of 0.25 AU and $\sim$0.02 AU, respectively (Backman & Paresce 1993).

We also observed broad, variable H$\alpha$ emission from both stars which exceeded the equivalent width (15–18 Å; Fang *et al.* 2009) and velocity width at 10% flux (270 km s$^{-1}$; White & Basri 2003) criteria for confirming circumstellar disc accretion. Multi-epoch H$\alpha$ velocity profiles are plotted in Figure 2. Equivalent widths for 2M1422–3623 ranged from $-91$ to $-33$ Å over the four epochs of medium-resolution ($R = 7,000$) Australian National University 2.3-m/WiFeS spectroscopy, with $v_{10}$ velocity widths of 240–340 km s$^{-1}$. For 2M1239–5702 these ranges were $-63$ to $-27$ Å and 240–330 km s$^{-1}$ across two epochs.

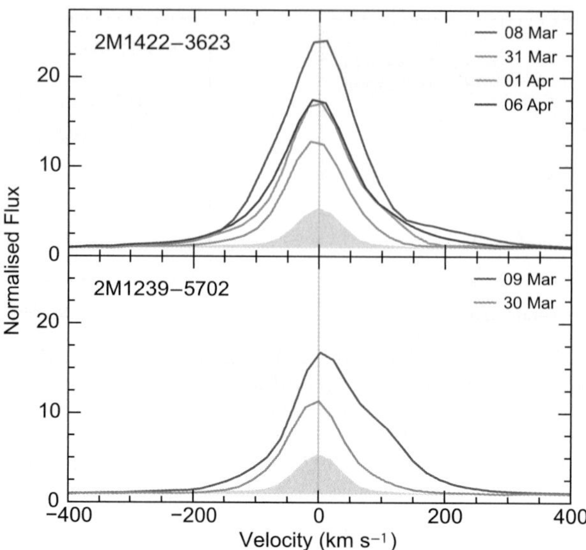


**Figure 2.** 2.3-m/WiFeS H$\alpha$ velocity profiles for 2M1422–3623 (top) and 2M1239–5702 (bottom). Each panel also shows the emission line profile of a typical active but non-accreting M-dwarf of similar spectral type (EW $= -9.5$ Å, $v_{10} = 169$ km s$^{-1}$).

Applying the $v_{10}$–$\dot{M}$ relation of Natta *et al.* (2004) yields maximal mass accretion rates of $\sim 10^{-9.5}$ $M_\odot$ yr$^{-1}$ with a variation of $\sim 1$ dex across epochs. 2M1239–5702 and 2M1422–3623 show near-IR excesses at 3–5 $\mu$m, indicating the presence of some dust in the inner disc. The variable accretion we observed is presumably driven by clumpy gas-rich material in this region.

In addition to H$\alpha$ we also observed Na I D and forbidden [O I] $\lambda$6300/6363 emission in 2M1422–3623 and strong He I $\lambda$5876/6678 emission from both stars. The Na I and He I emission are believed to originate with H$\alpha$ in magnetospheric infall regions (Muzerolle *et al.* 1998), whereas the forbidden emission is usually attributed to accretion-driven outflows or winds (Appenzeller & Mundt 1989).

## 4. Discussion

Outside of the younger ($\sim 10$ Myr) Upper Scorpius subgroup, 2M1239–5702 and 2M1422–3623 join a select group of Sco-Cen members known to be accreting. These include the M3.5 LCC member 2MASS J13373839–4736297 (Rodriguez *et al.* 2011) and the higher-mass accretors HD 139614, HD 135344 and AK Sco (Preibisch & Mamajek 2008). The rarity of such objects is unsurprising given the observed rapid decline of accretor fraction with age, with $e$-folding time-scales of 2–3 Myr (e.g. Fedele *et al.* 2010). For a putative population of $10^4$ low-mass Sco-Cen members we may naively expect only a handful of active accretors to have survived until the $\sim 16$ Myr median age of UCL and LCC. The serendipitous discovery of two such objects in a survey for Octans members therefore seems very unlikely. We propose two scenarios which may explain the presence of these supposedly rare objects — significant age spreads in the Sco-Cen subgroups, and/or increased circumstellar disc lifetimes around lower-mass stars.

Mamajek *et al.* (2002) and Pecaut *et al.* (2012) have investigated the issue of intrinsic age spreads in UCL and LCC. They found $1\sigma$ spreads of a few Myr up to 8 Myr depending on the subgroup, spectral type and evolutionary model grid considered. At face value, this would imply star formation in the subgroups ceased in the last 5–10 Myr, or even later if the larger age spreads are realistic. There is also evidence for spatial variation in ages within the subgroups (Pecaut & Mamajek, in prep; Preibisch & Mamajek 2008). Given these proposed age spreads, 2M1239–5702 and 2M1422–3623 may be as young as 5–10 Myr and represent some of the last stars to have formed in UCL and LCC. Outside of the presence of substantial circumstellar material and accretion, young ages for both stars are supported by their CMD positions and undepleted lithium ($EW_{\lambda 6708} > 600$ mÅ).

A second possibility is that the time-scale for circumstellar disc dispersal is stellar mass dependent, being less efficient around lower-mass stars. Such a trend has been observed in several young groups, including Upper Scorpius (Carpenter *et al.* 2006; Kennedy & Kenyon 2009; Luhman & Mamajek 2012) and has critical implications for the time available to form giant planets around low-mass stars. These studies indicate that a significant fraction ($\sim 25\%$) of M-dwarfs are able to retain inner discs at an age of 10 Myr. Extreme age spreads in Sco-Cen would thus not be required to explain the excesses and accretion observed in 2M1239–5702 and 2M1422–3623 . If such a trend is universal there may be hundreds more disc-bearing UCL and LCC members (out of a total low-mass population of several thousand stars) awaiting discovery within *WISE* and contemporary proper motion catalogues (e.g. UCAC4, SPM4), and soon, *Gaia*. These new members would also be very useful for lithium depletion (e.g. Song *et al.* 2012) and exoplanet direct imaging studies, among other works.

## Acknowledgement

SJM thanks the IAU and organisers of IAUS 314 for the generous award of a travel grant to attend the symposium and present this work.

## References

Appenzeller I. & Mundt R., 1989, *ARAA*, 1, 291

Backman D. E. & Paresce F., 1993, in E.H. Levy, J.I. Lunine, eds, Protostars and Planets III. pp. 1253–1304

Blaauw A., 1964, ARAA, 2, 213

Carpenter J. M., Mamajek E. E., Hillenbrand L. A., & Meyer M. R., 2006, *ApJL*, 651, L49

Chen C. H., Mamajek E. E., Bitner M. A., Pecaut M., Su K. Y. L., & Weinberger A. J., 2011, *ApJ*, 738, 122

D'Alessio P., Calvet N., Hartmann L., Lizano S., & Cantó J., 1999, *ApJ*, 527, 893

de Zeeuw P. T., Hoogerwerf R., de Bruijne J. H. J., Brown A. G. A., & Blaauw A., 1999, *AJ*, 117, 354

Espaillat C. *et al.*, 2012, *ApJ*, 747, 103

Fang M., van Boekel R., Wang W., Carmona A., Sicilia-Aguilar A., & Henning T., 2009, *A&A*, 504, 461

Fedele D., van den Ancker M. E., Henning T., Jayawardhana R., & Oliveira J. M., 2010, *A&A*, 510, A72

Galli P. A. B., Bertout C., Teixeira R., & Ducourant C., 2013, *A&A*, 558, A77

Kennedy G. M. & Kenyon S. J., 2009, *ApJ*, 695, 1210

Luhman K. L. & Mamajek E. E., 2012, *ApJ*, 758, 31

Mamajek E. E., Meyer M. R., Liebert J., 2002, *AJ*, 124, 1670

Muzerolle J., Hartmann L., & Calvet N., 1998, *AJ*, 116, 455

Natta A., Testi L., Muzerolle J., Randich S., Comerón F., & Persi P., 2004, *A&A*, 424, 603

Pecaut M. J., Mamajek E. E., & Bubar E. J., 2012, *ApJ*, 746, 154

Preibisch T. & Mamajek E. E., 2008, Handbook of Star Forming Regions, Vol. II. The Southern Sky, ASP Press. p. 235

Rodriguez D. R., Bessell M. S., Zuckerman B., & Kastner J. H., 2011, *ApJ*, 727, 62

Song I., Zuckerman B., & Bessell M. S., 2012, *AJ*, 144, 8
White R. J. & Basri G., 2003, *ApJ*, 582, 1109
Williams J. P. & Cieza L. A., 2011, *ARAA*, 49, 67
Wright E. L. *et al.*, 2010, *AJ*, 140, 1868
Wyatt M. C., 2008, *ARAA*, 46, 339

*Young Stars & Planets Near the Sun*
Proceedings IAU Symposium No. 314, 2015
J. H. Kastner, B. Stelzer, & S. A. Metchev, eds.

© International Astronomical Union 2016
doi:10.1017/S1743921315005992

# The WISE Census of Young Stellar Objects in Canis Major

**W. J. Fischer[1,2], D. L. Padgett[1], and K. R. Stapelfeldt[1]**

[1] NASA Goddard Space Flight Center
email: william.j.fischer@nasa.gov
[2] NASA Postdoctoral Program Fellow

**Abstract.** While searches for young stellar objects (YSOs) with the Spitzer Space Telescope focused on known molecular clouds, photometry from the Wide-field Infrared Survey Explorer (WISE) can be used to extend the search to the entire sky. As a precursor to more expansive searches, we present results for a 100 deg$^2$ region centered on the Canis Major clouds.

**Keywords.** protoplanetary disks – stars: formation – stars: protostars

## 1. Introduction

With its all-sky survey at 3.4, 4.6, 12, and 22 $\mu$m, the Wide-field Infrared Survey Explorer (WISE; Wright *et al.* 2010) can be used to identify young stellar objects (YSOs) with criteria similar to those established for the Spitzer Space Telescope but over the entire sky. Newly identified YSOs may refine the initial stellar mass function, allow a better characterization of star and planet formation in regions with low initial gas densities, and identify nearby targets for high-resolution follow-up imaging.

As a pilot study for our more expansive search, we present results for Canis Major. Star formation in the vicinity of Canis Major is centered amid the CMa OB1 association, 1–2° below the Galactic plane near a longitude of 224° and at a distance of $\sim$ 1000 pc (Gregorio-Hetem 2008). It is home to $5 \times 10^4$ $M_\odot$ of material distributed across 22 clouds as traced by $^{13}$CO gas (Kim *et al.* 2004). Although parts of it have recently been mapped by the outer Galactic plane surveys of Spitzer and the Herschel Space Observatory, there is no comprehensive study of star formation across the region. We searched the AllWISE catalog for Class I YSOs, in which a dusty protostellar envelope dominates the infrared emission, and more evolved Class II YSOs, in which a dusty circumstellar disk dominates.

## 2. YSO Selection Techniques

We targeted the 10° × 10° square centered at 106.67° right ascension and −11.29° declination. For inclusion in our initial catalog, we required detections in Bands 1 and 2, and we ignored sources flagged as artifacts or extended emission. To identify candidates, we adopted the WISE color-color criteria of Koenig & Lesiawitz (2014). These are based on the colors of known YSOs in Taurus, extragalactic sources, and galactic contaminants.

Most of the sources selected by these color criteria are faint and uniformly distributed across the region, unlike YSOs, which should be clustered near the sites of their formation. To deselect these likely extragalactic sources, we required a WISE band 1 magnitude $W1 < 12$ or a WISE band 4 magnitude $W4 < 5$. To remove bright galactic contaminants such as red giant stars, we required $W1 > 6$. Of the sources with the requisite colors, 155 Class I and 375 Class II candidates satisfy the magnitude requirements in the full 100 deg$^2$ search region.

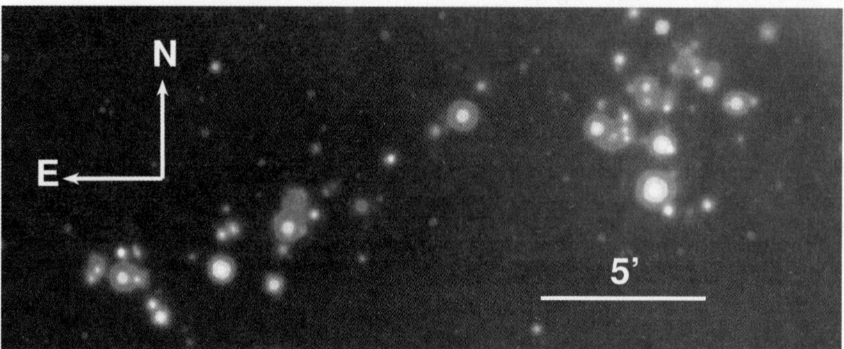


**Figure 1.** A three-color WISE image of a YSO cluster in Canis Major, centered at $7^h 10^m$ RA and $-10°30'$ Dec. Blue is 3.6 $\mu$m, green is 12 $\mu$m, and red is 22 $\mu$m. The 5' scale bar corresponds to 1.45 pc at a distance of 1000 pc. Generally, the bright red sources are the Class I candidates. *See the electronic edition for full color.*

## 3. A New Young Cluster of YSOs

We discovered a cluster of YSOs (Fig. 1) located 2° to the northeast of the well studied object Z CMa. Assuming a distance of 1000 pc, it covers an area of 9.3 pc$^2$ according to the method of Gutermuth *et al.* (2009). The cluster contains 31 Class I candidates based on the above criteria, 20% of the total for the 100 deg$^2$ search region. It also contains 12 Class II candidates, although this is a lower limit due to the insensitivity of WISE bands 3 and 4 to sources with Class II colors and magnitudes at 1000 pc. This cluster has more Class I sources and a larger physical area than about 80% of the 36 clusters within 1 kpc of the sun studied by Gutermuth *et al.* (2009), and the density of Class I sources is 3.3 pc$^{-2}$, larger than that of about two thirds of the Gutermuth sample.

The reddest source, just east of the center of Figure 1, has WISE magnitudes consistent with a 4.5 $L_\odot$ Class 0 protostar at 1000 pc. Its detection suggests that the WISE 22 $\mu$m channel is useful for identifying isolated, deeply embedded young protostars.

## 4. Conclusions

With WISE photometric criteria, we detected over 500 candidate YSOs in a 10° × 10° square centered on the Canis Major star forming region. The most populous cluster of candidates has 31 Class I candidates and an area of 9.3 pc$^2$, typical of the larger, more populous clusters in the nearest 1 kpc. Identification of analogs to the nearby star forming regions at larger distances provides the opportunity for efficient environmental studies with current and future long-wavelength observing facilities. Subsequent application of our YSO finding criteria to the entire sky may reveal isolated young stellar objects near the sun for high-resolution studies.

## References

Gregorio-Hetem, J. 2008, in *Handbook of Star Forming Regions*, B. Reipurth, ed., p. 1
Gutermuth, R. A., Megeath, S. T., & Myers, P. C., *et al.* 2009, *ApJS*, 184, 18
Kim, B. G., Kawamura, A., Yonekura, Y., & Fukui, Y. 2004, *PASJ*, 56, 313
Koenig, X. P. & Leisawitz, D. T. 2014, *ApJ*, 791, 131
Wright, E. L., Eisenhardt, P. R. M.., & Mainzer, A. K., *et al.* 2010, *AJ*, 140, 1868

*Young Stars & Planets Near the Sun*
Proceedings IAU Symposium No. 314, 2015
J. H. Kastner, B. Stelzer, & S. A. Metchev, eds.

© International Astronomical Union 2016
doi:10.1017/S1743921315006298

# A New, Young, Low-Mass Spectroscopic Binary Without a Home

**Laura S. Flagg[1,2], Evgenya L. Shkolnik[1], Alycia J. Weinberger[3], Brendan P. Bowler[4], Adam L. Kraus[5], and Michael C. Liu[6]**

[1]Lowell Observatory, Flagstaff, AZ 86001, USA, email: `lflagg@lowell.edu`
[2]Department of Physics and Astronomy, Northern Arizona University, Flagstaff, AZ 86011, USA
[3]Department of Terrestrial Magnetism, Carnegie Institution for Science, Washington, DC 20015 USA
[4]California Institute of Technology, Pasadena, CA 91125, USA
[5]Department of Astronomy, University of Texas at Austin, TX, USA
[6]Institute for Astronomy, University of Hawai'i at Manoa, Honolulu, HI 96822, USA

**Abstract.** We have discovered that 2MASS 08355977−3042306 is an accreting K7, double-lined, spectroscopic binary younger than ∼20 Myr. The age of a dispersed young star can best be determined if it is a member of a known young moving group. However, the three dimensional space velocities ($UVW$) we calculate using radial velocity measurements, proper motions, and plausible photometric distances make membership in any known young moving group unlikely.

**Keywords.** stars: low-mass, binaries: spectroscopic, stars: pre–main-sequence, stars: kinematics

---

The parameters of young, spectroscopic binaries are important for increasing our understanding of star formation and testing stellar evolutionary models. For stars that are 10 to 300 Myr old, the age can be determined by classifying the system into a young moving group (YMG). As part of our broader search for 20 Myr $\beta$ Pic YMG members, we acquired two epochs of high resolution (R≈45,000) optical spectra from the 2.5m Irénée du Pont telescope of 2MASS 08355977−3042306. The spectra revealed it to be a young, accreting, double-lined spectroscopic binary with a composite spectral type of K7.

The one dimensional cross-correlation function (CCF) from March 28, 2014, calculated with the IRAF task *fxcor* using the radial velocity (RV) standard GJ 653 (SpT=K5), showed two blended peaks in each order of the spectrum with the primary star blueshifted. The CCF has a different, asymmetric shape on January 13, 2015 where the primary peak is redshifted. By fitting two Gaussians to the March CCF (Fig. 1), we find a flux ratio of ∼0.7 in the V-band. We calculate an average RV for each component of $-9 \pm 2$ km/s and $+31 \pm 3$ km/s. The systemic RV, averaged over all epochs, including a recently acquired CHIRON spectrum, is $6 \pm 2$ km/s.

We measure a composite spectral type of K7 $+/-$ 1 subclass using the TiO-5 band index (Reid *et al.* 1995). Stars of this spectral type that have H$\alpha$ in emission and lithium absorption are typically less than 20 Myr old (Chabrier & Baraffe 1997; Barrado y Navascués & Martín 2003). We measure a lithium equivalent width of 0.25 Å and an H$\alpha$ equivalent width of $-3$ Å (Fig. 1). The H$\alpha$ emission has a 10% velocity width of ∼350 km/s, indicating that it is likely that one or both of the stars are still accreting (White & Basri 2003). Our measurements of the K I equivalent width and CaH index both indicate low surface gravity (Shkolnik *et al.* 2011). This also implies that the system is young, and the stars have not yet fully contracted onto the main sequence.

The age of a dispersed young star can best be determined if it is a member of a known YMG. Using the Baraffe et. al (1998) models and an assumed age of 15 Myr, we

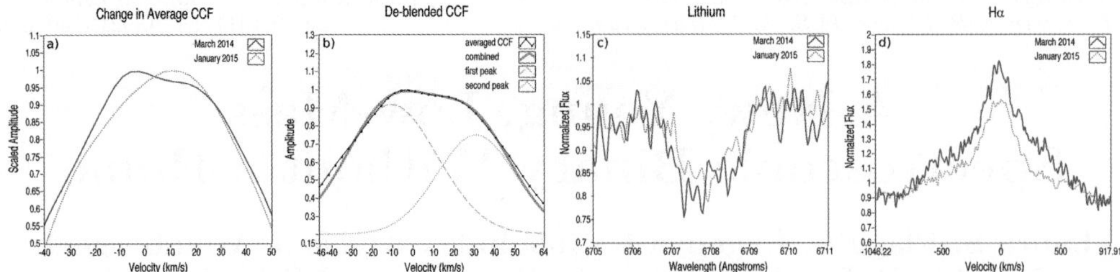

**Figure 1.** (a) The CCF of 2MASS 08355977−3042306 observed on two different nights. (b) The CCF from March 2014 is deblended into components. (c) The equivalent width of the lithium 6707.8 Å line is 0.25 Å. (d) The Hα emission has an equivalent width of -3 Å and the 10% velocity width is ∼350 km/s.

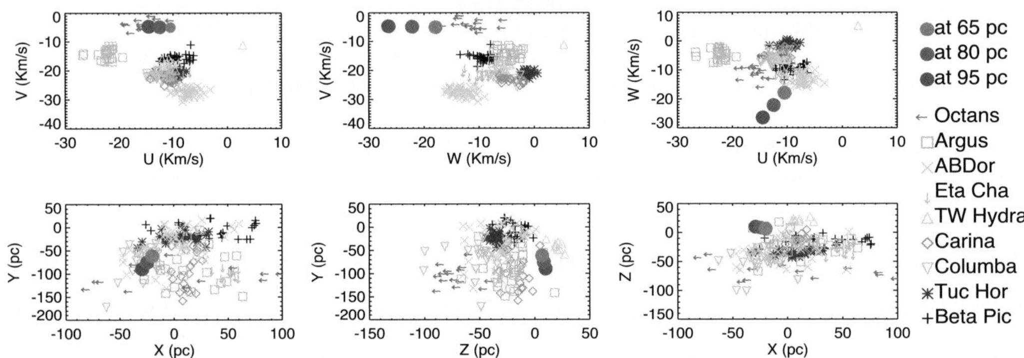

**Figure 2.** The $UVW$ and $XYZ$ values of 2MASS 08355977−3042306 for assumed photometric distances of 65, 80, and 95 pc (colored circles), compared with the same values for various nearby YMG members. Error bars are smaller than the symbols.

calculate a photometric distance of 80 pc. There are significant uncertainties ($\pm$ 15 pc) associated with this distance calculated based on the $1\sigma$ uncertainty in spectral type and the range of possible ages ($\leqslant$20 Myr). The derived proper motions are $-64.8$ mas/yr in R.A. and $-15.6$ mas/yr in declination following the methods of Kraus and Hillenbrand (2007). Using these, we calculate the $UVW$ velocities to be $[-12.5 \pm 1.8,\ -4.8 \pm 2.1,\ -22.2 \pm 2.1]$ km/s. The system's coordinates in $XYZ$ space are $[-24.5 \pm 1.5,\ -75.7 \pm 5.1,\ 8.3 \pm 0.5]$ pc. We also calculate these $UVW$ and $XYZ$ values using distances of 95 pc and 65 pc to account for uncertainties. The resulting range in $UVW$ and $XYZ$ values are indicated as circles in Fig. 2. The lack of consistent overlap with any YMG implies that membership in any of these—including $\beta$ Pic—is highly unlikely. Possible explanations for the system's apparent homelessness include that the system formed on its own, that it was kinematically ejected from a known YMG, or that it is a member of a yet to be discovered YMG.

## References


Baraffe, I., Chabrier, G., Allard, F., & Hauschildt, P. H. 1998, *A&A*, 337, 403

Barrado y Navascués, D. & Martín, E. L. 2003, *AJ*, 126, 2997

Chabrier, G. & Baraffe, I. 1997, *A&A*, 327, 1039

Kraus, A. L. & Hillenbrand, L., 2007, *ApJ*, 134, 2340

Reid, I. N., Hawley, S. L., & Gizis, J. E. 1995, *AJ*, 110, 1838

Shkolnik, E. L., Liu, M. C., Reid, I. N., Dupuy, T., & Weinberger, A. J., 2011, *ApJ*, 727, 6

White, R. J. & Basri, G. 2003, *ApJ*, 582, 1109

*Young Stars & Planets Near the Sun*
*Proceedings IAU Symposium No. 314, 2015*
*J. H. Kastner, B. Stelzer, & S. A. Metchev, eds.*

© International Astronomical Union 2016
doi:10.1017/S1743921315006341

# Effect of Prior Information on Bayesian Membership Calculations for Nearby Young Star Associations

## Jinhee Lee and Inseok Song

Department of Physics and Astronomy, The University of Georgia, Athens, GA 30602-2451
email: `jinhee@uga.edu, song@uga.edu`

**Abstract.** We present a refined moving group membership diagnostics scheme based on Bayesian inference. Compared to the BANYAN II method, we improved the calculation by updating bona fide members of a moving group, field star treatment, and uniform spatial distribution of moving group members. Here, we present the detailed description of our method and the new results for Bayesian membership calculation. Comparison of our method with BANYAN II shows probability differences up to $\sim$90%. We conclude that more cautious consideration is needed in moving group membership based on Bayesian inference.

**Keywords.** open clusters and associations, stars: kinematics and dynamics, stars: statistics, stars: pre-main-sequence

---

## 1. Introduction

It is difficult to determine the memberships of moving groups, i.e., stellar groups that are young ($< 100$ Myr), nearby ($< 100$ pc), and gravitationally unbound. However, the youth and proximity of their members provides the opportunity to understand the characteristics and evolution of young late type stars and planetary systems.

Malo *et al.* (2013) developed a tool to assess moving group memberships based on Bayesian inference, and Gagné *et al.* (2014) improved this tool (BANYAN II). Since the Bayesian probability is obtained based on prior information, one has to carefully select priors and understand the effect of chosen priors.

## 2. Method

Following Gagné *et al.* (2014), we consider kinematical observables such as RA, Dec, proper motion, distance, and radial velocity inside of the Bayesian inference scheme.

## 3. Improved priors and their effects

Here we suggest some improvements and investigate their effects. We test the effect of different priors using a set of bona fide moving group members as listed in Malo *et al.* (2013).

### 3.1. *Bona Fide Members for Group Properties*

Since bona fide members of moving groups represent 3-D ellipsoidal structures in $XYZ$ and $UVW$ space, different selections of "bona fide" stars change the group properties and have effects on membership probabilities of new stars. We improved on previous lists of bona fide $\beta$ Pic Moving Group (BPMG) members based on kinematics and unambiguous youth indicators. Our selection criteria and the list of selected stars will be presented in Lee & Song (in prep.). Figure 1 shows the effect of our new treatment on

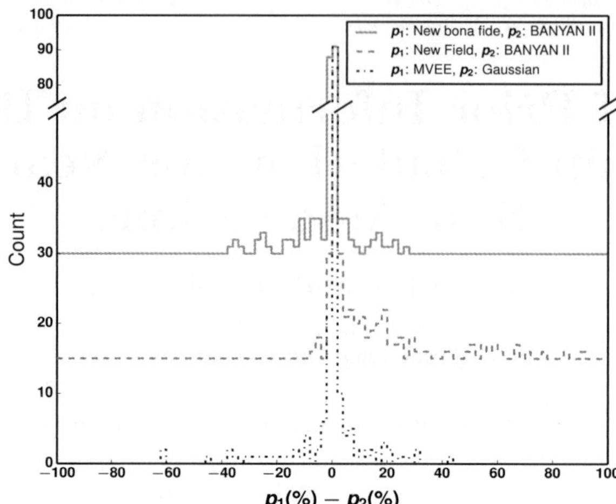


**Figure 1.** Effects of our new treatment on membership probabilities.

membership probabilities. Even using identical parameters for the BPMG stars included for consideration as members, the difference in membership probability produced using different assumed sets of bona fide members can be as large as ∼40%.

### 3.2. *Field Star Treatment*

Based on Skuljan *et al.* (1999), we populated four groups of field stars in $UVW$ space. Since we consider the local Galaxy ($< 100$ pc), we assumed that the four field star groups are distributed uniformly in $XYZ$ space. We compare the effect of our field star treatment to that of BANYAN II based on the Bensançon model. Figure 1 shows that the probability difference between the two calculations can be up to ∼90%.

### 3.3. *Minimum Volume Enclosing Ellipsoid (MVEE)*

The discussion in Ortega *et al.* (2002) - that the BPMG might have formed as an unbound star system with a large initial volume - indicates that assuming a Gaussian distribution around a spatial center in $XYZ$ for BPMG may be unrealistic. We compare the effect of the Gaussian distribution and the uniform distribution inside of the Minimum Volume Enclosing Ellipsoid (MVEE) of BPMG. The effect is up to ∼60% (Figure 1).

## 4. Conclusion

BANYAN II is a very useful and effective tool to quantitatively calculate the membership probability of candidate young star within a given moving group. However, we should be careful in using the calculated membership probability since the Bayesian probability is highly sensitive to prior information. Use of an updated list of bona fide members of BPMG, a uniform distribution inside of the MVEE, and a new field star treatment can change membership probabilities up to ∼90%. This indicates that we should be cautious when interpreting membership probabilities based on Bayesian inference.

### References


Gagné, J., Lafrenière, D., & Doyon, R., *et al.* 2014, *ApJ*, 783, 121
Malo, L., Doyon, R., & Lafrenière, D., *et al.* 2013, *ApJ*, 762, 88
Ortega, V. G., Reza, R., Jilinski, E., & Bazzanella, B. 2002, *ApJ*, 575, 75
Skuljan, J., Hearnshaw, J. B., & Cottrell, P. L. 1999, *MNRAS*, 308, 731

*Young Stars & Planets Near the Sun*
*Proceedings IAU Symposium No. 314, 2015*
*J. H. Kastner, B. Stelzer, & S. A. Metchev, eds.*

© International Astronomical Union 2016
doi:10.1017/S1743921315006365

# Where the Wild Young M Dwarfs Are: the SUPERBLINK Proper Motion Survey and a Search for Low-mass Moving Group Candidates

## Sébastien Lépine[1]

[1]Department of Physics and Astronomy, Georgia State University
email: slepine@astro.gsu.edu

**Abstract.** The SUPERBLINK survey catalogs all stars brighter than $R = 19$ mag and with proper motions larger than 40 mas yr$^{-1}$, down to a declination of $-33°$. The catalog inevitably includes a significant fraction of the presumed low-mass members of several nearby young moving groups (Beta Pic, AB Dor, Tuc-Hor, Argus), or low-mass escapees from the Hyades and Pleiades clusters. We discuss opportunities and challenges in identifying the missing M dwarf members of these moving groups. While rounding up the majority of the potential M dwarf members of these groups, such samples are significantly affected by co-moving field stars, both young and old, due to the heavy clumping of the local field population in velocity space.

**Keywords.** surveys, stars: kinematics, stars: low-mass, brown dwarfs

## 1. Searching for Young M dwarfs

The SUPERBLINK survey, is an all-sky census of $\sim 3$ million stars with very large proper motion, currently with a proper motion limit of $\mu > 40$ mas yr$^{-1}$ north of DEC $= -33°$, and $\mu > 150$ mas yr$^{-1}$ south of that limit (Lépine & Shara 2005, Lépine & Gaidos 2011). The survey reaches optical magnitude $R < 19$ mag, has a high completeness rate within the proper motion and magnitude limits, and also a very high ($> 99.5\%$) reliability, i.e. it is a 'verified' catalog, with virtually no spurious entries. The survey is accurately cross-matched to several photometric catalogs (ROSAT, GALEX, USNO-B1.0, SDSS, 2MASS). From a combination of proper motion and magnitude cuts, 233,727 M dwarfs estimated to be located within 100 parsecs of the Sun have been identified and their kinematics estimated (Fig. 1).

A kinematic search for young M dwarfs can be performed by searching for stars that appear to be co-moving with local clusters or moving groups. The search also takes the magnitude and colors of the stars into account, searching for objects that are consistently over-luminous, like the known nearby young stars (see e.g. Lépine & Simon 2009). The clear identification of hundreds of M dwarfs in the Pleiades and Hyades clusters show the method to be very efficient at recovering potential M dwarf members (Fig. 2), but the contamination from field stars is severe, especially for nearby groups that are significantly spread out. This happens because the nearby young clusters and moving groups occupy a heavily populated region in velocity space, and are coincidentally co-moving with numerous local field stars, old and young.

In the end the method most probably round up a significant fraction of the M dwarf members of those young moving groups and clusters, but various follow-up observations are required to search for signs of youth and confirm group membership.

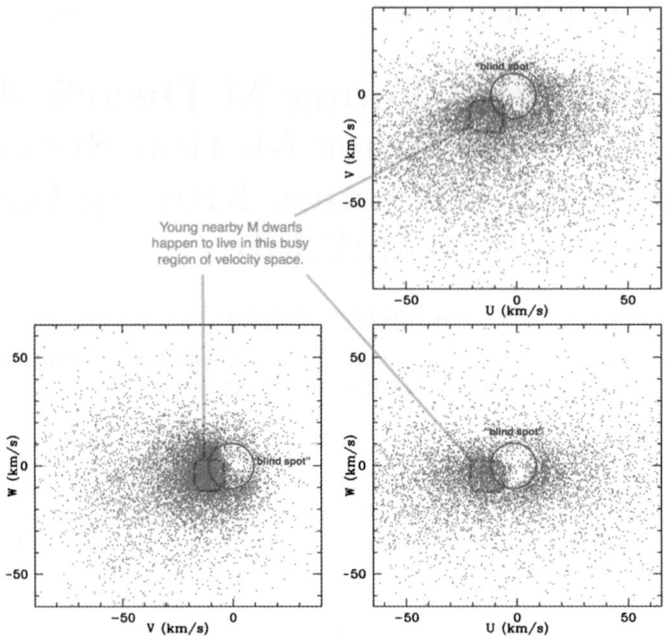



**Figure 1.** Kinematics from proper motion and photometric distances reveals that the velocity-space distribution of M dwarfs is very anisotropic and inhomogenous, consistent with the very same substructure or 'streams' observed for more massive G stars (Nordström *et al.* 2004).

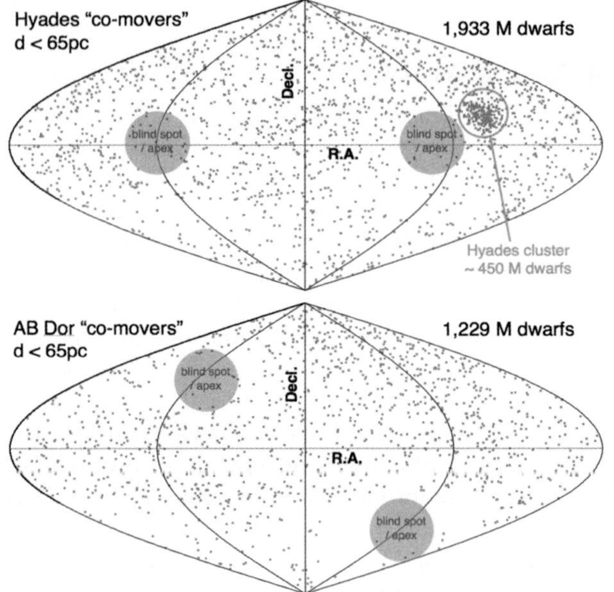



**Figure 2.** Kinematic selection: candidate stars from the Hyades and AB Dor moving group.

## References


Lépine, S. & Shara, M. M. 2005, *AJ*, 129, 13
Lépine, S. & Gaidos, E. 2011, *AJ*, 142, 138
Lépine, S. & Simon, M. 2009, *AJ*, 137, 383
Nordström, B., *et al.* 2004, *A&A*, 418, 989

*Young Stars & Planets Near the Sun*
Proceedings IAU Symposium No. 314, 2015
J. H. Kastner, B. Stelzer, & S. A. Metchev, eds.

© International Astronomical Union 2016
doi:10.1017/S1743921315006602

# Kinematics of M dwarfs in the CARMENES Input Catalogue: Membership in Young Moving Groups

**D. Montes[1], J. A. Caballero[2], I. Gallardo[1], M. Cortés-Contreras[1], and F. J. Alonso-Floriano[1]**

[1] Dpto. Astrofísica, Facultad de CC. Físicas, Universidad Complutense de Madrid, E-28040 Madrid, Spain
email: dmontes@ucm.es

[2] Centro de Astrobiologia (CSIC-INTA), Campus ESAC, PO Box 78, E-28691 Villanueva de la Cañada, Madrid, Spain

**Abstract.** We present a detailed study of the kinematics of M dwarfs in the CARMENES (Calar Alto high-Resolution search for M dwarfs with Exoearths with Near-infrared and optical Échelle Spectrographs) input catalog. We have selected all M dwarfs with known parallactic distance or a good photometric distance estimation, precise proper motion in the literature or as determined by us, and radial velocity measurements. Using these parameters, we computed the M dwarfs galactic space motions $(U, V, W)$. For the stars with $U$ and $V$ velocity components inside or near the boundaries that determine the young disk population, we have analyzed the possible membership in the classical moving groups and nearby loose associations with ages between 10 and 600 Myr. For the candidate members, we have compiled information available in the literature in order to constrain their membership by applying other age-dating methods.

**Keywords.** Galaxy: open clusters and associations, Stars: kinematics and dynamics, Stars: late-type, proper motions

## 1. Introduction

We are compiling the most comprehensive database of M dwarfs ever built, CARMENCITA, the CARMENES Cool dwarf Information and daTa Archive, which will be the CARMENES 'input catalog' (Quirrenbach *et al.* 2014). In addition to the science preparation with low- and high-resolution spectrographs and lucky imagers (see Alonso-Floriano *et al.* 2015; Cortés-Contreras *et al.* 2014; Passegger *et al.* 2015), we have compiled a large amount of public data on over 2100 M dwarfs, and we are analyzing them. Here we describe the preliminary results about the kinematics derived from these data.

## 2. Membership in young moving groups

From CARMENCITA we have now 1462 M dwarfs with all parameters (distance, radial velocity and proper motion) needed to determine the Galactic velocity components $(U, V, W)$. When parallactic distance is not available we have derived a photometric distance. For a large group of stars (in particular 214 stars in Lépine & Shara (2005), LSPM) with imprecise proper motions or missing uncertainties in the literature we have determined the proper motions using data from 2MASS, CMC14, CMC15, GSC2.3, USNO-A2, SDSS-DR9, ALLWISE and SuperCOSMOS. Figure 1 shows the $(U, V, W)$ diagrams for the 1462 M dwarfs in CARMENCITA with accumulated data and the possible candidate members of moving groups (Montes *et al.* 2001; Montes 2010, 2015; Klutsch *et al.* 2014).

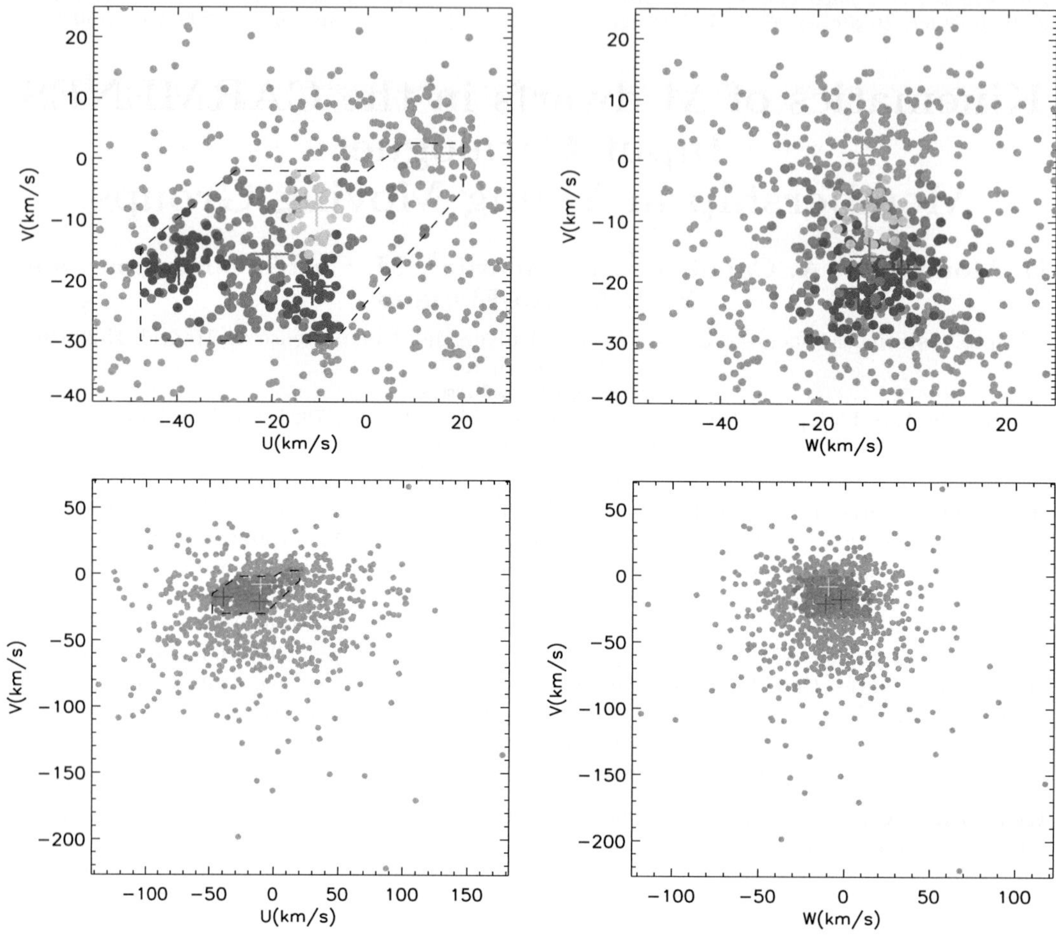

**Figure 1.** *Upper panel*: a zoom in *UV* and *WV* diagrams around the boundaries (dashed line) that determine the young disk population as defined by Eggen (1989) and the position of the classical moving groups (Local Association, Hyades Supercluster, Ursa Major, Castor, IC 2391 and Hercules-Lyra). The different colors indicate the possible candidates to each moving group. *Lower panel*: The full range of motions on *UV* and *WV* diagrams.

## Acknowledgements

This work was supported by the Univ. Complutense de Madrid (UCM) and the Ministry of Economy and Competitiveness (MINECO) under grant AYA2011-30147-C03-02.

## References

Alonso-Floriano, F. J., Morales, J. C., Caballero, J. A., *et al.* 2015, *A&A*, 577, A128
Cortés-Contreras, M., Caballero, J. A., & Montes, D. 2014, *The Observatory*, 134, 348
Eggen, O. J. 1989, *PASP*, 101, 366
Lépine, S. & Shara, M. M. 2005, *AJ*, 129, 1483
Klutsch, A., Freire Ferrero, R., Guillout, P., *et al.* 2014, *A&A*, 567, A52
Montes, D., López-Santiago, J., Gálvez, M. C., *et al.* 2001, *MNRAS*, 328, 45
Montes, D. 2010, *IAU Special Session SpS7, Highlights of Astronomy*, 15, E1
Montes, D. 2015, *Highlights of Spanish Astrophysics VIII*, p. 606
Passegger, V. M., Wende, S., Reiners, A. 2104, *Proceedings of Towards other Earths II*
Quirrenbach, A., Amado, P. J., Caballero, J. A., *et al.* 2014, *SPIE Proceeding*, 9147, 91471F

*Young Stars & Planets Near the Sun*
*Proceedings IAU Symposium No. 314, 2015*
*J. H. Kastner, B. Stelzer, & S. A. Metchev, eds.*

© International Astronomical Union 2016
doi:10.1017/S1743921315006183

# A Wide Angle Search for Hot Jupiters and Pre-Main Sequence Binaries in Young Stellar Associations

**Ryan J. Oelkers[1], Lucas M. Macri[1], Jennifer L. Marshall[1], Darren L. DePoy[1], Carlos Colazo[2], Pablo Guzzo[2], Diego G. Lambas[2,3], Ceci Quiñones[2], Katelyn Stringer[1], Luis Tapia[2], Colin Wisdom[1]**

[1]George P. and Cynthia Woods Mitchell Institute for Fundamental Physics and Astronomy,
Dept. of Physics and Astronomy, Texas A&M University, College Station, TX 77843, USA
[2]Observatorio Astronómico, Universidad Nacional de Córdoba
[3]Instituto de Astronomía Teórica y Experimental, IATE-CONICET
email: ryan.oelkers@physics.tamu.edu

**Abstract.** The past two decades have seen a significant advancement in the detection, classification and understanding of exoplanets and binary star systems. The vast majority of these systems consist of stars on the main sequence or on the giant branch, leading to a dearth of knowledge of properties at early times ($< 50$ Myr). Only one transiting planet candidate and a dozen eclipsing binaries are known among pre-main sequence objects, yet these are the systems that can provide the best constraints on stellar and planetary formation models. We have recently completed a photometric survey of 3 young ($< 50$ Myr), nearby (D$< 150$ pc) moving groups with a small-aperture instrument, nicknamed "AggieCam". We detected 7 candidate Hot Jupiters and over 200 likely pre-main sequence binaries, which are now being followed up photometrically and spectroscopically.

**Keywords.** stars: pre-main sequence, binaries: eclipsing, planets and satellites: detection

---

## 1. Introduction and Methods

AggieCam consists of a 16 Mpix CCD, a 54-mm aperture Mamiya lens and UV/IR cut filter (400-700 nm). The plate scale is 6.2″/pix and the field of view is 50 sq. degrees. AggieCam was deployed at two locations in Argentina through a collaboration with the Observatory of Córdoba University and IATE, UNC-CONICET.

We surveyed 3 young southern stellar associations, listed in Table 1, from June 2013 to September 2014. We used the difference imaging pipeline of Oelkers *et al.* (2015) for data reduction and analysis and achieved a photometric precision of 2 mmag at V=10 in 30 minutes (Fig. 1). Hot Jupiter (HJ) candidates were identified using the Box Least Squares algorithm (Kovács *et al.* 2002), requiring signal-to-pink noise $> 7$, 3+ transits and $\Delta\chi^2_+/\Delta\chi^2_- > 1$. Pre-main sequence binary candidates (PMB) were selected using a Lomb-Scargle periodogram (Lomb 1976, Scargle 1982) by searching for periods with false alarm probability $< 5\%$ falling outside of known aliases. Nearly 300,000 stars were monitored across all fields, yielding over 200 likely binaries, 7 possible transiting planets and a variety of transient and variable objects (Fig. 2)

## 2. Ongoing and Future Work

Higher-precision follow-up photometry is being obtained using telescopes at McDonald Observatory, Las Cumbres Observatory Global Telescope Network, Bosque Alegre Astro-

**Table 1.** Young stellar associations observed with AggieCam

| Association Name | Coordinates RA | Dec | Age[1] [Myr] | D[1] [pc] | Candidates PMB | HJ | Hours Observed |
|---|---|---|---|---|---|---|---|
| Upper Scorpius | 16h30m | −24° | 5-11 | 145 | 161 | 6 | ∼100 |
| $\eta$ Chamaeleontis | 08h45m | −79° | 2-18 | 97 | 67 | 1 | ∼75 |
| IC 2391 | 08h40m | −53° | 30-50 | 147 | 145 | 0 | ∼40 |

*Notes:* [1] As detailed in Pettersson 2008.

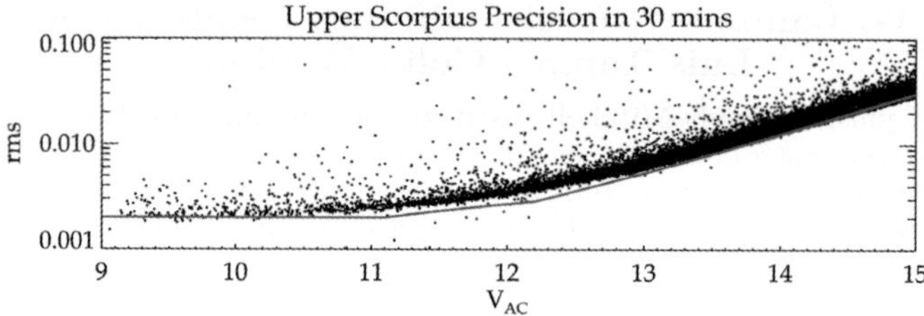

**Figure 1.** Expected (line; photon, sky and scintillation noise) and obtained (dots) photometric precision for AggieCam in 30 minutes of observation for the Upper Scorpius association.

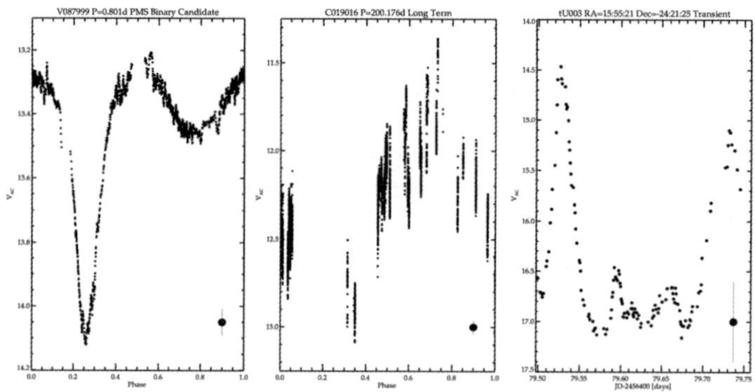

**Figure 2.** AggieCam light curves (smoothed over 10 minutes) of a sample of variable objects. Left: PMB candidate in IC 2391; Middle: long-term variable (P∼200d) in $\eta$ Chamaeleontis; Right: a transient in Upper Scorpius.

physical Station and the Texas A&M campus observatory. Initial photometric follow up has ruled out all the HJ candidates. We are also using the SES instrument (McCarthy *et al.* 1993) at the McDonald 2.1-m telescope for initial radial velocity measurements and to search for the Li I feature at 6708 Å as an indicator of young age. We expect to complete the initial follow up by the end of summer 2015.

## References


Kovács, G., Zucker, S., & Mazeh, T. 2002, *A&A*, 391, 369

Lomb, N. R. 1976, *Ap&SS*, 39, 447

McCarthy, J. K., Sandiford, B. A., Boyd, D., & Booth, J. 1993, *PASP*, 105, 881

Oelkers, R. J., Macri, L. M., Wang, L., Ashley, M. C. B., Cui, X., Feng, L.-L., Gong, X., Lawrence, J. S., Qiang, L., Luong-Van, D., Pennypacker, C. R., Yang, H., Yuan, X., York, D. G., Zhou, X., & Zhu, Z. 2015, *AJ*, 149, 50

Pettersson, B., 2008, in Handbook of Star Forming Regions, Volume II, 43, ed. Reipurth, B. (San Francisco: Astronomical Society of the Pacific)

Scargle, J. D. 1982, *ApJ*, 263, 835

*Young Stars & Planets Near the Sun*
*Proceedings IAU Symposium No. 314, 2015*
J. H. Kastner, B. Stelzer, & S. A. Metchev, eds.

© International Astronomical Union 2016
doi:10.1017/S1743921315006651

# Searching for Young M Dwarfs Near the Sun: Fast Rotators in the K2 Field

**Dicy Saylor[1] and Lepine[1]**

[1]Georgia State University
email: dsaylor@astro.gsu.edu, slepine@astro.gsu.edu

**Abstract.** We present a novel way to mine the Kepler data for young M dwarfs.

**Keywords.** stars: rotation, methods: data analysis

## 1. Introduction

The Kepler 2 (K2) mission is targeting large numbers of nearby K and M dwarfs selected from the SUPERBLINK proper motion survey. Kepler K2 Campaign 0 and 1 monitored a total of 8607 of these cool main-sequence stars. We used the Auto-Correlation Function to search for fast rotators by examining photometric modulation in the light curves due to star spots. We identified dozens of candidate fast rotators with rotation periods less than 3 days. We investigate the possibility that they are members of nearby young moving groups. We also discuss the potential of the K2 mission to identify new nearby young K & M dwarfs.

## 2. Methods

We cross-matched the SUPERBLINK Catalog with the Ecliptic Plane Input Catalog (EPIC) to find high proper motion ($> 40mas/yr$) K and M stars that were observed by Kepler K2 Campaign 0 and 1. We then ran an Auto-Correlation Function (ACF) period finding routine on those resulting 8607 stars. This ACF routine is based on the work in McQuillan *et al.* (2013). We preferentially selected for stars with rotation rates $0.25d < P < 4d$, which for K-M dwarfs is typical of stars with ages $< 150Myr$ (Mamajek & Hillenbrand 2008). Our search identified 46 candidate fast rotators in Campaign 0 and 214 fast rotators in Campaign 1 (due to page limit, email the authors for an electronic copy of the candidates). An example light curve and corresponding ACF is plotted in Figure 1. We estimated photometric distances for all the stars, and calculated their transverse components of motion, equivalent to [V,W] for campaign 0 stars, which are near the Galactic anti-center, and [U,V] for campaign 1 stars, which are near the Galactic pole.

## 3. Conclusions

Figure 2 plots the UVW space velocities of the Kepler K2 fast rotators found by this analysis. A table listing the results of this work can be supplied by the author by request. The fast rotators generally have similar kinematics to known nearby moving groups. There is a bias towards slightly lower velocities as the distances are underestimated for these young targets. Our analysis indicates that a measured fast rotation period can be an indicator of youth. Before the Kepler mission is was not feasible to measure rotation periods simply in the interest of diagnosing age. Our work implies that rotation rate

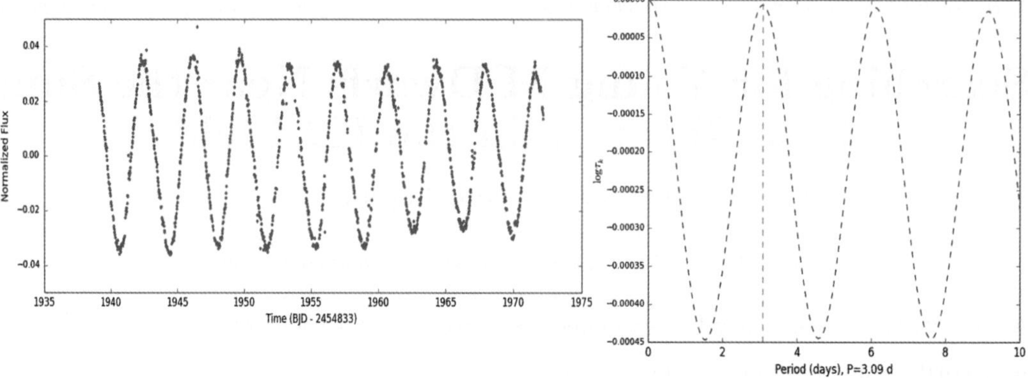

Figure 1: Example light curve of fast rotator EPIC 202068580 discovered in our search of Kepler K2 campaign 0 (left), and the corresponding auto-correlation function (right). The rotation period was calculated using an Auto-Correlation Function (ACF) developed by the authors. The resulting ACF is shown to the right of the light curve, with the detected period ($3.09 days$) indicated by a dashed red line.

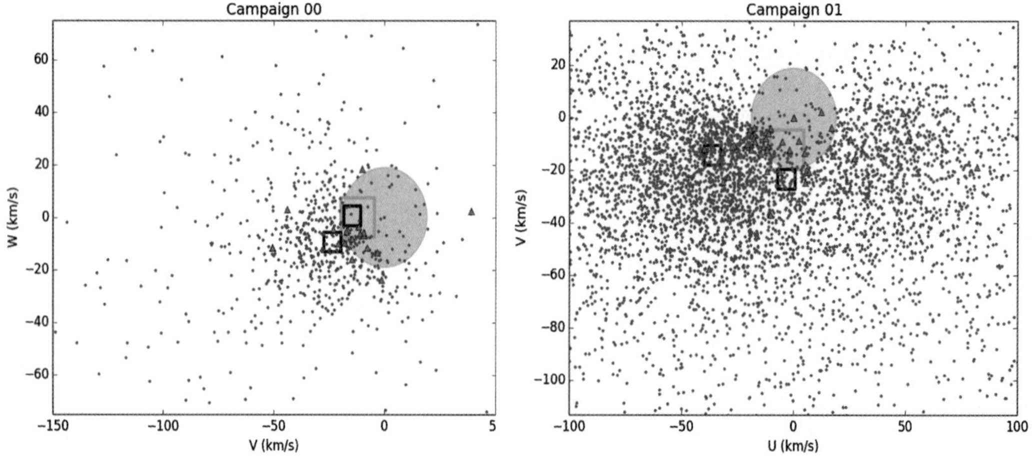

Figure 2: The derived UVW space velocities of SUPERBLINK high proper motion stars in Kepler K2 campaign 0 and 1. The blue points are all SUPERBLINK stars in the field. The fast rotators are overplotted in red triangles. The red circle represents the incompletely sampled area of SUPERBLINK. The green square indicates the average UVW velocities of the nearby kinematic moving groups while the black squares indicate the Hyades and Pleiades motions. The SUPERBLINK fast rotators show similar space kinematics to the nearby young moving groups, indicating youth.

could be an independent indicator of youth. Future work will include searching upcoming Kepler K2 campaigns for fast rotators and computing important statistics.

## References

Mamajek, E. E. & Hillenbrand, L. A. 2008, *ApJ*, 687, 1264
McQuillan, A., Aigrain, S., & Mazeh, T. 2013, *MNRAS*, 432, 1203

*Young Stars & Planets Near the Sun*
*Proceedings IAU Symposium No. 314, 2015*
*J. H. Kastner, B. Stelzer, & S. A. Metchev, eds.*

© International Astronomical Union 2016
doi:10.1017/S1743921315006596

# The All Sky Young Association (ASYA): a New Young Association

## C. A. O. Torres[1], G. R. Quast[1], and D. Montes[2],

[1] Laboratório Nacional de Astrofísica/ MCT, Rua Estados Unidos 154, 37504-364, Itajubá, MG, Brazil
email: beto@lna.br

[2] Dpto. Astrofísica, Facultad de CC. Físicas, Universidad Complutense de Madrid, E-28040 Madrid, Spain
email: dmontes@ucm.es

**Abstract.** To analyze the SACY (Search for Associations Containing Young stars) survey we developed a method to find young associations and to define their high probability members. These bona fide members enable to obtain the kinematical and the physical properties of each association in a proper way. Recently we noted a concentration in the $UV$ plane and we found a new association we are calling ASYA (All Sky Young Association) for its overall distribution in the sky with a total of 38 bonafide members and an estimated age of 110 Myr, the oldest young association found in the SACY survey. We present here its kinematical, space and Li distributions and its HR diagram.

**Keywords.** Galaxy: open clusters and associations, Stars: kinematics and dynamics, Stars: late-type

## 1. Introduction

The SACY (Search for Associations Containing Young Stars) survey was a spectroscopic effort to find southern nearby associations using as targets the possible Tycho–2/HIPPARCOS stars counterparts of the ROSAT All-Sky Bright Sources Catalogue (Torres *et al.* 2006; 2008). To analyze the SACY survey we developed a method to find young associations and to define their high probability members. These bonafide members enable to obtain the kinematical and the physical properties of each association in a proper way.

## 2. ASYA (All Sky Young Association)

Recently we noted a concentration in the $UV$ plane and we found a new association we are calling ASYA (All Sky Young Association) for its overall distribution in the sky. We also search HIPPARCOS catalogue for other possible members and we found a total of 38 bonafide members (including those from HIPPARCOS and SACY). Although kinematically ($U = -15.2, V = -26.9, W = -2.8$ km/s) near the Her-Lyr moving group, ASYA is definitively distinct from it and younger - we estimated an age of 110 Myr, the oldest of the young associations found in the SACY survey. A weak expansion in the X direction is present $U = <U> +k(X- <X>)$ and the solution was obtained with a $k = 0.02$. We present its kinematical ($UVW$) and space ($XYZ$) distributions in Fig. 1 and its Li distribution ($EW$(Li) vs $V - I$) and its HR diagram ($M_V$ vs. $V - I$) in Fig. 2.

## Acknowledgements

This work was supported by the Univ. Complutense de Madrid (UCM) and the Ministry of Economy and Competitiveness (MINECO) under grant AYA2011-30147-C03-02.

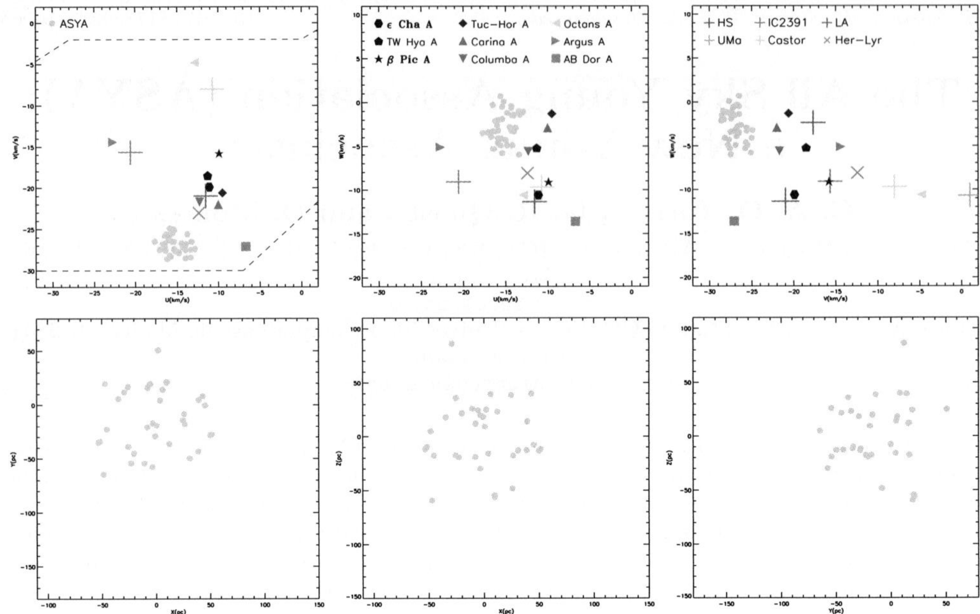

**Figure 1.** *Upper panel*: $UVW$ space for ASYA showing the well defined kinematical clustering. The position of the classical moving groups (Hyades Supercluster (HS), Ursa Major (UMa), IC2391, Castor, Local Association (LA) and Hercules-Lyra, Montes *et al.* 2001 and Montes 2010) and the young associations of SACY ($\epsilon$ Cha, TW Hya, $\beta$ Pic, Tuc-Hor, Carina, Columba, Octans, Argus and AB Dor) are also presented for comparison. Note that ASYA is clearly separated from Her-Lyr and the other younger associations. *Lower panel*: $XYZ$ space for ASYA.

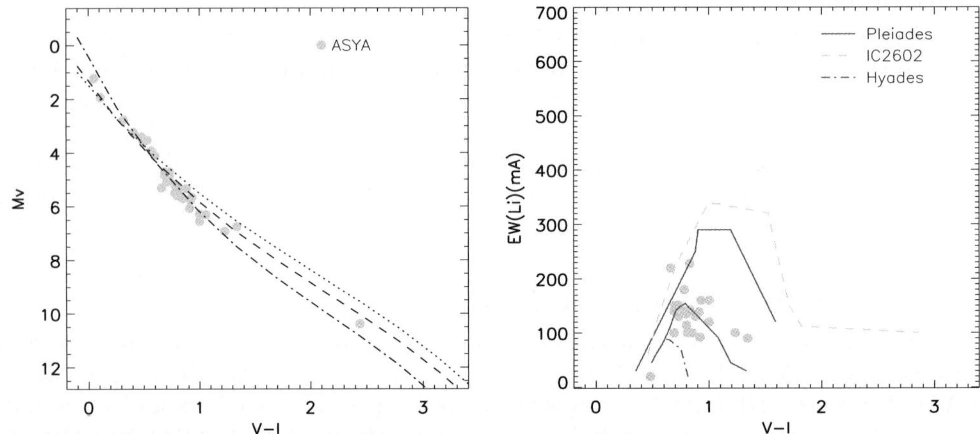

**Figure 2.** *Left panel*: The HR diagram for the proposed members of ASYA. The over-plotted curves are our ad-hoc isochrones. The upper curve is estimated as about 20 Myr and the lower curve 110 Myr. *Right panel*: Lithium distribution ($EW$(Li) vs $V - I$) of ASYA compared with the Li distribution of IC 2602, Pleiades and Hyades open clusters.

### References


Montes, D., López-Santiago, J., Gálvez, M. C., *et al.* 2001, *MNRAS*, 328, 45

Montes, D. 2010, *IAU Special Session SpS7, Highlights of Astronomy*, 15, E1

Torres, C. A. O., Quast, G. R., da Silva, L., *et al.* 2006, *A&A*, 460, 695

Torres, C. A. O., Quast, G. R., Melo, C. H. F., & Sterzik, M. F. 2008, *Handbook of Star Forming Regions*, Volume II, 757

*Young Stars & Planets Near the Sun*
Proceedings IAU Symposium No. 314, 2015
J. H. Kastner, B. Stelzer, & S. A. Metchev, eds.

© International Astronomical Union 2016
doi:10.1017/S1743921315006067

# Stellar Evolution Models of Young Stars: Progress and Limitations

**Gregory A. Feiden**[1]

[1]Department of Physics & Astronomy, Uppsala University, SE-751 20 Uppsala, Sweden
email: `gregory.feiden@physics.uu.se`

**Abstract.** Stellar evolution models are a cornerstone of young star astrophysics, which necessitates that they yield accurate and reliable predictions of stellar properties. Here, I review the current performance of stellar evolution models against young astrophysical benchmarks and highlight recent progress incorporating non-standard physics, such as magnetic field and starspots, to explain observed deficiencies. While addition of these physical processes leads to improved agreement between models and observations, there are several fundamental limitations in our understanding about how these physical processes operate. These limitations inhibit our ability to form a coherent picture of the essential physics needed to accurately compute young stellar models, but provide rich avenues for further exploration.

**Keywords.** stars: evolution, stars: fundamental parameters, stars: interiors, stars: low-mass, stars: magnetic fields, stars: pre-main-sequence, stars: spots

## 1. Introduction

Stellar evolution models are essential for understanding the time dependence of astrophysical phenomena. Theory provides absolute ages of stellar systems, which are particularly consequential for young populations, where a variety of processes are on-going that lead directly to observed properties of older stars and stellar systems. Ages provided by young stellar models inform our understanding about the lifetimes of protoplanetary disks, timescales for the formation of giant planets, timescales for stellar angular momentum evolution, and the time evolution of stellar magnetic activity. Furthermore, young stellar ages provide information regarding the formation and thermal evolution of stars and sub-stellar objects. Given their role in providing constraints on a number of astrophysical processes, it is critical that stellar models provide accurate characterizations of stellar populations, further necessitating that models yield accurate properties of individual stars.

This review attempts to consolidate our current understanding about the performance of young stellar evolution models and highlight current efforts to improve their accuracy. Development of young star models has a rich history; however, it will not be possible to provide a comprehensive historical review. Several shorter reviews exist for the interested reader (e.g., Stahler 1988). Nevertheless, much of what we know today is the result of over half a century of hard work and dedication of theorists and observers, alike.

## 2. Current Performance

Performance of stellar evolution models can be assessed using two complementary astrophysical benchmarks: color-magnitude diagrams and touchstone stars. Each reveals different information about the successes and shortcomings of stellar evolution models.

## 2.1. *Color-Magnitude Diagrams*

Color-magnitude diagrams provide extensive diagnostic information about the validity of stellar evolution models across a range of effective temperatures and luminosities (see Bell, this volume). The current state of modeling color-magnitude diagram morphologies is conflicting, but can be roughly divided into three sub-categories encompassing clusters or stellar associations younger than 20 Myr, those between 20 and 100 Myr, and those that are older than 100 Myr. For the youngest populations, there is evidence of a significant luminosity spread at constant color in color-magnitude diagrams, a spread that cannot be reproduced by a single standard stellar model isochrone (e.g., Hillenbrand 1997; Da Rio *et al.* 2010). Observational errors do not appear to be the source of the spread. Instead, spreads are indicative of either genuine age spreads of several Myr resulting from extended star formation processes or they are highlighting effects that result from physics not presently included in standard stellar models. The precise origin of luminosity spreads is unresolved (e.g., Jeffries 2012), but multiple mechanisms have been identified that are able to generate intrinsic spreads in color-magnitude diagrams without the need to invoke extended star formation events. These include mechanisms like starspots (Somers, this volume) and episodic accretion (Baraffe *et al.* 2010).

Beyond about 20 Myr, the luminosity scatter in observed color-magnitude diagrams starts to decrease with single standard stellar model isochrones providing more accurate representations of the data. This may simply be due to the fact that relative age differences of several Myr grow increasingly irrelevant after about 20 Myr or it may be the result of non-standard physics playing less of a role in governing the observed properties of stars. For example, the effects of early episodic accretion are only expected to be noticeable up to ages of around 20 Myr before theoretical predictions converge with standard stellar model results. Alternatively, it may be a combination of the two effects. However, this is contrasted by the fact that, for ages below about 100 Myr, there are noticeable disagreements between ages of young stellar populations derived from cool low-mass stars and those from hotter intermediate-mass stars that have reached the main sequence, with ages decreasing as stellar mass decreases (e.g., Hillenbrand 1997; Mamajek & Bell 2014; Herczeg & Hillenbrand 2015). So, while apparent age scatters at a given color decrease, standard models are unable to provide consistent age estimates across the color-magnitude diagram for young stellar populations, strongly suggesting that there are missing physics in stellar models.

For ages older than 100 Myr, a majority of the stars in young stellar populations are either on the main-sequence or their position in color-magnitude diagrams is indistinguishable from the zero-age main-sequence. The Pleiades provides a characteristic cluster sequence upon which to assess model validity for ages of approximately 100 Myr. The Pleiades has long been a source of trouble for stellar models of low-mass stars, with cooler K-dwarfs appearing to be significantly bluer than model and empirical predictions (Herbig 1962; Stauffer *et al.* 2003). However, current generations of stellar models provide reasonable agreement across various color-magnitude diagrams, as demonstrated in Figure 1. Models yield excellent agreement in near-infrared (NIR) and optical color-magnitude diagrams until the onset of strong TiO absorption, signalling the beginning of the M dwarf sequence around $M_V = 8.0$ mag. Notably, a single standard model isochrone provides a consistent fit to the color-magnitude diagrams, in contrast with results from younger populations.

Furthermore, agreement is consistent across different flavors of models, as seen in Figure 1. This improvement appears to be largely related to the adoption of updated model atmospheres used to prescribe bolometric corrections, which adopt recent solar

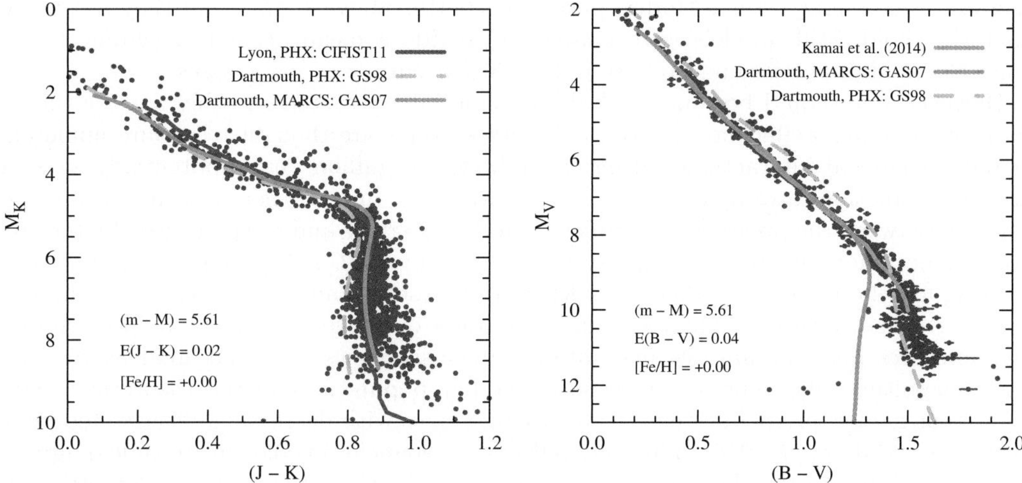

**Figure 1.** Color-magnitude diagrams for the Pleiades. Data are from Stauffer *et al.* (2007) and Kamai *et al.* (2014). (*left*) 2MASS $(J - K)$-$M_K$. (*right*) Johnson $(B - V)$-$M_V$. Dartmouth (Feiden *et al.*, in prep) and Lyon (Baraffe *et al.* 2015) stellar evolution isochrones are overplotted. Dartmouth isochrones are shown with different solar abundances and bolometric corrections. An empirical isochrone from Kamai *et al.* (2014) is also shown in the right panel.

abundances (Baraffe, this volume; Grevesse *et al.* 2007; Caffau *et al.* 2011). In particular, lower oxygen abundances decrease the amount of water absorption in the NIR, yielding better agreement with NIR colors and providing support for their adoption in the face of known problems with helioseismology (Basu & Antia 2004). Line lists and oscillator strengths for TiO and various monohydrides appear to still be a problem for the coolest models in optical passbands. As may be expected, optical colors are still fairly unreliable when it comes to drawing conclusions about young M stars.

Unfortunately, despite the utility of color-magnitude diagrams, they can mask significant errors present in stellar models. Bolometric corrections and photometric colors derived from either theoretical model atmosphere calculations or empirical methods can provide a counter-balance against more fundamental model errors.

### 2.2. *Stellar Fundamental Properties*

Observations of touchstone (or benchmark) stars allow tests of stellar evolution theory at the most fundamental level by providing direct measurements of stellar masses, radii, effective temperatures, and luminosities. Dynamical systems with measured stellar masses are of greatest interest, as mass is a direct input parameter in stellar models, largely determining the entire evolution of a star. Over the past decade, several studies have leveraged dynamical information from young binary systems to confront stellar model predictions (e.g., Hillenbrand & White 2004; Mathieu *et al.* 2007; Stassun *et al.* 2014). Each study has invariably led to the same result: modern stellar evolution models are unable to accurately predict the properties of young, low-mass stars in binary systems.

Hillenbrand & White (2004) demonstrated that young stellar models systematically predict masses that are too small for a given stellar effective temperature and luminosity (radius). Their study relied on a large collection of predominantly astrometric binaries, with a few known eclipsing binary systems. Recently, Stassun *et al.* (2014) showed that the latest generations of stellar models fair no better than those adopted by Hillenbrand & White (2004). Using a smaller sample of carefully selected eclipsing binaries, stellar models were shown to overestimate observed masses for primaries, but underestimate masses for secondary stars. Models performed better above 1 $M_\odot$ than they did below that threshold. Still, models were unable to provide a mean absolute accuracy better than 5% at high masses and no better than 20% accuracy at low masses.

However, one should be concerned about using binary systems to characterize the reliability of young stellar models. Young binary systems are thought to endure numerous strong dynamical interactions during their first few million years, potentially slowing stellar contraction rates by pumping energy into stars. In fact, there is a tantalizing correlation between the presence of significant modeling errors and the presence of a tertiary companion among eclipsing binary systems (Stassun *et al.* 2014), suggesting that tidal heating may be important when considering the structure and evolution of young stars.

Nevertheless, investigations focusing on samples of single stars with well determined temperatures and luminosities find internal inconsistencies in stellar models. As with color-magnitude diagrams, comparisons of stellar populations in the theoretical plane show that models predict systematically younger ages with decreasing effective temperatures (e.g., Malo *et al.* 2014). This may reflect discrepancies in color-magnitude diagrams being translated to the theoretical plane with the adoption of empirical bolometric corrections. However, Malo *et al.* (2014) relied on an independent method to arrive at the same conclusion, suggesting the result is a general feature of stellar model predictions.

## 3. Progress & Fundamental Limitations

Efforts are currently underway to mitigate the aforementioned discrepancies between young stellar models and observations. These efforts have largely, though not exclusively, focused on magnetic fields and starspots.

### 3.1. *Magnetic Fields*

Stellar evolution models that include effects related to the presence of globally pervasive magnetic fields (D'Antona *et al.* 2000; Mullan & MacDonald 2001; Feiden & Chaboyer 2012) were recently applied to young stellar systems (MacDonald & Mullan 2010; Malo *et al.* 2014). Magnetic fields can inhibit large-scale convective flows, decreasing a star's net outward flux of energy. For young stars that are powered by the release of gravitational potential energy, decreasing their convective energy flux cools their effective temperature and traps energy within their interiors, slowing their contraction. Magnetic young stars are therefore cooler and larger at a given age than their non-magnetic counterparts, but note that radii of young stars are not "inflated" in the same sense as radii of low-mass main sequence stars; magnetic young stars simply undergo a more gradual contraction.

Magnetic inhibition of convection provides a natural explanation for observed age differences as a function of effective temperature in young stellar associations (see, e.g., MacDonald & Mullan 2010; Malo *et al.* 2014). At the same time, decreasing the effective temperature of young stars leads to an overall decrease in internal temperatures owing to their largely adiabatic stratification. Lithium depletion timescales are extended as a result, since it takes longer for stellar interiors to reach temperatures required for lithium destruction ($\sim$ 2.5 MK; Malo *et al.* 2014), leading to better agreement between ages derived from the location of the lithium depletion boundary and other methods. Notably, MacDonald & Mullan (2010; and later Malo *et al.* 2014) showed that the apparent age of 5 Myr derived for the low-mass stars in the $\beta$-Pic moving group can be reconciled with the 20 – 30 Myr age estimate from the higher mass population, lending some support to the idea that magnetic fields are important in early phases of stellar evolution.

It is encouraging to note that magnetic models of young stars can reconcile ages of low-mass stars with the higher mass population using seemingly reasonable magnetic field properties. Surface magnetic field strengths are $\sim 3$ kG (equipartition values in low-mass K and M stars), while interior field strengths are $\sim 50$ kG. As tantalizing as magnetic fields may be, their adoption must be approached with caution. Significant uncertainties exist about what constitutes realistic interior and surface magnetic field strengths for young stars, particularly as a function of age and effective temperature. Nevertheless, magnetic models provide firm predictions about surface magnetic field strengths that require validation.

### 3.2. *Starspots*

An alternative to global magnetic inhibition of convection is that starspots block outgoing flux, causing stars to thermally restructure (Spruit & Weiss 1986; Jackson *et al.* 2009). While starspots are, in part, the physical manifestation of inhibited convection on the surface of stars, their effects on stellar positions in color-magnitude diagrams are distinct. Global inhibition of convection leads to an overall shift of theoretical predictions toward redder colors, primarily due to their cooler effective temperatures. In contrast, starspots create a scenario where there are regions of different temperatures on the stellar surface, leading to emergent spectra and colors that are a combination of flux from various surfaces. Starspots can shift stars toward either bluer or redder colors as a result, depending on the precise areal coverages, temperature contrasts, and photometric colors under consideration (Jackson & Jeffries 2014).

Jackson *et al.* (2009) demonstrate that spotted models can match observed morphologies of color-magnitude diagrams and fundamental properties of young low-mass stars. In particular, redistribution of flux throughout the stellar convection zone leads to unspotted regions of a star appearing hotter than for the same star without spots. This provides a natural explanation for the observed blue color of Pleiades K dwarfs and excess flux in blue regions of K dwarf spectra (Stauffer *et al.* 2003). Since spots also block outgoing flux, they can slow the contraction rate of young stars in much the same way as magnetic inhibition of convection (Jackson & Jeffries 2014). Finally, starspots may provide an explanation of observed spreads in color-magnitude diagrams and spreads in measured lithium abundances in young stellar populations (see Somers, this volume).

However, as with magnetic fields, spotted models must be approached with caution. There is some empirical data yielding estimates of starspot areal coverages and temperature contrasts, but only for a few isolated cases, making the process of validating spotted model predictions difficult. Synthesizing data from several color-magnitude diagrams and photometric lightcurves is currently the best means of constraining these models. Still, starspot physics is not very well understood. It is not clear that flux trapped by any given starspot is redistributed throughout the convection zone before the star is able to radiate away that excess energy (e.g., Spruit 1982). Nor is it understood how this redistribution occurs: is flux completely redistributed without a need for the star to restructure, or must a star thermally restructure? These are interesting questions that we are now beginning to address, at least in the context of stellar structure and evolution. Often ignored for simplicity, starspots may come to play an important role in our understanding of young, low-mass stars.

## References


Baraffe, I. & Chabrier, G. 2010, *A&A*, 521, A44
Baraffe, I., Homeier, D., Allard, F., & Chabrier, G. 2015, *A&A*, 577, A42
Basu, S. & Antia, H. M. 2004, *ApJL*, 606, L85
Caffau, E., Ludwig, H.-G., Steffen, M., Freytag, B., & Bonifacio, P. 2011, *SoPh*, 268, 255

Da Rio, N., Robberto, M., & Soderblom, D. R., et al. 2010, *ApJ*, 722, 1092

D'Antona, F., Ventura, P., & Mazzitelli, I. 2000, *ApJ*, 543, L77

Feiden, G. A. & Chaboyer, B. 2012, *ApJ*, 761, 30

Grevesse, N., Asplund, M., & Sauval, A. J. 2007, *SSRv*, 130, 105

Herbig, G. H. 1962, *ApJ*, 135, 736

Herczeg, G. J. & Hillenbrand, L. A. 2015, *arXiv*: 1505.06518

Hillenbrand, L. A. 1997, *AJ*, 113, 1733

Hillenbrand, L. A. & White, R. J. 2004, *ApJ*, 604, 741

Jackson, R. J. & Jeffries, R. D. 2014, *MNRAS*, 441, 2111

Jackson, R. J., Jeffries, R. D., & Maxted, P. F. L. 2009, *MNRAS*, 339, L89

Jeffries, R. D. 2012, in Star Clusters in the Era of Large Surveys, ed. A. Moitinho & J. Alves, 163

Kamai, B. L., Vrba, F. J., Stauffer, J. R., & Stassun, K. G. 2014, *AJ*, 148, 30

MacDonald, J. & Mullan, D. J. 2010, *ApJ*, 723, 1599

Malo, L., Doyon, R., & Feiden, G. A., et al. 2014, *ApJ*, 792, 37

Mamajek, E. E. & Bell, C. P. M. 2014, *MNRAS*, 445, 2169

Mathieu, R. D., Baraffe, I., Simon, M., Stassun, K. G., & White, R. 2007, in Protostars & Planets V, 411

Mullan, D. J. & MacDonald, J. 2001, *ApJ*, 559, 353

Spruit, H. C. 1982, *A&A*, 108, 348

Spruit, H. C. & Weiss, A. 1986, *A&A*, 166, 167

Stahler, S. W. 1988, *ApJ*, 332, 804

Stassun, K. G., Feiden, G. A., & Torres, G. 2014, *New Astronomy Reviews*, 60, 1

Stauffer, J. R., Hartmann, L. W., & Fazio, G. G., et al. 2007, *ApJS*, 172, 663

Stauffer, J. R., Jones, B. F., & Backman, D., et al. 2003, *AJ*, 126, 833


## Discussion

CHABRIER: My concern with invoking magnetic fields to explain shortcomings of fully convective stellar models is that interior magnetic field strengths need to be of order 1 MG or greater. Magnetic fields of such a magnitude will be buoyantly unstable and cannot be maintained by dynamo action. What are your thoughts about this?

FEIDEN: I agree that 1 MG or stronger magnetic fields deep within fully convective stars are probably not realistic and should be taken as a serious flaw in the models. However, this requirement is only necessary for fully convective *main sequence* stars. Typical interior magnetic field strengths invoked to reconcile models with observations of low-mass pre-main-sequence stars are on the order of 50 kG, quite consistent with estimates of dynamo generated field strengths from 3D MHD simulations.

PINSONNEAULT: Two points. First, adoption of the revised solar abundances not only introduces disagreement with the solar convection zone boundary, but predictions of neutrino rates are lower due to the low Fe abundance, although still formally consistent with observations. Second, we showed that not only did the Pleiades K dwarfs appear bluer in the CMD, but their spectra showed noticeable excess emission at bluer wavelengths. How would you explain this excess emission if theoretical colors are correct?

FEIDEN: Interesting question. Some of this may be the result of the metallicity difference between stars in Praesepe and the Pleiades. Praesepe stars may exhibit more significant line blanketing in the blue, leading to the appearance of sizable continuum excess at blue wavelengths for Pleiades stars. Other emission, say from Balmer lines, may just not contribute significantly to the integrated flux, thus playing a negligible role in governing broadband optical colors. This is testable with synthetic spectra.

*Young Stars & Planets Near the Sun*
Proceedings IAU Symposium No. 314, 2015
J. H. Kastner, B. Stelzer, & S. A. Metchev, eds.

© International Astronomical Union 2016
doi:10.1017/S1743921315006079

# Anomalous Spectral Types and Intrinsic Colors of Young Stars

## Mark J. Pecaut

Rockhurst University, 1100 Rockhurst Rd., Kansas City, MO 64110
email: mark.pecaut@rockhurst.edu

**Abstract.** We highlight differences in spectral types and intrinsic colors observed in pre-main sequence (pre-MS) stars. Spectral types of pre-MS stars are wavelength-dependent, with near-infrared spectra being 3-5 spectral sub-classes later than the spectral types determined from optical spectra. In addition, the intrinsic colors of young stars differ from that of main-sequence stars at a given spectral type. We caution observers to adopt optical spectral types over near-infrared types, since Hertzsprung-Russell (H-R) diagram positions derived from optical spectral types provide consistency between dynamical masses and theoretical evolutionary tracks. We also urge observers to deredden pre-MS stars with tabulations of intrinsic colors specifically constructed for young stars, since their unreddened colors differ from that of main sequence dwarfs. Otherwise, $V$-band extinctions as much as $\sim$0.6 mag erroneously higher than the true extinction may result, which would introduce systematic errors in the H-R diagram positions and thus bias the inferred ages.

**Keywords.** stars: pre-main sequence, stars: fundamental parameters, starspots, open clusters and associations: individual ($\eta$ Cha Cluster, TW Hydra Association, $\beta$ Pictoris moving group, Tucana-Horologium moving group)

---

## 1. Introduction

Two of the most fundamental parameters of a star – the effective temperature ($T_{\rm eff}$) and luminosity – are based on simple, easy-to understand data such as the spectral type, extinction, and bolometric corrections. Determining these should be relatively error-free, right? Experience has taught observers that special care must be taken when characterizing pre-main sequence (pre-MS) stars, as these young stars require more attention than that of main-sequence dwarfs.

When observers desire to characterize a star's properties, they normally start with the spectral type. The spectral type of the star is determined by comparing characteristics of the spectrum with spectral standard stars. In doing this we can obtain an estimate of the temperature and a gross estimate of the surface gravity of the star. To quantify the extinction and reddening, the observer will compare the target star's observed colors to tabulated intrinsic colors of stars of the same spectral type to determine a color excess and use a total-to-selective extinction ratio value to estimate the extinction. Once the extinction has been quantified, a distance can be estimated (if not known), by assuming an age and consulting a theoretical isochrone, or assuming the star is on the main-sequence and calculating a main-sequence distance. Alternatively, if a trigonometric or kinematic parallax is known, we may compute the luminosity of the star. Normally, the point of these calculations is to place the star on the Hertzsprung-Russell (H-R) diagram and compare it to theoretical evolutionary models to estimate an age and mass.

Though this process seems fairly straightforward, there are many assumptions and systematic effects which can creep in and result in systematic errors in the fundamental parameters, such as temperature and luminosity (which are relatively model-free), and therefore the parameters derived from the evolutionary models. Assuming we are able to determine the spectral type with perfect fidelity, there are uncertainties in the tables which relate the spectral type, $T_{\mathrm{eff}}$, intrinsic color and bolometric correction. There are also many different varieties of such tables, with slight variations in their intrinsic colors, underlying temperature scale, and bolometric corrections. Some tables even contain self-inconsistent values for the bolometric corrections and the bolometric luminosity of the Sun (Torres 2010)†!

For populations of young stars this presents many problems because systematic errors in the fundamental properties of young, pre-MS stars will propagate into systematic errors in masses and ages (see Soderblom *et al.* 2014 for a full discussion of ages of young stars). This can skew the inferred evolutionary lifetime of gas-rich disks. Since gas giant planets can only form when gas is present in the circumstellar disk, these ages are also used to constrain giant planet formation timescales. Problems are also present when individual stars are mischaracterized. If a young star hosts a directly-imaged substellar object, the mass of the substellar object is estimated by comparing the luminosity and assumed age of the object with evolutionary models. Systematic effects in assumed ages can then propagate to wrong assumptions about the model-derived masses (e.g., $\kappa$ And b; Hinkley *et al.* 2013; Bonnefoy *et al.* 2014), which may misdirect planet formation theories. Thus, systematic errors in individual parameters, such as spectral types and intrinsic colors, can propagate down and fundamentally limit our ability to test star and planet formation theories.

## 2. Spectral Types

The spectral type of a target is one of the most useful measurements, since many other stellar properties are usually derived with some dependence on the spectral type. Young stars, like most stars, are typed by comparing their spectra with that of spectral standards, and this has historically been performed using optical spectra. However, many low-mass stars are brighter in the near-infrared (NIR) and so it seems completely reasonable to perform this same measurement with NIR spectra as well. Two interesting cases are that of TW Hya and V4046 Sgr.

TW Hya, one of the most well-studied classical T-Tauri stars and a member of the youngest nearby moving group that bears its name (the TW Hydra Association), has typically been assigned a temperature type of K7 using optical spectra (K8IVe, Pecaut & Mamajek 2013; K6Ve, Torres *et al.* 2006; K6e, Hoff *et al.* 1998; K7e, de la Reza *et al.* 1989; K7 Ve, Herbig 1978). However, Vacca & Sandell (2011) assigned a type of M2.5V using NIR spectra from SpeX, which implied a very young age of $\sim$3 Myr instead of the much older, more often quoted age of $\sim$ 10 Myr (Barrado Y Navascués 2006). This discrepancy is a source of great confusion – which spectral type should one adopt when characterizing young stars?

V4046 Sgr, a young binary member of the $\beta$ Pictoris moving group harboring a gas-rich disk of its own, has also been typed in both the optical and NIR and the same effect is observed - the NIR spectral type is about 3-5 subtypes later than the optical spectral type (Kastner *et al.* 2015). However, unlike TW Hya, V4046 Sgr has dynamical mass constraints from radial velocities and gas dynamics (Rosenfeld *et al.* 2012). Kastner *et al.* (2015) have placed these two young stars on the H-R diagram assuming the optical spectral types in one case and the NIR spectral types in another case. Comparing these

---

† There are even a variety of values used for the bolometric magnitude of the Sun. Here we adopt $M_{bol,\odot} = 4.7554 \pm 0.0004$, as advocated in Mamajek (2012).

two sets of H-R diagram positions with theoretical evolutionary models, the NIR spectral types are inconsistent with the dynamical mass constraints and Kastner *et al.* (2015) thus urged caution against using the NIR spectral types on young stars.

Stauffer *et al.* (2003) have studied this wavelength-dependent spectral type effect among the zero-age main sequence K dwarfs in the Pleiades ($\sim$135 Myr; Bell *et al.* 2014). The Stauffer *et al.* (2003) study found that the spectral type of Pleiades K-type stars were systematically $\sim$1 subtype later in the red optical spectra than the blue optical spectra. Furthermore, they did not observe this spectral type anomaly in members of the older Praesepe cluster ($\sim$650-800 Myr; Gáspár *et al.* 2009; Bell *et al.* 2014; Brandt & Huang 2015). Stauffer *et al.* (2003) argued that spots were a major factor in this effect, and concluded that there must be more than one photospheric temperature present in the Pleiades K-dwarfs. Pre-main sequence stars are magnetically very active as well, and observations indicate large filling factors on their surfaces (Berdyugina 2005), consistent with this effect.

## 3. Intrinsic Colors

Many studies in the past two decades have pointed out that stellar intrinsic colors of young stars are different than that of typical main sequence dwarfs. Gullbring *et al.* (1998) had noted this for both classical and weak-lined T-Tauri stars in Taurus. Da Rio *et al.* (2010) have noted this in the Orion Nebula Cluster (ONC) and dereddened the very young ONC cluster members using a specially constructed spectral-type color sequence for young stars. However, their spectral type-color sequence was not published as part of their study. Luhman (1999) and Luhman *et al.* (2010b,a) went much further and constructed a color-spectral type sequence for pre-MS late K- and M-type stars and brown dwarfs. Most recently, the Pecaut & Mamajek (2013) study released a comprehensive tabulation of the intrinsic colors for pre-MS stars from F-type to late M-type with the most popular photometric bands – Johnson–Cousins $BVI_C$, 2MASS $JHK_S$ and the recently available WISE $W1$, $W2$, $W3$ and $W4$ bands. In addition, we fit the observed spectral energy distributions to Phoenix–NextGen synthetic spectra (Allard *et al.* 2012) to infer an effective temperature ($T_{\rm eff}$) and bolometric correction (BC) scale for pre-MS stars. This young spectral type–color–$T_{\rm eff}$–BC scale was constructed using all the known (as of July 2013) members of the nearest moving groups – the $\eta$ Cha Cluster, the TW Hydra Association (TWA), the $\beta$ Pictoris moving group and the Tucana–Horologium (Tuc-Hor) moving group.

Two color-color plots for young stars are shown in Figure 1. We plot $V-K_S$ on the horizontal axis as a proxy for $T_{\rm eff}$ against $J-H$ (left) and $K_S-W1$ (right). The stars are predominantly clustered around the dwarf and giant sequence until the two diverge around $V-K_S$ of $\sim$4, where they lie between the dwarf and giant locus. This strongly suggests surface gravity is a major factor in the deviation of intrinsic colors from that of dwarfs, since we expect that pre-MS stars will have surface gravities somewhere between that of dwarfs and giants.

An indication of the importance of accounting for the difference between pre-MS colors and dwarf colors is shown in Figure 2a. The $J-H$ colors of young stars are significantly redder than the dwarf colors at a given spectral type. If $J-H$ dwarf colors were used to estimate extinction for an unreddened M0 star, it would erroneously appear to have an $A_V \simeq 0.6$ mag of extinction! If dwarf colors are used to deredden a population of young stars, a systematic bias may be introduced, depending on the particular color used, which may cause their luminosities to be systematically overestimated and thus their ages would be systematically underestimated.

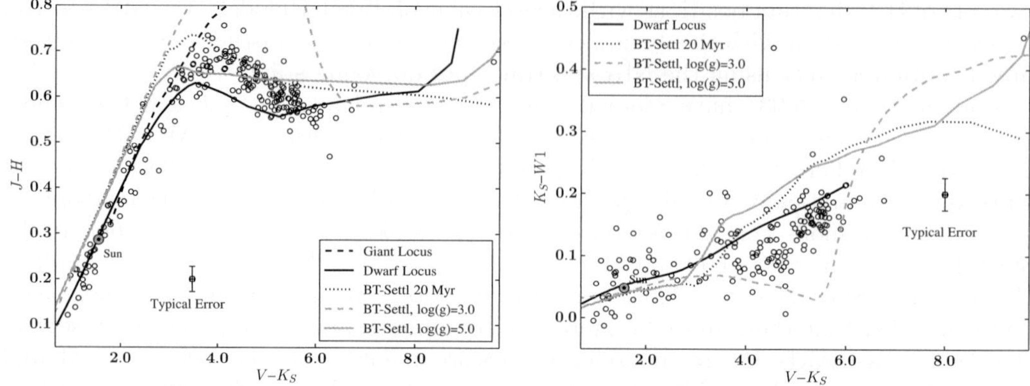

**Figure 1.** $V–K_S$ versus $J–H$ (left) and $V–K_S$ versus $K_S–W1$ (right) for young stars in the $\eta$ Cha cluster, the TW Hydra Association, the $\beta$ Pictoris moving group and the Tucana–Horologium moving group from the sample in Pecaut & Mamajek (2013), with the addition of 129 new Tuc-Hor members from Kraus *et al.* (2014). The dwarf locus is adopted from Pecaut & Mamajek (2013) and the giant locus is adopted from Bessell *et al.* (1998). 20 Myr isochronal colors are constructed by adopting surface gravities from a Baraffe *et al.* (1998) 20 Myr isochrone with synthetic colors from the BT-Settl models of Allard *et al.* (2012).

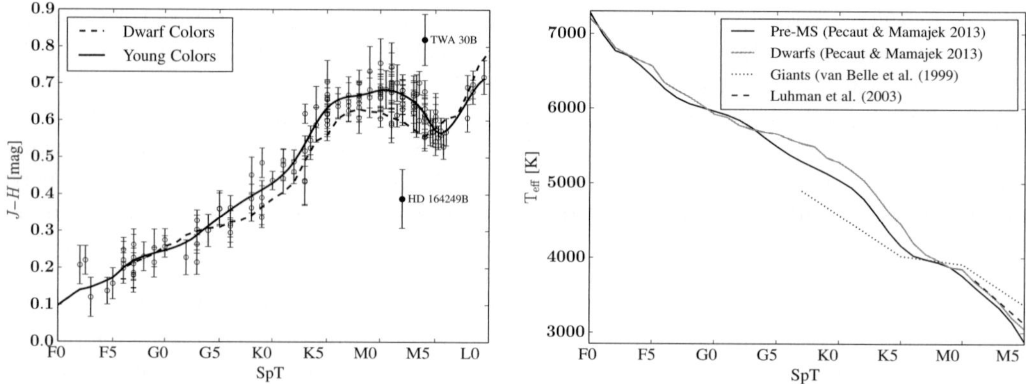

**Figure 2.** Left: Spectral type versus $J–H$ for young stars in the $\eta$ Cha cluster, the TW Hydra Association, the $\beta$ Pictoris moving group and the Tucana–Horologium moving group from the sample in Pecaut & Mamajek (2013), with the addition of 129 new Tuc-Hor members from Kraus *et al.* (2014). Right: Spectral type versus effective temperature for pre-MS stars (Pecaut & Mamajek 2013), dwarf stars (Pecaut & Mamajek 2013), giants (van Belle *et al.* 1999), and the M-type $T_{\mathrm{eff}}$ scale from Luhman *et al.* (2003). The pre-MS temperature scale shown is $\sim$200K cooler than the dwarf scale for types $\sim$G5–K5.

A careful look at the temperature scale of pre-MS stars is warranted, since a star's adopted spectral type is used to infer the $T_{\mathrm{eff}}$. A frequently used temperature scale for M-type stars is the scale of Luhman *et al.* (2003), which is intermediate between dwarfs and giants. However, in the past few years two important developments have been made available: (1) with the release of the *2MASS* and *WISE* catalogs (Skrutskie *et al.* 2006; Cutri *et al.* 2012), photometry is available which covers large sections of the star's spectral energy distributions (SED) and (2) high-quality synthetic spectra are available from the Phoenix-NextGen group (Allard *et al.* 2012) which include updated molecular opacities and model low-temperature stars and brown dwarfs more successfully than ever possible previously. Thus it is possible to obtain tight constraints on the $T_{\mathrm{eff}}$ of an individual unreddened pre-MS member of one of the nearby, young moving groups through fitting the observed SED with synthetic models. Pecaut & Mamajek (2013) fit Phoenix-NextGen BT-Settl models of Allard *et al.* (2012) to the observed SEDs of the members of $\eta$ Cha, TWA, $\beta$ Pic and Tuc-Hor to tie their updated spectral type intrinsic color tabulation to a $T_{eff}$-BC system. The results of this $T_{eff}$ scale are shown in Figure 2b. Pre-MS stars are $\sim$200K systematically cooler than their main sequence counterparts at a given spectral type, for spectral types $\sim$G5–K5.

## 4. Conclusions

When characterizing the basic observations and properties of young stars, such as spectral type, $T_{eff}$, reddening and extinction, care must be exercised to avoid systematic errors and biases. It is important that optical spectra be used to constrain the spectral type of the star, since these seem to best represent the effective temperature of the stars. Some evidence points to spots as a factor in wavelength-dependent spectral types, discussed in detail by Gullbring *et al.* (1998) and Stauffer *et al.* (2003).

When considering or using tabulations of intrinsic colors, it is important to adopt intrinsic colors for pre-MS stars when the stars are still contracting to the main sequence. Comparisons with dwarf and giant colors as well as theoretical synthetic spectra of different surface gravity points to surface gravity as an important effect altering the intrinsic colors of pre-MS stars. If individual extinctions are mis-estimated using dwarf colors, stars placed on the H-R diagram may have their luminosities systematically over-estimated and the population may appear younger simply because observers are unable to account for reddening and extinction properly. This clearly inhibits our ability to accurately test theoretical evolutionary models and may systematically bias ages, which has consequences reaching down to mis-estimating age spreads in star-forming regions, underestimating the evolutionary timescales of disks and planet formation timescales, and systematic errors when inferring the initial mass function (IMF) in pre-MS stellar populations.

## References

Allard, F., Homeier, D., & Freytag, B. 2012, Royal Society of London Philosophical Transactions Series A, 370, 2765

Baraffe, I., Chabrier, G., Allard, F., & Hauschildt, P. H. 1998, *A&A*, 337, 403

Barrado Y Navascués, D. 2006, A&A, 459, 511

Bell, C. P. M., Rees, J. M., Naylor, T., *et al.* 2014, *MNRAS*, 445, 3496

Berdyugina, S. V. 2005, Living Reviews in Solar Physics, 2, 8

Bessell, M. S., Castelli, F., & Plez, B. 1998, A&A, 333, 231

Bonnefoy, M., Currie, T., Marleau, G.-D., *et al.* 2014, A&A, 562, A111

Brandt, T. D., & Huang, C. X. 2015, ArXiv e-prints, arXiv:1504.00004

Cutri, R. M., Skrutskie, M. F., van Dyk, S., *et al.* 2012, VizieR Online Data Catalog, 2281, 0

Da Rio, N., Robberto, M., Soderblom, D. R., *et al.* 2010, *ApJ*, 722, 1092

de la Reza, R., Torres, C. A. O., Quast, G., Castilho, B. V., & Vieira, G. L. 1989, *ApJ*, 343, L61

Gáspár, A., Rieke, G. H., Su, K. Y. L., *et al.* 2009, *ApJ*, 697, 1578

Gullbring, E., Hartmann, L., Briceno, C., & Calvet, N. 1998, *ApJ*, 492, 323

Herbig, G. H. 1978, Can Post-T Tauri Stars Be Found?, ed. L. V. Mirzoyan, 171

Hinkley, S., Pueyo, L., Faherty, J. K., *et al.* 2013, *ApJ*, 779, 153

Hoff, W., Henning, T., & Pfau, W. 1998, A&A, 336, 242

Kastner, J. H., Rapson, V., Sargent, B., Smith, C. T., & Rayner, J. 2015, in Cambridge Workshop on Cool Stars, Stellar Systems, and the Sun, Vol. 18, Cambridge Workshop on Cool Stars, Stellar Systems, and the Sun, ed. G. T. van Belle & H. C. Harris, 313–320

Kraus, A. L., Shkolnik, E. L., Allers, K. N., & Liu, M. C. 2014, *AJ*, 147, 146

Luhman, K. L. 1999, *ApJ*, 525, 466

Luhman, K. L., Allen, P. R., Espaillat, C., Hartmann, L., & Calvet, N. 2010a, *ApJS*, 189, 353

—. 2010b, *ApJS*, 186, 111

Luhman, K. L., Stauffer, J. R., Muench, A. A., *et al.* 2003, *ApJ*, 593, 1093

Mamajek, E. E. 2012, *ApJ*, 754, L20

Pecaut, M. J., & Mamajek, E. E. 2013, *ApJS*, 208, 9

Rosenfeld, K. A., Andrews, S. M., Wilner, D. J., & Stempels, H. C. 2012, *ApJ*, 759, 119

Skrutskie, M. F., Cutri, R. M., Stiening, R., *et al.* 2006, *AJ*, 131, 1163

Soderblom, D. R., Hillenbrand, L. A., Jeffries, R. D., Mamajek, E. E.,& Naylor, T. 2014, Protostars and Planets VI, 219

Stauffer, J. R., Jones, B. F., Backman, D., *et al.* 2003, *AJ*, 126, 833

Torres, C. A. O., Quast, G. R., da Silva, L., *et al.* 2006, A&A, 460, 695

Torres, G. 2010, *AJ*, 140, 1158

Vacca, W. D. & Sandell, G. 2011, *ApJ*, 732, 8

van Belle, G. T., Lane, B. F., Thompson, R. R., *et al.* 1999, *AJ*, 117, 521

*Young Stars & Planets Near the Sun*
*Proceedings IAU Symposium No. 314, 2015*
*J. H. Kastner, B. Stelzer, & S. A. Metchev, eds.*

© International Astronomical Union 2016
doi:10.1017/S1743921315006092

# The Impact of Starspots on Mass and Age Estimates for Pre-main Sequence Stars

**Garrett Somers and Marc H. Pinsonneault**

The Ohio State University Astronomy Department
email: `somers@astronomy.ohio-state.edu`

**Abstract.** We investigate the impact of starspots on the evolution of late-type stars during the pre-main sequence (pre-MS). We find that heavy spot coverage increases the radii of stars by 4-10%, consistent with inflation factors in eclipsing binary systems, and suppresses the rate of pre-MS lithium depletion, leading to a dispersion in zero-age MS Li abundance (comparable to observed spreads) if a range of spot properties exist within clusters from 3-10 Myr. This concordance with data implies that spots induce a range of radii at fixed mass during the pre-MS. These spots decrease the luminosity and $T_{\rm eff}$ of stars, leading to a displacement on the HR diagram. This displacement causes isochrone derived masses and ages to be systematically underestimated, and can lead to the spurious appearance of an age spread in a co-eval population.

**Keywords.** stars: starspots − stars: pre-main sequence − stars: fundamental parameters − stars: activity

## 1. Introduction

Despite recent improvements in stellar modeling, several notable discrepancies remain between theoretical predictions of low mass stellar properties and high precision measurements from techniques such as eclipsing binary (EB) analysis and interferometry. One such discrepancy is the inflated radius problem, where young, low-mass stellar radii are observed to be larger by $5 - 10\%$ than theoretical predictions. A second discrepancy is between the observed lithium patterns of young clusters such as the Pleiades, which host abundance dispersions in excess of an order-of-magnitude at fixed $T_{\rm eff}$, and standard model theoretical predictions, which anticipate no dispersion at fixed $T_{\rm eff}$. Notably, the most Li rich stars in the Pleiades are also the most rapidly rotating, the opposite sense of the correlation expected from rotational mixing.

In two recent papers, we suggested that if some mechanism induced inflated radii in rapidly rotating, low-mass stars during the pre-MS, then the Li destruction rate in the inflated stars would be suppressed, and one could explain both the inflated radius problem and the Li-rotation correlation in the Pleiades, with a single mechanism (Somers & Pinsonneault 2014; Somers & Pinsonneault 2015a). In this proceedings, we summarize results from our recent work which explores this possibility using a specific inflation mechanism, starspots.

## 2. Starspots

Starspots are the visual manifestation of concentrated magnetic fields near the surfaces of stars. Spots inhibit energy transport through convection, and alter the pressure conditions at the photosphere, inducing a global structural response that can be modeled (Gough & Taylor 1966). Spots have long been argued to impact the luminosities, temperatures, and radii of stars (e.g. Spruit & Weiss 1986), but in only a few cases has

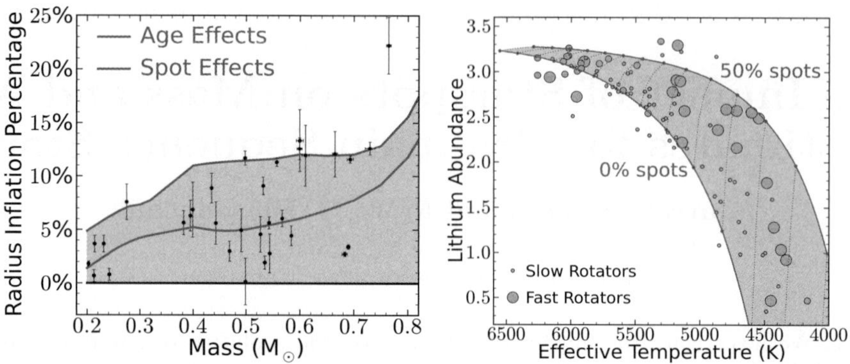

**Figure 1.** *Left:* The impact of spot and age effects on stellar radii. Each point shows the radius inflation inferred for an eclipsing binary member relative to our models (data from Torres *et al.* 2010 and Feiden & Chaboyer 2012). The red region shows the radius increase incurred over 10 Gyr, and the blue region shows the combined impact of heavy spot coverage and age. With both effects considered, nearly all the data can be explained. *Right:* The impact of spots on Li destruction during the pre-MS. Spot coverage up to 50% can suppress Li destruction enough to explain the Li spread at fixed $T_{\rm eff}$ observed in the Pleiades (grey circles; Li from Soderblom *et al.* 1993, rotation from Hartman *et al.* 2010).

their effect on the pre-main sequence (pre-MS) been examined (e.g. Jackson & Jeffries 2014), and never with a modern stellar evolution code.

Young, active stars are typically the most heavily spotted, so a systematic study of the structural impact of spots during the pre-MS and early MS may prove crucial to our understanding of early stellar evolution. To this end, we have incorporated in our stellar evolution code a treatment of spots which accounts for two relevant physical processes: 1) the redistribution of flux in the interior, modeled by explicitly altering the radiative gradient; 2) the altered boundary conditions at the surface, modeled by self-consistently solving for the pressure in the un-spotted regions (the full details are presented in §2.2 of Somers & Pinsonneault 2015b).

Using this method, we calculated a grid of low mass ($0.1$-$0.8 M_\odot$) and solar mass ($0.8$-$1.2 M_\odot$) stars, with various spot properties. Our spotted models each have a spot$-$photosphere temperature contrast of 0.8, and host a variety of spot filling factors up to 50% surface coverage, the upper limit of claimed properties in the literature.

## 3. Impact on Stellar Radii and Lithium Abundances

We first explored the impact of spots on stellar radii. We find that heavy spot coverage leads to a substantial enhancement at all ages, developing on the early pre-MS and persisting onto the MS. For all masses in our range, radius anomalies peak at 8-12% during the pre-MS, and converge to a mass dependent pattern at the ZAMS, where fully convective stars are inflated by $\sim 4\%$, and solar mass stars are inflated by up to 10%. The larger anomalies on the pre-MS result primarily from the reduced Hayashi contraction rate of spotted stars, a consequence of their suppressed luminosity (§4). By contrast, MS anomalies result predominately from the impact of spots on the pressure boundary conditions at the surface, and at the transport of flux in the upper envelope.

In the left panel of Fig. 1, we compare the radius predictions of our spotted models to EB masses and radii. Each data point shows the fractional inflation percentage relative to our zero-age MS predictions. The red region shows the impact of 10 Gyrs of stellar evolution on the radius, indicating that a subset of the data could be explained solely by age effects, though more than half cannot. The blue region shows the area permitted when including both spot and age effects, which could both be at work in short-period, (presumably) tidally-locked EBs. With only one exception, our spot models can account for the anomalous radii of the comparison sample. This demonstrates that our models predict reasonable inflation factors for observationally motivated spot properties.

Next, we explore the destruction of lithium in our models. Li burns at $T \sim 2.5 \times 10^6$ K, and is depleted from stellar surfaces during the pre-MS until a substantial radiative core has formed. Standard models (red line in the right panel of Fig. 1) predict that the mass and composition of a star uniquely determine its Li destruction during the MS, and consequently predict no spread in Li at fixed $T_{\mathrm{eff}}$ within clusters. By contrast, a large dispersion in Li abundance at fixed $T_{\mathrm{eff}}$ (grey points) is observed in the 120 Myr Pleiades cluster, suggesting additional mechanisms are at play.

Spotted models are physically larger during the pre-MS, leading to a reduction in the temperature at the base of the surface convection zone, and thus a suppressed rate of Li destruction. The blue line in the right panel of Fig. 1 shows the Li abundances resulting from a constant filling factor of 50% during the pre-MS, and the blue shaded regions show abundances produced by intermediate spot properties. The majority of the data fall within this range, suggesting that stars of equal mass must have dissimilar radii during the pre-MS to explain the data. Furthermore, the most Li-rich stars are the fastest spinning, suggesting that this radius dispersion arises from a physical process linked to rotation − spots are a prime candidate.

Collectively, Li and radius data suggest that a range of radii at fixed mass, driven by magnetic activity, emerge in open clusters during the pre-MS and persist onto the MS. We now discuss an important implication of this conclusion.


## 4. Consequences for Inferred Masses and Ages

The left panel of Fig. 2 shows standard (red) and 50% spotted (blue) isochrones at 3 Myr in the HR diagram. The blue isochrone is displaced downward and to the right, indicating that spots reduce the luminosity and $T_{\mathrm{eff}}$ of their host stars. Now consider the location of the black cross. If a pre-MS stars is observed at this HR-diagram location, one might infer a mass of $0.5 M_\odot$ and an age of 3 Myr through comparison with the standard isochrones. However, if this star is actually spotted, then the blue isochrones should be used to derive its properties, in which case one finds $M \sim 0.75 M_\odot$, and $t \sim 8$ Myr. This demonstrates that spotted stars are *systematically older and more massive than often measured*.

To estimate the magnitude of this effect, we return to the Pleiades Li abundances in Fig. 1. If the cause of the Li dispersion is indeed pre-MS spots, then the current Li abundance of each individual star tells us what its spot properties must have been on the pre-MS. We can therefore derive the distribution of pre-MS spot properties that the Pleiades *must have had*, in order to produce the current $T_{\mathrm{eff}}$ vs. Li distribution. We can then ask what masses and ages *would we have measured* from this distribution, using standard isochrones. The results are in the right panel of Fig. 2. The blue points show the masses derived from the Pleiades, backwards modeled to a coeval age of 10 Myr. The red points show the masses and ages that isochrone fitting gives for this population. As can be seen, there is a small systematic trend towards lower inferred masses, and a very strong trend towards lower ages, with errors reaching a factor of $3\times$ for the most vulnerable stars. This is not a subtle effect! We further note that the lowest mass stars appear to be the youngest, a feature which has been observed in some young clusters (e.g. Herczeg & Hillenbrand 2015). This suggests that many young associations may in fact

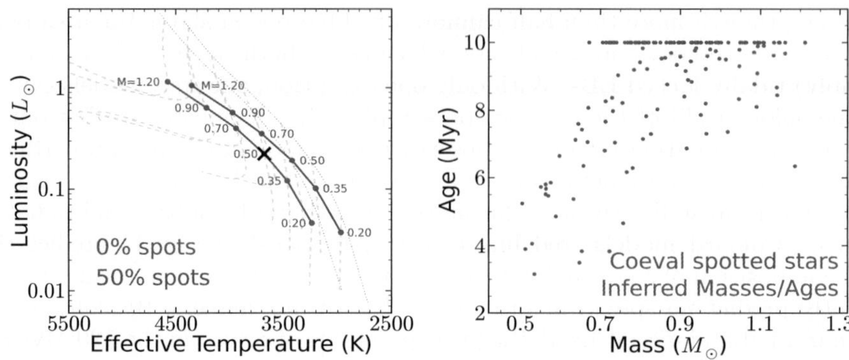


**Figure 2.** *Left:* 3 Myr isochrones for un-spotted (red) and spotted (blue) models. A pre-MS star at the location of the black cross will be incorrectly interpreted if its spot properties are not accounted for. *Right:* Mass and age errors incurred when inferring properties of a spotted population with standard isochrones.

be older than often quoted, that age spreads claimed in some clusters may be spurious, and that age errors are largest for the lowest mass stars.

## 5. Conclusions

We have found that the impact of spots on stellar models is to increase their radii, and to lower their luminosities and effective temperatures, in a mass dependent fashion. The increased radii due to spots can explain the observed radius inflation of some low mass stars, and can account for the suppressed Li depletion in rapid rotators during the pre-MS. These data collectively suggest that stars of equal mass have different physical sizes at fixed age around 3-10 Myr. As a consequence, masses and ages derived with isochrones will be erroneously low and young, respectively, and may create the appearance of age spreads in co-eval populations.

## References

Feiden, G. A. & Chaboyer, B. 2012, *ApJ*, 757, 42
Gough, D. O. & Tayler, R. J. 1966, *MNRAS*, 133, 85
Hartman, J. D., Bakos, G. Á., Kovács, G., & Noyes, R. W. 2010, *MNRAS*, 408, 475
Herczeg, G. J. & Hillenbrand, L. A. 2015, arXiv:1505.06518
Jackson, R. J. & Jeffries, R. D. 2014, *MNRAS*, 441, 2111
Soderblom, D. R., Jones, B. F., & Balachandran, S., *et al.* 1993, *AJ*, 106, 1059
Somers, G. & Pinsonneault, M. H. 2014, *ApJ*, 790, 72
Somers, G. & Pinsonneault, M. H. 2015a, *MNRAS*, 449, 4131
Somers, G. & Pinsonneault, M. H. 2015b, arXiv:1506.01393
Spruit, H. C. & Weiss, A. 1986, *AAP*, 166, 167
Torres, G., Andersen, J., & Giménez, A. 2010, *AAPR*, 18, 67

*Young Stars & Planets Near the Sun*
Proceedings IAU Symposium No. 314, 2015
J. H. Kastner, B. Stelzer, & S. A. Metchev, eds.

© International Astronomical Union 2016
doi:10.1017/S174392131500664X

# Lithium Depletion Boundary Ages of Young Stars: Inconsistencies in Pre-Main Sequence Models

Inseok Song[1]

[1]Department of Physics & Astronomy, University of Georgia, Athens, GA 30602
email: song@uga.edu

**Abstract.** For proper interpretations of various phenomena in young stars and planetary systems, knowledge of accurate stellar ages is very important. Among a handful of age dating methods commonly used for young ($\lesssim 500\,\mathrm{Myr}$) stars, lithium depletion boundary (LDB) ages have recently become the most cited and accepted age estimates. However, because of inconsistencies in theoretical evolutionary models, especially for lithium depletion calculations, one has to be cautious in using LDB ages. For a given luminosity, the lithium depletion process is too slow, causing LDB ages to appear older. Various stellar processes affect the surface lithium abundance, and these effects include star spots, accretion history, and magnetic fields. Until we have a self-consistent theoretical evolutionary model for young stars including all relevant stellar effects, caution should be taken when LDB ages are used.

**Keywords.** stars: fundamental parameters (ages), stars: pre–main-sequence, stars: abundances

## 1. Introduction

Obtaining an accurate age estimate of a young star is important because ages provide the basis for proper interpretations of various aspects of the formation and evolution of young stars and planets. It is especially important to precisely age-date nearby young moving groups because their ages overlap with important planet formation epochs over the range of 10-100 Myr.

Ages of nearby young moving groups (NYMGs) can be estimated by comparing various activity indicators against patterns from the members of open clusters with well determined ages (see Zuckerman & Song 2004). The Pleiades ($\sim 100\,\mathrm{Myr}$) and Hyades ($\sim 650\,\mathrm{Myr}$) clusters have been used as the best age anchors in such stellar age-dating. Ages of NYMGs younger than the Pleiades have been estimated using sparser clusters, such as IC 2602 and $\eta$ Cha, as age calibrators. Because this age-dating is essentially a relative age ordering, uncertainties in age dating are limited by the available age calibrators. Typical uncertainties in commonly used age determination methods for $\sim$10 Myr old stars are $^{+10}_{-5}$ Myr. Surface lithium abundance can be effectively used as a clock because Li burning (1) is fast, (2) starts at temperature of about 2 million Kelvin, and (3) is sensitive to the internal convective structure. At a given age beyond $\sim 8 - 10\,\mathrm{Myr}$, the Li $\lambda$6708 feature disappears among early M-type stars and suddenly reappears for types around mid-M; this is the so-called lithium depletion boundary (LDB). Because young stars around this LDB are fully convective, the precise location of the LDB in $T_{eff}$ and luminosity can be readily transformed to a mass (hence an age) by using theoretical pre-main sequence (PMS) evolutionary models. LDB ages obtained for open clusters are typically 50% older than upper main sequence fitting ages, and this LDB-based age dating technique has been used for some NYMGs. The age of the $\beta$ Pictoris moving group (BPMG) has been estimated to be $\sim$12 Myr based on various combinations of age-dating

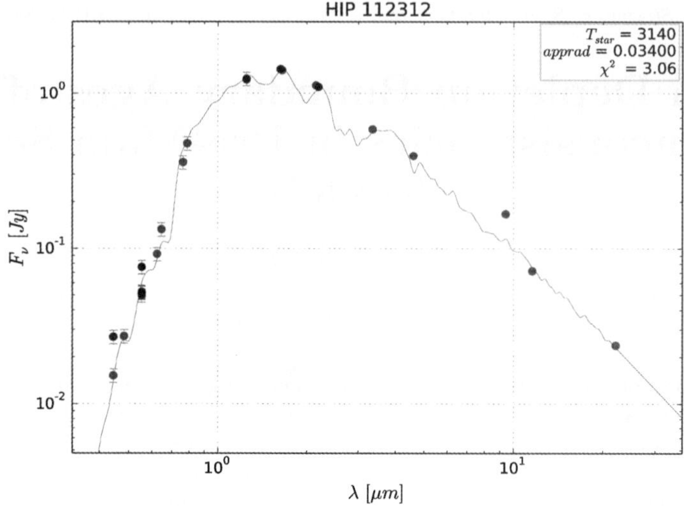



**Figure 1.** An example of SED fitting, here for BPMG member HIP 112312A.

methods (Song, Zuckerman, & Bessel 2003; Ortega *et al.* 2002; Zuckerman & Song 2004), while older BPMG ages, based on the LDB technique, have been reported more recently (Binks & Jeffries 2014; Mamajek & Bell 2014; Malo *et al.* 2014).

However, as shown in Song *et al.* (2002), Li depletion calculations from various PMS models are inconsistent. For a given PMS model, there exists a significant discrepancy between the observed LDB luminosity and the expected luminosity. Also, various processes commonly seen among young stars, such as accretion, magnetic fields, and starspots, can affect surface Li depletion. In this paper, I will re-evaluate the LDB age of BPMG using $T_{eff}$ and luminosity values obtained from spectral energy distribution (SED) fitting, so as to demonstrate the inconsistencies among PMS model calculations.

## 2. The LDB Age of the $\beta$ Pictoris Moving Group

The location of the LDB in terms of temperature and luminosity (hence mass) is a sensitive function of age, so precise measurement of the LDB is important. Observationally, an LDB position is measured in terms of photometric magnitudes and/or colors. Therefore, transforming these observational parameters into temperature and luminosity is necessary, and this has frequently been done using a temperature scale for young stars (e.g., Pecaut & Mamajek 2013). In this paper, an SED-fitting approach is used instead and, because of the covariance nature of several photometric magnitudes for a given source, this method is known to be very robust (Kraus *et al.* 2014). For known BPMG members, I have collected 7 or more photometric measurements for each star and these measurements were fit against a set of NextGen synthetic stellar spectra (see Rhee *et al.* 2007 for details on the SED fit). Best fit parameters ($T_{eff}$ and $\theta_{app}$, where $\theta_{app}$ is the apparent stellar radius in arcseconds) are translated to $T_{eff}$ and luminosity ($L$) using the trigonometric parallax distance for each star.

In Figure 1, an example SED fit is displayed; best fit parameters were obtained for other BPMG stars using SED fits of similar quality. Then, estimated temperatures and luminosities were plotted against a set of commonly used pre-main sequence evolutionary models (Figure 2). As shown in Figure 2, the position of the LDB for the BPMG is estimated to be around $T_{eff} = 3050\,\mathrm{K}$ and $L = 0.033L_\odot$. This LDB is interpreted as the

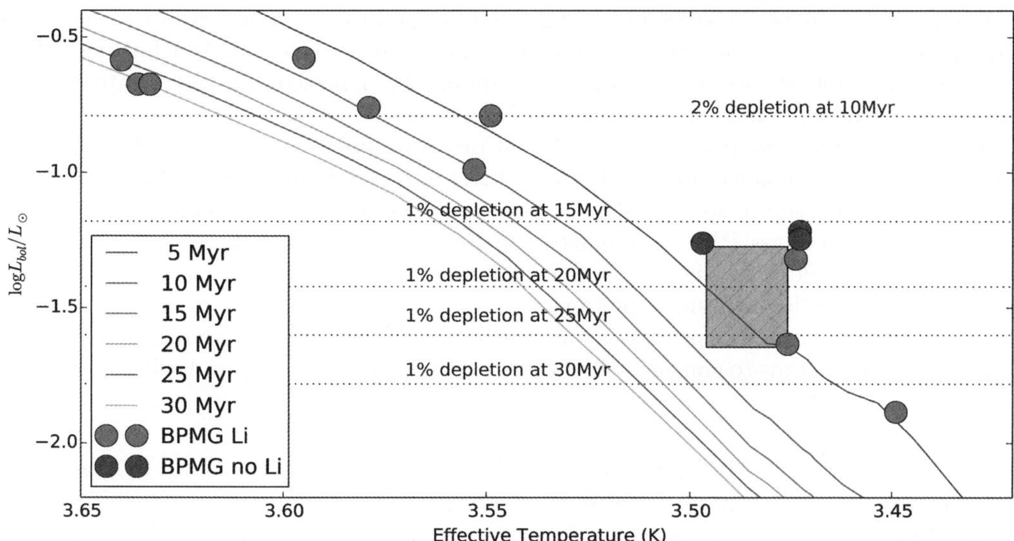


**Figure 2.** BPMG LDB compared to model lithium depletion for various ages. Model calculations are from Baraffe *et al.* (2015).

depletion of Li by more than a factor 100 relative to its initial abundance. In the same figure, several horizontal lines are plotted indicating a factor 100 depletion of lithium for various ages. For PMS models that are consistent, the LDB should appear where a 1% depletion horizontal line intersects with an appropriate age isochrone. However, the shaded box denoting the observed LDB for the BPMG is not consistent with the lithium depletion pattern, because it is located at a 10 Myr or younger isochrone, where the maximum lithium depletion cannot reach the 100× level.

Even for fully convective PMS stars (i.e., all stars around the LDB point), temperature is very sensitive to the degree of superadiabadicity in the interior, which in turn depends on the treatment of the convective temperature gradient (Siess *et al.* 2000). For such cases, an LDB analysis using luminosity has been regarded as more trustworthy than an analysis using temperature (Burrows *et al.* 2001). Therefore, in LDB analyses, generally only the luminosity is considered. As evident in the Figure, the observed location of the LDB luminosity and the predicted LDB luminosity are very different.

In Song *et al.* (2002), we used the M4.5+M4.5 BPMG binary system HIP 112312 A&B, for which a LDB was detected between the two binary components, to illustrate such a PMS model discrepancy, and we advised caution in interpreting the LDB. After ten years, although several more BPMG members have been added, the BPMG LDB is still essentially defined by this same binary system. As shown in the next section, because surface lithium abundance can be affected by stellar evolutionary histories, the LDB detected in a binary should be regarded as more trustworthy than an LDB defined by members of moving groups or clusters. Open cluster (or moving group) members might have been exposed to more diverse evolutionary environments than the two components in a single binary.

## 3. Physical Processes Affecting Lithium Depletion in PMS Stars

It has been shown that a star's previous accretion history can affect its surface lithium abundance at later stage (Baraffe *et al.* 2002). Also, starspots and magnetic fields can also affect the surface lithium content.

A high accretion rate results in a smaller stellar radius compared to a non-accreting object, which in turn leads to a hotter core temperature (Baraffe & Chabrier 2010). Because of the hotter core temperature, lithium can be depleted at a faster rate, in good agreement with the results shown in our LDB analysis.

On the contrary, starspots impact internal pressure and temperature, inhibiting convection. If this effect is significant, then all LDB ages are younger than they should be (Somers & Pinsonneault 2015). Likewise, the presence of a strong magnetic field in the evolution of a low mass PMS star makes the star appear brighter at a given age (Malo et al. 2014). This effect mitigates the aforementioned apparent discrepancy in the PMS model calculations for lithium depletion.

Because several physical processes can affect surface lithium abundance in a complex manner, it is important to consider all of these effects simultaneously and self-consistently.

## 4. Conclusions

Recently, older ($\gtrsim 20\,\mathrm{Myr}$) ages for the BPMG that are based on various LDB analyses have become widely accepted, because of a perception that the LDB age-dating method is based on simpler physics and the fact that multiple research groups have reported similarly old LDB ages for the BPMG. However, as shown earlier in Song, Bessel, & Zuckerman (2002) and in this paper, there are significant model inconsistencies in PMS calculations. The observed luminosity of the BPMG LDB is different from the expected luminosity obtained from the same model and the inferred LDB age. Until we have self-consistent PMS model calculations, LDB ages have to viewed with caution. Furthermore, there are several important physical processes (accretion, magnetic fields, and starspots) that can change the surface lithium abundance in complex ways. Considering these problems, a new BPMG age of $15 \pm 5\,\mathrm{Myr}$ — which spans the previous younger, chromospheric ages and the recent older, LDB ages — is suggested to be used, until a community consensus can be reached.

### References


Baraffe, I., Chabrier, G., Allard, F., & Hauschildt, P. H. 2002, *A&A*, 382, 563
Baraffe, I. & Chabrier, G. 2010, *A&A*, 521, A44
Baraffe, I., Homeier, D., Allard, F., & Chabrier, G. 2015, *A&A*, 577, A42
Binks, A. S. & Jeffries, R. D. 2014, *MNRAS*, 438, L11
Burrows, A., Hubbard, W. B., Lunine, J. I., & Liebert, J. 2001, *RvMP*, 73, 719
Kraus, A. L., Shkolnik, E. L., Allers, K. N., & Liu, M. C. 2014, *AJ*, 147, 146
Malo, L., Doyon, R., Feiden, G. A., *et al.* 2014, *ApJ*, 792, 37
Mamajek, E. E. & Bell, C. P. M. 2014, *MNRAS*, 445, 2169
Ortega, V. G., de la Reza, R., Jilinski, E., & Bazzanella, B. 2002, *ApJ*, 575, L75
Pecaut, M. J. & Mamajek, E. E. 2013, *ApJS*, 208, 9
Rhee, J. H., Song, I., Zuckerman, B., & McElwain, M. 2007, *ApJ*, 660, 1556
Siess, L., Dufour, E., & Forestini, M. 2000, *A&A*, 358, 593
Somers, G. & Pinsonneault, M. H. 2015, *MNRAS*, 449, 4131
Song, I., Zuckerman, B., & Bessell, M. S. 2003, *ApJ*, 599, 342
Song, I., Bessell, M. S., & Zuckerman, B. 2002, *ApJ*, 581, L43
Zuckerman, B. & Song, I. 2004, *ARA&A*, 42, 685


### Discussion

ZUCKERMAN: What about LDB ages for other moving groups and open clusters? Would the discrepancy equally affect these other LDB ages?

AUTHOR: Yes. The discrepancy will affect all other LDB ages in the same manner.

*Young Stars & Planets Near the Sun*
*Proceedings IAU Symposium No. 314, 2015*
*J. H. Kastner, B. Stelzer, & S. A. Metchev, eds.*

© International Astronomical Union 2016
doi:10.1017/S1743921315006626

# Cloud Driven Variability on Young Brown Dwarfs and Giant Exoplanets

## Beth Biller[1]

[1]Institute for Astronomy, University of Edinburgh
email: **bb@roe.ac.uk**

**Abstract.** Variability has now been robustly observed in a range of L and T type field brown dwarfs, primarily at near-IR and mid-IR wavelengths. The probable cause of this variability is surface inhomogeneities in the clouds of these objects, causing a semi-periodic variability signal when combined with the rotational modulation from the 3-12 hour period expected for these objects. Variability at similar or even higher amplitudes may be expected for young brown dwarfs and giant exoplanets, which share similar $T_{\mathrm{eff}}$ as field brown dwarfs, but have considerably lower surface gravities. Variability studies of these objects relative to old field objects is then a direct probe of the effects of surface gravity on atmospheric structure. Here I discuss ongoing efforts to detect variability from these young objects, both for free-floating objects and companions to stars, including preliminary results from an ongoing survey of young, low surface gravity objects with NTT SOFI.

**Keywords.** infrared: planetary systems, (stars:) brown dwarfs, planets and satellites: atmospheres

## 1. Introduction

Numerous young exoplanets have been directly imaged in the infrared (c.f. among others, Marois *et al.* 2008; Marois et al. 2010; Lagrange *et al.* 2010; Rameau et al. 2013a) Initial observations of directly imaged planets have yielded some surprises. Young directly imaged exoplanets were expected to share similar atmospheric properties with the well-studied population of brown dwarfs, which have comparable temperatures albeit higher masses. However, most young directly imaged exoplanets are much redder in the near-IR than their brown dwarf counterparts at similar $T_{\mathrm{eff}}$. In fact, only two directly imaged exoplanets (GJ 504b, Kuzuhara et al. 2013; Janson *et al.* 2013, HD 95086b, Rameau *et al.* 2013a; Rameau *et al.* 2013b) to date possesses the methane absorption feature robustly observed in brown dwarfs with similar $T_{\mathrm{eff}}$. This discrepancy may be explained by the persistence of dusty clouds at lower temperatures as well as non-equilibrium chemistry (e.g., Barman et al. 2011; Marley *et al.* 2012), likely to result from the lower surface gravity of these objects. Furthermore, patchy cloud coverage may be required to model observations across a wide wavelength range (e.g., HR 8799b; Bowler *et al.* 2010; Skemer *et al.* 2012). Like their higher mass brown dwarf counterparts (Zapatero Osorio *et al.* 2006), planets are expected to rotate on hour to day timescales. Thus, a key probe of cloud properties in exoplanet atmospheres is time-variability (brightness as a function of phase), which is sensitive to the spatial distribution of condensates as the planet rotates.

## 2. Conditions for Variability

Detectable rotationally modulated variability in a specific class of object requires the presence of: 1) relatively short rotation periods (i.e. observable on the timescales of a night or two) and 2) surface inhomogeneities.

It is already clear that field brown dwarfs possess quite short rotational periods. (Zapatero Osorio *et al.* 2006) obtained projected rotational velocities (*vsini*) for a sample of 19 M6.5-T8 very low mass stars and brown dwarfs using KECK NIRSPEC (see Fig. 1). They found that the T dwarfs observed were generally faster rotators than the M stars in the sample. Assuming brown dwarfs are around $\sim$1 $R_{Jup}$ in size (a reasonable assumption, given electron degeneracy in their cores, Burrows *et al.* 2003), they find a maximum rotational period for L and T brown dwarfs of 12.5 hours. No similar study has been done to date for young brown dwarfs, but considerations from conservation of momentum as these objects form and collapse suggest their periods may be somewhat longer.

Recently, Snellen *et al.* (2014) measured a rotational velocity using VLT CRIRES for the 8-10 $M_{Jup}$ planet $\beta$ Pic b (Lagrange et al. 2009; Lagrange *et al.* 2010). This is the first rotational velocity measured for any extrasolar planet. In fact, $\beta$ Pic b rotates faster than any of the planets in the Solar system with $v_{rotation} = 25\pm3$ km/s$^{-1}$. Using spectral modeling from broadband photometry to infer a radius of $1.65 \pm 0.06 R_{jup}$, this gives a rotational period of $\sim 8.1\pm1.0$ hours. Thus, at least one exoplanet is also a rapid rotator.

In the case of objects with spectral types >L2, the expected source of surface inhomogeneities is patchiness in cloud cover or even multiple cloud components, especially at the L/T spectral type transition. Marley *et al.* (2010) find that patchy cloud models best bridge the transition between the L and T spectral types, with entirely clear models best describing the later T spectral types and single component, entirely cloudy models best describing the earlier L spectral types (see Fig. 1). Observationally, Skemer *et al.* (2012) find that patchy cloud coverage may be required to model observations across a wide wavelength range.

Earlier spectral types (e.g., M type brown dwarfs in young star-forming regions such as Taurus) may have variability dominated by mechanisms other than clouds: e.g., accretion from a circum-brown-dwarf or circum-planetary disk or magnetic phenomena such as starspots. In contrast, L and T type brown dwarfs are too cool to have significant magnetically-driven starspots.

## 3. Variability in Brown Dwarfs

Variability likely due to cloud structure is already found in numerous L and T type field brown dwarfs. Indeed, for the closest known L/T transition brown dwarf binary WISE J104915.57531906.1AB (generally referred to as Luhman 16AB, consisting of T0.5 and L7.5 components, Luhman 2013, Burgasser *et al.* 2013) resolved MPG/ESO 2.2 m GROND simultaneous six-band (r'i'z' JHK) photometric monitoring yielded detections of periodic variability in all 6 bands for Luhman 16B, with amplitudes from 5-15% (see Fig. 2, Biller *et al.* 2013, see also Gillon *et al.* 2013) and in i'z' for Luhman 16A with amplitudes from 1-3% (Biller *et al.* 2013). This variability has a measured period of 4.9 hours, evolves on daily to weekly timescales (Gillon *et al.* 2013), and is likely due to changing cloud structures on these objects. Recent large-scale surveys of brown dwarf variability with Spitzer have revealed mid-IR variability of up to a few percent in >50% of L and T type brown dwarfs (Metchev *et al.* 2015, see Fig. 3). (Buenzli *et al.* 2014) find that $\sim$30% of the L5-T6 objects surveyed in their HST SNAP survey show variability trends and large ground-based surveys also find ubiquitous variability (Radigan *et al.* 2014,

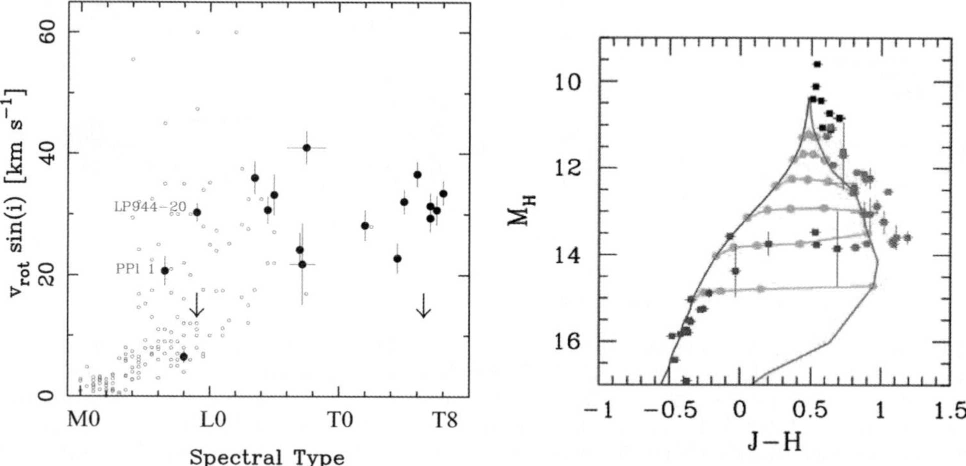

**Figure 1.** Left: Figure from (Zapatero Osorio *et al.* 2006). Keck NIRSPEC measurements of *vsini* for 19 M to T objects show that T-type objects are generally faster rotators than M-type objects and set a maximum rotation period of 12.5 hours for the sample. Right: figure from (Marley *et al.* 2010). Black points are M stars, red points are L dwarfs, and blue points are T dwarfs. The solid red line is a single-component, completely cloudy model, the solid blue line is a completely clear model, and the solid green lines are patchy cloud models with different values of the cloud sedimentation parameter $f_{sed}$. The patchy cloud models best reproduce the notable shift from red to blue colors seen at the L/T spectral type transition.

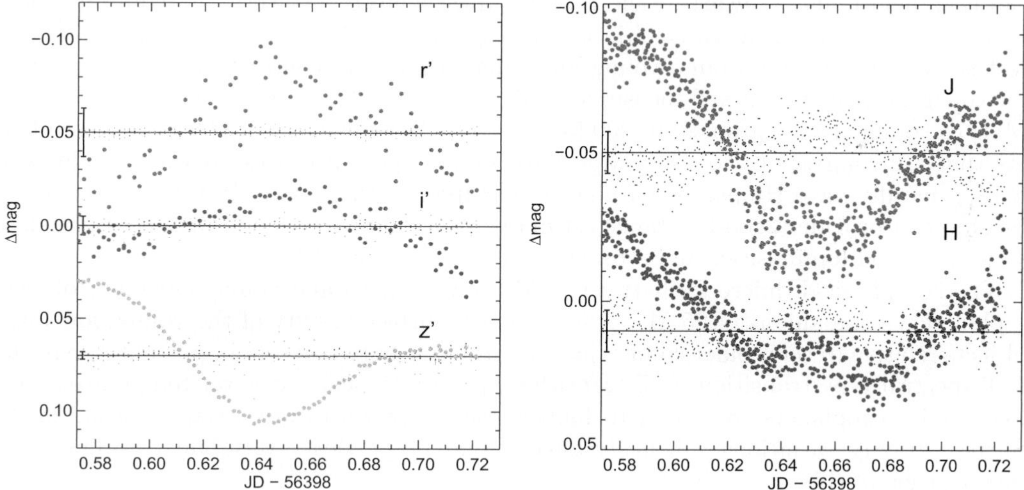

**Figure 2.** Figure from (Biller *et al.* 2013). Multiwavelength $rizJHK_S$ unresolved variability monitoring of the nearest brown dwarfs, the L/T transition binary Luhman 16AB using the GROND imager at the ESO/MPG 2.2 m telescope. Variability is seen from the few percent level up to the 15% level in all 6 bands.

Wilson *et al.* 2014, see Fig. 3). While variability amplitude may be increased across the L/T transition (Radigan *et al.* 2014), variability is now robustly observed across the full range of L and T spectral types. Thus, variability may be expected for young extrasolar planets, which share similar $T_{\rm eff}$ and L and T spectral types. Indeed, the spectrum of the young planet HR 8799b is nearly identical to that of the most variable brown dwarf known, 2M 2139 (27% variable, Radigan *et al.* 2012).

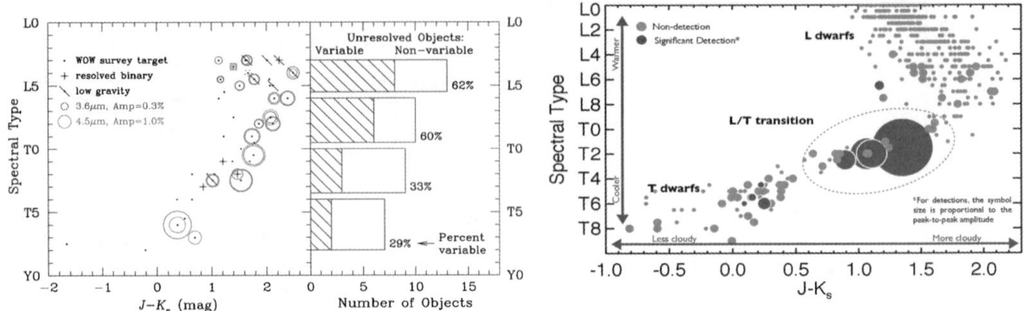

**Figure 3.** Left: Figure from Metchev *et al.* (2015), from a Spitzer mid-IR variability survey of 44 brown dwarfs. Metchev *et al.* (2015) find ubiquitous variability at low amplitudes as well as a tentative association (92% confidence) between low surface gravity and high-amplitude variability among L3L5.5 dwarfs. Right: Figure from Radigan *et al.* (2014), from a variability survey of 62 brown dwarfs. Of the 57 brown dwarfs included in their final analysis, Radigan *et al.* (2014) found that 9 were significantly variable with >99% confidence, with higher variability amplitudes found for L/T transition objects.

## 4. Variability in Planetary Mass Objects

Young directly imaged extrasolar giant planets and very low mass young brown dwarfs / planetary mass objects (estimated mass <25 $M_{Jup}$ using the conservative planet definition of Schneider et al. 2011, henceforth "young planetary mass objects" or PMOs) are in fact much better targets for variability monitoring (and hence, testing atmospheric models) than field L and T dwarfs. Field L and T dwarfs often have very uncertain distances and no age constraints, as these objects are degenerate with age – an old higher-mass object is indistinguishable from a young lower-mass object. In contrast, distances and ages for young exoplanets are well-measured. PMOs that are members of young clusters (1-10 Myr) or young moving groups (10-200 Myr) have well-defined ages. Extrasolar giant planets that are companions to stars often have well-defined distances, as their bright host stars generally have Hipparcos parallaxes. For free-floating young PMOs, cluster membership can provide a good distance estimate. Well-defined ages and distances translate to well-defined luminosities and stronger constraints on models.

The key physical difference between field brown dwarfs and young substellar objects / planetary companions (<25 $M_{Jup}$) is the lower surface gravity of the young, low mass objects relative to their older counterparts. Surface gravity plays an important role in the L-T spectral type transition; L-T transition appears to occur at lower temperatures for low-gravity exoplanets compared to higher gravity brown dwarfs (Marley *et al.* 2012). Surface gravity variations will likely affect cloud properties – variability studies are a direct probe of this.

### 4.1. *Variability in Free-floating Planetary Mass Objects*

While instruments such as GPI (Macintosh *et al.* 2014) and SPHERE (Beuzit *et al.* 2008) achieve contrasts suitable to enable variability studies for a handful of exoplanet companions, it is currently prohibitively difficult to study variability for a statistically significant sample of directly imaged planets within 1" of their star. However, we can circumvent the contrast issue and probe variability for young 3-25 $M_{Jup}$ objects by observing widely (>1") separated companions and isolated very low mass young brown dwarfs / PMOs.

I (along with collaborators at Edinburgh, MPIA, IPAG, University of Hawaii, and elsewhere) am currently conducting the first statistically significant survey of weather

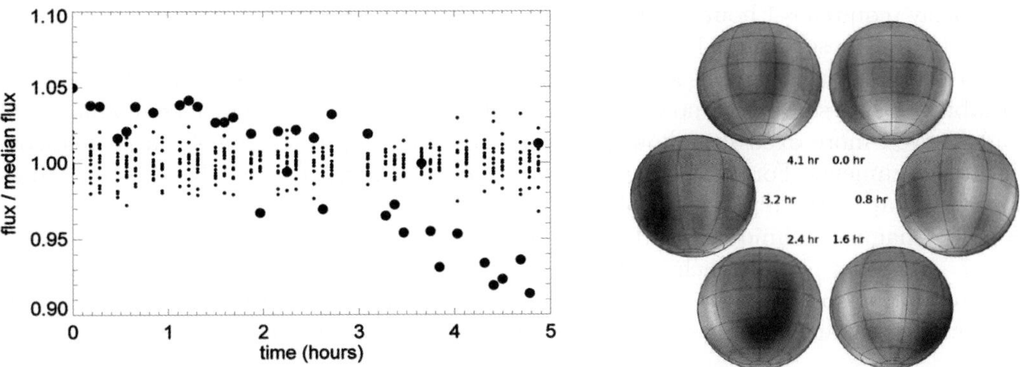



**Figure 4.** Left: First tentative detection from our SOFI survey of 22 young brown dwarfs and planetary mass objects. Lightcurves are shown after detrending using a calibration curve built from 8 carefully selected reference stars. The variable target is plotted with large filled circles; small circles show detrended lightcurves for the 8 reference stars. Comparing the peak-to-valley amplitude of the variability with the rms attained on similar brightness reference stars, variability is detected in this case at the 10-$\sigma$ level. Right: Figure from Crossfield *et al.* (2014). Doppler imaging map of the brown dwarf Luhman 16B. It will eventually be possible to make similar maps for exoplanet companions using high resolution spectrographs at extremely large telescopes, such as METIS, planned for the E-ELT.

patterns on young planetary mass objects using NTT SOFI. This project is the very first targetted survey of exoplanet variability. We have obtained to date $J$-band photometric variability monitoring observations for a sample of 22 young PMOs or very low mass brown dwarfs (mass estimates $<25\,\mathrm{M}_{Jup}$ according to the conservative "planet" definition of Schneider *et al.* 2011) which are either known members of young star-forming clusters (age $< 8$ Myr) or of nearby moving groups (age $< 100$ Myr). Our carefully chosen sample are all: 1) either high-probability cluster or moving group members with well-estimated ages and distances or with other indicators of young age (e.g., signs of low surface gravity), 2) with spectral type L1 or greater, e.g., predicted $T_{\mathrm{eff}} <2000$ K, and, 3) if a companion, with contrast low enough or at wide enough separation from its primary to easily be retrieved in a single SOFI exposure. The vast majority of our carefully selected sample were identified as young objects only recently, (see e.g., Gagné *et al.* 2014, 2015). We avoid earlier spectral types (e.g., M type brown dwarfs in young star-forming regions such as Taurus) as these objects may have variability dominated by mechanisms other than clouds: e.g., accretion from a circum-brown-dwarf or circum-planetary disk or magnetic phenomena such as starspots. In contrast, L and T type brown dwarfs are too cool to have significant magnetically-driven starspots. We also avoid field late T and Y brown dwarfs. While some of these objects likely have planetary masses, their ages are unconstrained, leading to large uncertainties on their mass estimates from models.

Buenzli *et al.* (2014) find that $\sim$30% of the L5-T6 objects surveyed in their HST SNAP survey show variability trends. For our 22 targets, we thus expect to find variability in up to $\sim$6 of our targets in the most optimistic case that young, planetary mass objects share identical statistical properties to these older populations. Data analysis is still in progress; however, one tentative detection is presented in Fig. 4. Followup of tentative detections in our October / November 2014 data will take place in August 2015.

### 4.2. *Variability in Exoplanet Companions*

Searches for variability in exoplanet companions have only recently become possible with the advent of next generation planet imagers such as GPI and SPHERE. Planets which previously required ∼1 hour of on-sky time to image can be detected even in 5-10 minutes of data from these imagers! Thus, multiple groups are pursuing variability studies for bright exoplanet companions such as $\beta$ Pic b and HR8799b. It is important to constrain variability of these objects now, as significant variability in exoplanet companions will lead to even more exciting followup possibilities with future extremely large telescope (ELT) instruments. For instance, Crossfield *et al.* (2014) used VLT CRIRES to produce a surface map of the highly variable L/T transition brown dwarf Luhman 16B using the doppler imaging technique (see Fig. 4). This is the very first surface map of a brown dwarf – if young planets such as $\beta$ Pic b are similarly variable in the L′ bandpass, the high resolution spectroscopy mode anticipated for the E-ELT mid-IR instrument METIS will enable similar mapping of exoplanet surfaces (see e.g., Snellen *et al.* 2014).

## 5. Conclusions

The first surveys of variability in young free-floating PMOs as well as the first searches for variability in exoplanet companions are currently underway. In the next few years, these studies will provide a key probe of cloud properties in PMO atmospheres. This is already the case for field L and T brown dwarfs, where variability studies indicate that cloud inhomogeneity must be quite common for these objects.

## References

Barman, T. S., Macintosh, B., Konopacky, Q. M., & Marois, C. 2011, *ApJ*, 733, 65
Beuzit, J.-L., Feldt, M., Dohlen, K., *et al.* 2008, *proc. SPIE*, 7014,
Biller, B. A., Crossfield, I. J. M., Mancini, L., *et al.* 2013, *ApJ Letters*, 778, L10
Bowler, B. P., Liu, M. C., Dupuy, T. J., & Cushing, M. C. 2010, *ApJ*, 723, 850
Buenzli, E., Apai, D., Radigan, J., Reid, I. N., & Flateau, D. 2014, *ApJ*, 782, 77
Burgasser, A. J., Sheppard, S. S., & Luhman, K. L. 2013, *ApJ*, 772, 129
Burrows, A., Sudarsky, D., & Lunine, J. I. 2003, *ApJ*, 596, 587
Crossfield, I. J. M., Biller, B., Schlieder, J. E., *et al.* 2014, *Nature*, 505, 654
Gagné, J., Lafrenière, D., Doyon, R., Malo, L., & Artigau, É. 2014, *ApJ*, 783, 121
Gagné, J., Lafrenière, D., Doyon, R., Malo, L., & Artigau, É. 2015, *ApJ*, 798, 73
Gillon, M., Triaud, A. H. M. J., Jehin, E., *et al.* 2013, *A&A*, 555, L5
Heinze, A. N., Metchev, S., Apai, D., *et al.* 2013, *ApJ*, 767, 173
Janson, M., Brandt, T. D., Kuzuhara, M., *et al.* 2013, *ApJ letters*, 778, L4
Kuzuhara, M., Tamura, M., Kudo, T., *et al.* 2013, *ApJ*, 774, 11
Lagrange, A.-M., Gratadour, D., Chauvin, G., *et al.* 2009, *A&A*, 493, L21
Lagrange, A.-M., Bonnefoy, M., Chauvin, G., *et al.* 2010, *Science*, 329, 57
Luhman, K. L. 2013, *ApJ letters*, 767, L1
Macintosh, B., Graham, J. R., Ingraham, P., *et al.* 2014, *PNAS*, 111, 12661
Marley, M. S., Saumon, D., & Goldblatt, C. 2010, *ApJ letters*, 723, L117
Marley, M. S., Saumon, D., Cushing, M., *et al.* 2012, *ApJ*, 754, 135
Marois, C., Macintosh, B., Barman, T., *et al.* 2008, *Science*, 322, 1348
Marois, C., Zuckerman, B., Konopacky, Q. M., *et al.* 2010, *Nature*, 468, 1080
Metchev, S. A., Heinze, A., Apai, D., *et al.* 2015, *ApJ*, 799, 154
Radigan, J., Jayawardhana, R., Lafrenière, D., *et al.* 2012, *ApJ*, 750, 105
Radigan, J., Lafrenière, D., Jayawardhana, R., & Artigau, E. 2014, *ApJ*, 793, 75
Rameau, J., Chauvin, G., Lagrange, A.-M., *et al.* 2013, *A&A*, 553, A60
Rameau, J., Chauvin, G., Lagrange, A.-M., *et al.* 2013, arXiv:1305.7428
Schneider, J., Dedieu, C., Le Sidaner, P., Savalle, R., & Zolotukhin, I. 2011, *A&A*, 532, A79
Skemer, A. J., Hinz, P. M., Esposito, S., *et al.* 2012, *ApJ*, 753, 14
Snellen, I. A. G., Brandl, B. R., de Kok, R. J., *et al.* 2014, *Nature*, 509, 63
Wilson, P. A., Rajan, A., & Patience, J. 2014, *A&A*, 566, A111
Zapatero Osorio, M. R., Martín, E. L., Bouy, H., *et al.* 2006, *ApJ*, 647, 1405

*Young Stars & Planets Near the Sun*
*Proceedings IAU Symposium No. 314, 2015*
J. H. Kastner, B. Stelzer, & S. A. Metchev, eds.

© International Astronomical Union 2016
doi:10.1017/S174392131500633X

# Accretion, Disks, and Magnetic Activity in the TW Hya Association

## B. Stelzer[1], A. Frasca[2] and J.M. Alcalà[3]

[1] INAF - Osservatorio Astronomico di Palermo, Piazza del Parlamento 1, 90134 Palermo, Italy
email: `stelzer@astropa.inaf.it`

[2] INAF - Osservatorio Astrofisico di Catania, Via Santa Sofia 78, 95123 Catania, Italy
[3] INAF - Osservatorio Astronomico di Capodimonte, Via Moiariello, 16, 80131 Napoli, Italy

**Abstract.** We present new photometric and spectroscopic data for the M-type members of the TW Hya association with the aim of a comprehensive study of accretion, disks and magnetic activity at the critical age of $\sim 10\,\mathrm{Myr}$ where circumstellar matter disappears.

**Keywords.** stars: low-mass, brown dwarfs, stars: chromospheres, accretion, accretion disks

## 1. Introduction

Accretion, outflows and magnetic activity are key features driving young stellar evolution. All these phenomena are linked in a complex feedback mechanism where the activity from the stellar chromosphere and corona influences the evolution of disks and planet formation while mass accretion from the disk to the star in turn may act as a heating agent on the structure of the outer stellar atmosphere. The details of this mechanism, how it depends on stellar mass, and how it evolves with time is, thus, of utmost importance for an understanding of pre-main sequence evolution.

At a distance of $\sim 50\,\mathrm{pc}$ the TW Hya association (TWA) represents one of the most easily accessible laboratories for low-mass star formation. Moreover, its age of $\sim 8\,\mathrm{Myr}$ places the TWA at a crucial evolutionary phase where disks dissipate and the accretion/outflow process comes to a halt.

We are performing a broad-band (350–2500 nm) mid-resolution spectroscopic survey with X-Shooter on the VLT. These observations give access to a rich database of emission lines which probe accretion, outflows and magnetic activity. In addition, we have carried out a optical/near-infrared (NIR) monitoring campaign of both stars in the wide binary system TWA 30 to study photometric variability related to their circumstellar environments.

## 2. Chromospheric properties of TWA stars

We have analyzed X-Shooter/VLT spectra of 24 non-accreting pre-main sequence stars, so-called Class III sources, from three nearby star-forming regions ($\sigma$ Orionis, Lupus III, and the TWA). Of particular interest here are flux-flux relations between activity diagnostics that probe different atmospheric layers (from the lower chromosphere to the corona). We examine such relations for the first time for pre-main sequence stars. For main-sequence late-type stars the fluxes of chromospheric emission lines follow power-law relations, with the exception of a subgroup of M-type stars which define a separate "active" branch (Martinez-Arnáiz *et al.* 2010). Figure 1 (left) shows that all the Class III stars from our sample lie on this "active" branch, and—spanning a range of masses throughout the full M spectral sequence—they extend it to lower fluxes. We conclude

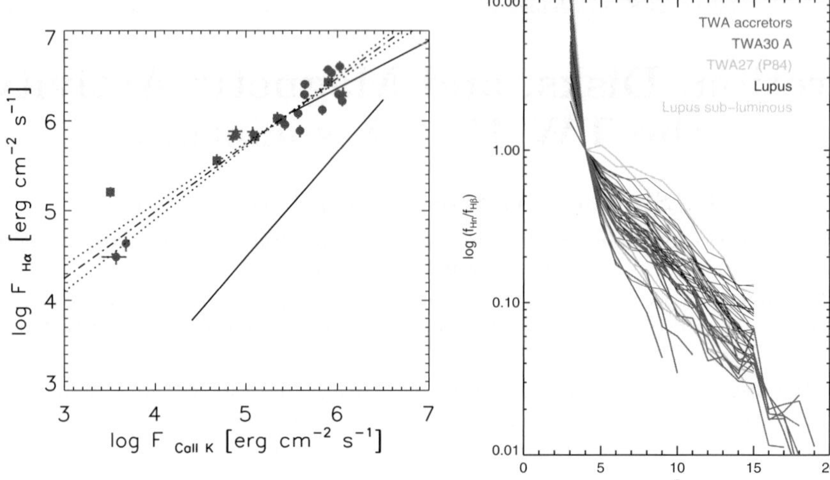



**Figure 1.** *(Left)* Flux-flux relation between chromospheric H$\alpha$ and Ca II K emission for Class III stars; upper/lower black lines denote the "active"/"inactive" branches defined by the main-sequence sample of Martínez-Arnáiz *et al.* (2010) and the blue line is the best-fit to the pre-main sequence sample. *(Right)* Balmer decrements for accretor candidates in TWA (colored curves) compared to those for accretors in Lupus (black curves; Antoniucci *et al.* 2015, in prep).

that chromospheric flux-flux relations can serve as a tool for confirming the youth of low-mass stars. More detailed investigations covering ages between the $\sim 10\,\mathrm{Myr}$ of TWA and the $\geqslant 1\,\mathrm{Gyr}$ of main-sequence field stars are required to explore this new age diagnostic.

Flux ratios between individual emission lines of the Class III sources show a smooth dependence on the effective temperature. Meunier & Delfosse (2009) found that in the solar chromosphere the different behavior of H$\alpha$ and Ca II H&K is due to their link to different structures, so-called filaments and plages. However, the observed spread in Figure 1 and in the H$\alpha$/Ca II K flux ratio for given $T_{\mathrm{eff}}$ suggests that a third parameter, in addition to age and $T_{\mathrm{eff}}$, determines the chromospheric structure of late-type stars.

H$\alpha$ and Ca II H&K are the best-studied emission lines in chromospheres of late-type stars. Our knowledge on the whole emission line spectrum is more elusive as a result of limited spectral range and sensitivity of most available spectrographs. This gap can be closed with wide-band spectroscopic studies, such as our comprehensive X-Shooter study of Class III sources (see Stelzer *et al.* 2013).

## 3. Accretion measurements for TWA members

We have obtained X-Shooter spectra for 15 "potentially accreting" TWA members. Here we concentrate on measurements of mass accretion rates ($\dot{M}_{\mathrm{acc}}$). We consider as "potential accretors" all TWA stars previously known to be accreting and further TWA members with evidence of IR (disk) excess in published photometry or a Balmer jump in the X-Shooter spectra. This sample excludes the Class III stars discussed in Sect. 2.

Two methods are used to measure mass accretion rates from the X-Shooter spectra. First, we determine the Balmer jump—if any—by modelling the UVB spectrum with a combination of a non-accreting Class III template and a hydrogen slab model representing emission from the accretion column as described by Manara *et al.* (2013). This yields the accretion luminosity, $L_{\mathrm{acc}}$. Secondly, we use the empirical calibrations between fluxes of individual emission lines and the $L_{\mathrm{acc}}$ values measured with the slab modelling derived by Alcalá *et al.* (2014) to obtain $\dot{M}_{\mathrm{acc}}$ from observed line fluxes.

We compare our preliminary results for $\dot{M}_{\mathrm{acc}}$ as a function of stellar mass ($M_*$) to the analogous study for the Lupus star forming region by Alcalá *et al.* (2014), and find that the TWA stars display smaller $\dot{M}_{\mathrm{acc}}$ at given mass. However, the slope in the $M_* - \dot{M}_{\mathrm{acc}}$ diagram is the same as for Lupus. This indicates that the TWA disks represent an evolved version of the same type of disks that are present in Lupus.

In Figure 1 (right) we compare the Balmer decrements of the TWA accretor sample to those of the accretors in Lupus (see Antoniucci *et al.* 2015, in prep.). The TWA stars tend to have slightly steeper decrements than the younger stars in Lupus. The physical conditions determining the shape of the decrements will be investigated by comparison to decrements predicted from the radiative transfer models by Kwan & Fischer (2011).

## 4. Multiband photometric monitoring of the TWA 30 binary

TWA 30A and 30B are nearly equal-mass components of a wide binary (spectral types M5 and M4; $80''$ separation) discovered and confirmed as kinematic TWA members by Looper *et al.* (2010a) and Looper *et al.* (2010b). They are among the nearest ($\sim 50\,\mathrm{pc}$) stars with signatures of disks, mass accretion and outflow activity. The disks of both stars are likely seen (nearly) edge-on: (i) The forbidden emission lines (FELs) have only small radial velocity shifts indicating that their jets are oriented in the plane of the sky; (ii) the FELs are much stronger than the accretion signatures suggesting that the accretion columns are partially hidden by parts of the circumstellar disk.

In the discovery papers, Looper *et al.* (2010a) and Looper *et al.* (2010b) showed with multi-epoch low-cadence spectroscopy that both stars have strongly variable NIR colors. To constrain the time-scales of the variability over its full optical/NIR spectral energy distribution we have carried out a photometric monitoring campaign with the robotic 60 cm telescope REM on La Silla operated/owned by the Italian Istituto Nazionale di Astrofisica (INAF). The REM is equipped with a NIR camera, REMIR, and an optical camera, ROSS 2, fed by the same telescope and providing simultaneous photometry in seven filters, $g'r'i'z'JHK'$. The REM lightcurves of TWA 30A and 30B are shown in Figure 2.

TWA 30A shows quasi-periodic dimmings throughout the monitoring. The lightcurves observed in the seven filters are highly correlated and the depth increases systematically for bluer bands. The NIR colors show that the data points are roughly aligned along the reddening vector. However, during the whole 3-month monitoring campaign the extinction did not reach the high ($\sim 10\,\mathrm{mag}$) values inferred from previous spectroscopy. Similar lightcurves have been observed from 9 stars in the NGC 2264 star forming region based on a long CoRoT and Spitzer monitoring campaign (Stauffer *et al.* 2015). The occurrence of such periodic dimmings can be interpreted as dust structures levitated above the disk and temporarily occulting the star light as they pass in front of the line of sight.

TWA 30B was for most of the time during our REM monitoring observed in a faint state with occasional "bursts" of several magnitudes. Contrary to TWA 30A, the brightness variations of TWA 30B are much stronger in the NIR than at optical bands. The "bursts" observed in the REM lightcurve are likely responsible for the NIR excess seen in the previous spectroscopic data, which was modelled by Looper *et al.* (2010b) as cool ($\sim 700\,\mathrm{K}$) blackbody emission. A physical interpretation for this excess NIR emission could be starlight reprocessed from a small portion of the disk that rotates in and out of the line-of-sight. For TWA 30B the NIR photometry has a much flatter slope in the $J - H$ vs $H - K_s$ diagram than TWA 30A, and NIR colors measured for the "bursts" seen during

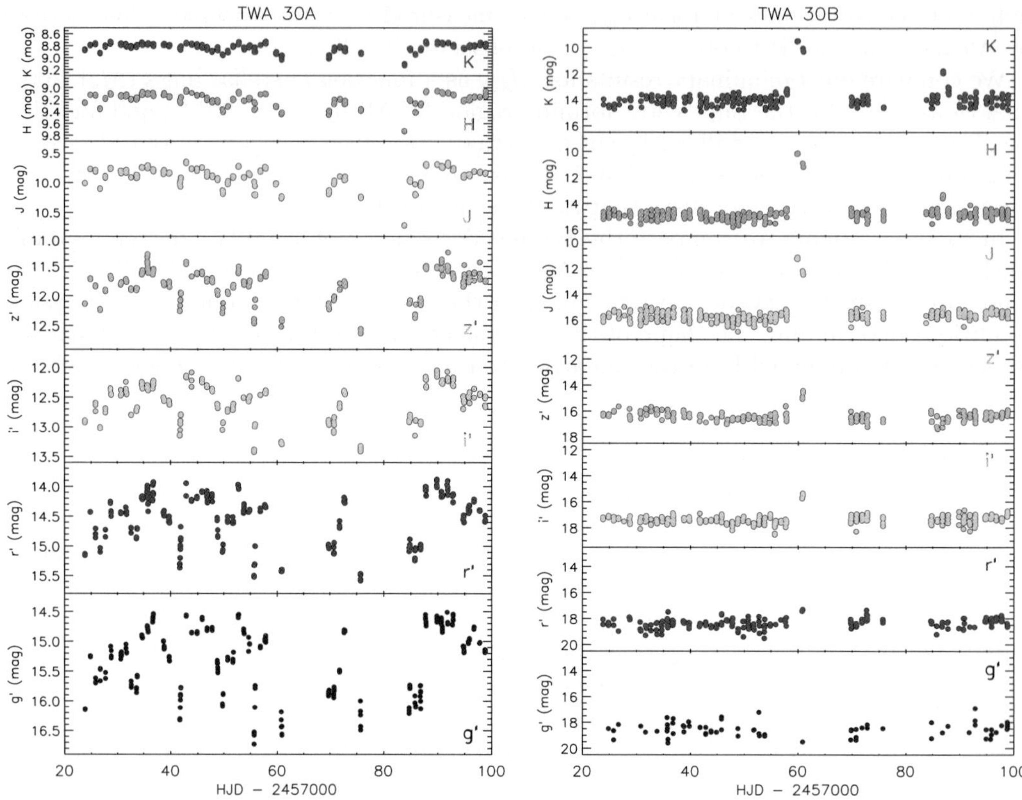

**Figure 2.** REM multi-band optical/NIR lightcurves of TWA 30A (left) and TWA 30B (right).

our REM monitoring are different from those inferred from previous spectrophotometry by Looper *et al.* (2010b).

## References


Alcalá, J. M., Natta, A., Manara, C. F., *et al.*, 2014, *A&A*, 561A, 2
Kwan, J. & Fischer, W., 2011, *MNRAS*, 411, 2383
Looper, D. L., Mohanty, S., Bochanski, J. J., Burgasser, A. J., *et al.*, 2010, *ApJL*, 714, 45
Looper, D. L., Bochanski, J. J., Burgasser, A. J., Mohanty, S., *et al.*, 2010, *ApJS*, 140, 1486
Manara, C.F., Beccari, G., Da Rio, N., *et al.*, *A&A*, 558A, 114
Martínez-Arnáiz, R., López-Santiago, J., Crespo-Chacón, I., & Montes, D., 2010, *MNRAS*, 414, 2629
Meunier, N. & Delfosse, X., 2009, *A&A* 501, 1103
Stauffer, J., Cody, A. M., McGinnis, P., *et al.*, 2015, *AJ*, 149, 130
Stelzer, B., Frasca, A., Alcalà, J. M., *et al.*, 2013, *A&A*, 558A, 141


## Discussion

RODRIGUEZ: Did you find periodicities in the REM lightcurves of TWA 30 ?

STELZER: Yes, indeed. The dips of TWA 30A are quasi-periodic with $P \sim 8\,$d. Assuming corotation, the observed period implies that the dust is located at a distance of $\sim 0.2\,$AU from the star (i.e. in the inner disk) and that its azimuthal position is roughly consistent over time. The variable depth of the dips indicates that the amount or optical depth of the occulting material is changing in time.

*Young Stars & Planets Near the Sun*
Proceedings IAU Symposium No. 314, 2015
J. H. Kastner, B. Stelzer, & S. A. Metchev, eds.

© International Astronomical Union 2016
doi:10.1017/S174392131500647X

# Angular Momentum Evolution
# of Young Solar-type Stars

**Louis Amard[1,2], Ana Palacios[1], Corinne Charbonnel[2,3]**

[1]LUPM,Université Montpellier, CNRS, Place Eugène Bataillon, 34095 Montpellier France
email: `louis.amard@univ-montp2.fr`
[2]Geneva Observatory, University of Geneva, Chemin des Maillettes, 51, 1290 Versoix, Switzerland
[3]IRAP departement of astronomy, CNRS UMR 5277, Université de Toulouse, 14, Av. E.Belin, 31400 Toulouse, France

**Abstract.** We present stellar evolution models of young solar-type stars including self consistent treatment of rotational mixing and extraction of angular momentum (AM) by magnetized wind including the most up-to-date physic of AM transport.

**Keywords.** stars: evolution, stars: rotation, stars: wind-braking, hydrodynamics

## 1. Introduction

During their early evolution, solar-type stars undergo a considerable modification of their AM distribution due mainly to interaction with their environment. As the still-accreting star is surrounded by its disc, the extraction of AM is mainly driven by jets and accretion; but once the star is spinning freely, the major part of the AM is removed at the surface by stellar winds outflowing along magnetic field lines while the rest is redistributed in the radiative region by hydrodynamical instabilities.

Here we use two specific prescriptions with different initial conditions to describe the impact of both phenomena (torque and internal transport of AM) on the evolution of surface and core rotation. We show in particular that rotation periods observed in open clusters over a broad age range can be reproduced with a weak core-envelope coupling using the most recent braking law available in the literature.

## 2. Physics input to the stellar models

We compute stellar models ($1M_\odot$, $Z = 0.0134$) with the STAREVOL code (Siess *et al.* 2000, Palacios *et al.* 2006, Lagarde *et al.* 2012) including suitable physics for this range of mass. We account for an atmosphere with a $T - \tau$ relation following Krishna-Swamy (1966) that has been extensively tested on stellar models and gives good results (e.g. Vandenberg *et al.* 2007).

The treatment of convection is based on the classical mixing-length formalism with the parameter $\alpha_{MLT} = l/H_P = 1.607$ calibrated on a solar model. Convective areas are instantaneously mixed and assumed to rotate as solid bodies. Moreover, we account for hydrostatic effects through centrifugal forces entering the hydrostatic equation following Endal & Sofia (1976).

We follow the formalism of Zahn (1992), Maeder & Zahn (1998) and Mathis & Zahn (2004) to compute the secular transport of AM in the radiative zone.

$$\rho\frac{\mathrm{d}}{\mathrm{d}t}\left(r^2\Omega\right) = \frac{1}{5r^2}\frac{\partial}{\partial r}\left(\rho r^4\Omega U_r\right) + \frac{1}{r}\frac{\partial}{\partial r}\left(r^4\rho\nu_v\frac{\partial\Omega}{\partial r}\right) + \tau_W \qquad (2.1)$$

with the vertical shear diffusion coefficient $\nu_v$ explicited in Talon & Zahn (1997) and the horizontal diffusion coefficient from Mathis *et al.* (2004). An extensive study of internal mixing formalism is done in Amard *et al.* (in prep.) with other prescriptions. We do not account for additional processes such as magnetic fields or internal gravity waves, which are also expected to transport AM very efficiently in the radiative zone.

The last term $\tau_W$ represents the AM loss due to the stellar-wind torque and is only applied at the external boundary layer. Here we present model predictions using Matt *et al.* (2015) prescription (other braking laws studied in Amard *et al.* in prep.) :

$$\tau_W = -\mathcal{T}_0 \left( \frac{\tau_{CZ}}{\tau_{CZ\odot}} \right)^p \left( \frac{\Omega_\star}{\Omega_\odot} \right)^{p+1} \geqslant \text{unsaturated,} \tag{2.2}$$

$$\tau_W = -\mathcal{T}_0 \chi^p \left( \frac{\Omega_\star}{\Omega_\odot} \right) \geqslant \text{saturated} \tag{2.3}$$

with $p = 1.7$ and the empirically calibrated scaling factor $\mathcal{T}_0$ as in Matt *et al.* (2015) (see also Amard *et al.* in prep.). This prescription accounts for a magnetic field saturation reached for $\chi = \frac{Ro_\odot}{Ro_{\rm crit}} = 10$ with $Ro_\odot = (\tau_{CZ\odot}\Omega_\odot)^{-1}$ the solar Rossby number and $Ro_{\rm crit}$ the critical one, $\tau_{CZ}$ being the convective turnover timescale in the stellar envelope.

## 3. Rotation distribution

In order to constrain the evolution, we need some quite accurate ages and only associations or open clusters can provide this accuracy. Young associations have usually too few members in a specific mass range, so we only use open cluster members with a mass between 0.9 $M_\odot$ and 1.1 $M_\odot$. We use the sample gathered by Gallet & Bouvier (2015 and references therein) which includes clusters with ages between 2 Myr (ONC, NGC6530,...) and 2.5 Gyr (NGC6819). We also include the sun to constrain long-term evolution of the rotation of solar-type stars. Observations are shown in Fig. 1, where we also indicate the 90th, 50th and 25th percentiles for each selected cluster computed using the rejection method (see Gallet & Bouvier 2013). We assume that these three statistical samples represent typical evolutions of the surface rotation of young solar-type stars and these are the data we aim to reproduce with our models in the present study.

## 4. Evolution of surface rotation

We follow the evolution as soon as the star has shrinked enough to ensure a structural cohesion (surface velocity below the break-up speed). The one solar mass model is computed with three different initial velocities corresponding to the three evolutions (fast-median-slow).

Following the observations, we assume a constant angular velocity during the disc-coupling phase which has a duration which is adapted accordingly: 3 Myr for the fast rotating model and 5 Myr for the median and slow ones.

We account for AM extraction as soon as the stars are decoupled from their disc. Between the end of the disc-coupling phase and the zero-age main sequence (ZAMS), the stars are contracting and contraction dominates the evolution of surface angular velocity for the first tens of Myr. At the arrival on the ZAMS, the stars stop contracting and the surface velocities reach a maximum before the torque becomes dominant and starts spinning down the star.

The spin-down phase is driven by the internal transport of AM. AM redistribution depends on the spinning rate and on the evolution of the magnetic field strength which itself is related to the Rossby number which itself depends on the stellar surface velocity.

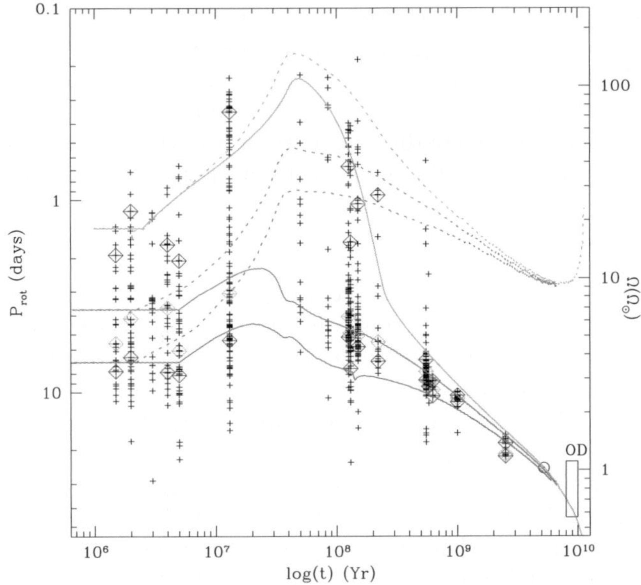

**Figure 1.** Evolution of angular velocities (right) and rotational periods (left) as a function of time. Crosses show stars belonging to open clusters at various ages (Gallet & Bouvier 2015). Red, green and blue tilted squares represent respectively the 25th, 50th and 90th percentile associated to each type of rotator. Solid lines show the evolution of surface velocity for slow (magenta), median (green) and fast rotators (blue). Dashed lines represent the angular velocity integrated over the whole radiative region of each model. The open circle is the angular velocity of the present Sun and the OD black rectangle shows the angular velocity dispersion of old disk field stars (McQuillan *et al.* 2013).

Figure 1 presents the theoretical evolution of surface rotation for our three models (slow, moderate, and fast rotation). Predictions do well reproduce the observed rotational periods at each cluster. We can highlight two important points:

(1) After the ZAMS, surface rotation velocity is dominated by the wind-induced torque, which drives internal transport of AM. The effects of internal transport of AM on surface rotation are erased by the wind-induced torque. The latter is indeed completely dominating the evolution of the rotational period at evolutionary time-scale. But during the PMS, fast rotation and contraction induce a very strong meridional circulation spinning up the star and counter-balancing the torque.

(2) The coupling between the radiative core and the envelope of the star depends here on the surface angular velocity of the star: the fast rotator is almost in solid-body rotation until it reaches the ZAMS while slower rotators are fully decoupled, their core spinning more than 10 times faster than the surface at the arrival on the ZAMS. This can be explained by two factors: AM internal transport efficiency depends on the velocity of the star and the torque is also stronger (proportionally to the AM content) when the stellar dynamo is in a non saturated regime. Therefore, slow rotators are below the saturation limit and the surface experiences a relatively strong braking while the radiative part is accelerating freely. We can probe these effects by looking at the maximal rotation rate reached at the ZAMS compared to the initial velocity: in the case of both slow and median rotators, this maximum happens a few ten Myr earlier than for the fast rotator case. This is due to cumulative effects of weak core-envelope coupling and strong torque. In parallel, the radiative core acceleration is comparable in every case and the mean velocity is proportional to the initial velocity.

Finally, the surface velocities of the three models converge beyond 500 Myr to reach the solar value at the solar age following a Skumanich-like trend. In parallel, we also observe the convergence of the integrated angular velocities in the radiative zone. This indicates that at this phase, all our models experience the same efficiency for the transport of AM independently of the initial conditions. Turning to other indicators such as the light elements surface abundances would help to disentangle between these three different evolutions.

## 5. Conclusions

We present stellar evolutionary models including a self-consistent treatment of rotation-induced mixing and realistic AM extraction allows us to study the effect of different physical processes on stellar rotation along the evolution. We show that including only hydrodynamical mechanisms we can reach a unique rotation profile at the age of the Sun independent of the initial angular momentum content, losing memory of all previous evolution.

The surface angular velocity evolution essentially constrains the torque. The latter dominates the integrated surface rotation evolution but is not constraining the internal AM transport. Indeed, asteroseismology (e.g. Garcia *et al.* 2011) and stellar chemical properties (e.g. Bouvier 2008) reveal a flattened profile for the actual Sun in opposition to the important core-envelope decoupling we obtain in our model. This can be explained by additional processes such as internal gravity waves (Charbonnel *et al.* 2013, Charbonnel & Talon 2005) or magnetic fields (Eggenberger *et al.* 2005), which appear to be very good candidates to transport efficiently AM in the radiative zone. The next step will be to include those processes in the computation to test their efficiency with a realistic torque.

## References

Amard, L., Palacios, A., Charbonnel, C. in prep
Bouvier, J. 2008, *A&A*, 489, 53
Charbonnel, C., Decressin, T., Amard, L., Palacios, A., & Talon, S. 2013, *A&A*, 554, A40
Charbonnel, C. & Talon, S. 2005, *Science*, 309, 2189
Eggenberger, P., Maeder, A., & Meynet, G. 2005, *A&A*, 440, L9
Endal, A. S. & Sofia, S. 1976, *ApJ*, 210, 184
Gallet, F. & Bouvier, J. 2013, *A&A*, 556, A36
Gallet, F. & Bouvier, J. 2015, *A&A*, 577, A98
García, R.-A., Salabert, D., & Ballot, J. et al. 2011, *JPhCS*, 271
Krishna-Swamy, K. S. 1966, *ApJ* 145,174
Lagarde, N. & Decressin, T., Charbonnel C. et al. 2012, *A&A*, 543, 108
Maeder, A. & Zahn, J.-P. 1998, *A&A*, 334, 1000
Mathis, S., Palacios, A., & Zahn, J.-P. 2004, *A&A*, 425, 243
Mathis, S. & Zahn, J.-P. 2004, *A&A*, 425, 229
Matt, S., Brun, A. S., Baraffe, I., Bouvier, J., & Chabrier, G. 2015, *ApJ*, 799, L23
McQuillan, A., Mazeh, T., & Aigrain, S. 2013, *ApJL*, 775, L11
Palacios, A., Charbonnel, C., Talon, S., & Siess, L. 2006, *A&A*, 453, 261
Siess, L., Dufour, E., & Forestini, M. 2000, *A&A*, 358, 593
Talon, S. & Zahn, J.-P. 1997, *A&A*, 317, 749
Vandenberg, D. A., Edvardsson, B., Eriksson, K., Gustafsson, B., & Ferguson, J. W. 2007, in
    IAU Symp. 241, *Stellar Populations as Building Blocks of Galaxies*, p. 23
Zahn, J.-P. 1992, *A&A*, 265, 115

## Discussion

PINSONNEAULT: Comment: K2 mission is about to get asteroseismic data from young stars around 30 to 100 Myr allowing to potentially probe internal rotation during this crucial phase.

*Young Stars & Planets Near the Sun*
*Proceedings IAU Symposium No. 314, 2015*
*J. H. Kastner, B. Stelzer, & S. A. Metchev, eds.*

© International Astronomical Union 2016
doi:10.1017/S1743921315006705

# The Evolution of Surface Magnetic Fields in Young Solar-type Stars

**Colin P. Folsom[1,2], Pascal Petit[3], Jérôme Bouvier[1,2], Julien Morin[4], Agnès Lèbre[4], and Jean-François Donati[3]**

[1]Université Grenoble Alpes, IPAG, F-38000 Grenoble, France
[2]CNRS, IPAG, F-38000 Grenoble, France
email: `colin.folsom@obs.ujf-grenoble.fr`
[3]IRAP, CNRS and Université de Toulouse, 14 avenue Édouard Belin, 31400, Toulouse, France
[4]LUPM, UMR 5299, CNRS and Université Montpellier II - Place E. Bataillon, 34090 Montpellier, France

**Abstract.** Surface rotation rates of young solar-type stars display drastic changes at the end of the pre-main sequence through the early main sequence. This may trigger corresponding changes in the magnetic dynamos operating in these stars, which ought to be observable in their surface magnetic fields. We present here the first results of an observational effort aimed at characterizing the evolution of stellar magnetic fields through this critical phase. We observed stars from open clusters and associations, which range from 20 to 600 Myr, and used Zeeman Doppler Imaging to characterize their complex magnetic fields. We find a clear trend towards weaker magnetic fields for older ages, as well as a tight correlation between magnetic field strength and Rossby number over this age range. Comparing to results for younger T Tauri stars, we observe a very significant change in magnetic strength and geometry, as the radiative core develops during the late pre-main sequence.

**Keywords.** stars: magnetic field, stars: evolution, stars: rotation, stars: pre-main sequence, stars: solar-type

---

## 1. Introduction

Solar type stars display a major change in their rotation rates as they cross the pre-main sequence (PMS) and settle on to the main sequence (e.g., Gallet & Bouvier 2013, 2015). Early on the PMS, stars strongly interact with their disks, and this regulates their rotation rates. After a few Myr, a star decouples from its disk and is still contracting, which results in a rapid spin up. On a longer timescale, the star loses angular momentum through its magnetized wind, and starts to spin down as it settles onto the main sequence.

Magnetic fields in solar-type stars are thought to be generated by a dynamo process, possibly an $\alpha$-$\Omega$ dynamo. Thus, the large change in rotation rates around the zero-age main sequence (ZAMS) should have a large impact on magnetic properties. Conversely, the rotational spin down is controlled by magnetic fields (e.g. Matt et al. 2012). It is thus crucial to understand the connection between magnetism, rotation, and age in order to to fully decipher early stellar evolution and its potential impact on, e.g., the early evolution of planetary systems and their habitability.

We present here the first results from an investigation of the magnetic properties of young solar-type stars as a function of rotation rate and age.

## 2. Observations

Targets for our study are members of known moving groups and clusters, in the age range of $\sim$20 to $\sim$600 Myr, when most of the rotational evolution occurs (cf. Bouvier et al. 2014). We exclude T Tauri stars, to avoid stars whose rotation rate is still regulated by disk interactions, and that have been studied in previous projects (e.g. Donati et al. 2011). We also exclude older stars from our selection, which have converged onto the gyrochronologic sequence and whose magnetic properties have been already characterized (e.g. Petit et al. 2008). Indeed, our sample was devised to fill the evolutionary gap between early PMS stars and mature main sequence stars. We focus on 0.7-1.0$M_\odot$ stars with known rotational periods, usually derived from photometric monitoring.

In order to characterize their magnetic fields, the targets were observed with high-resolution spectropolarimetry. We used ESPaDOnS at the Canada France Hawaii Telescope and Narval at the Télescope Bernard Lyot in France. These are essentially identical spectropolarimeters, with a resolution of 65000, and a wavelength coverage of 3700 to 10500 Å. Reduced observations contain a circularly polarized Stokes $V$ spectrum as well as a total intensity Stokes $I$ spectrum (see Donati et al. 1997 for data reduction). Observations of a target consist of a series of typically 15 observations, spread over the rotation cycle of the star. This phase-resolved series of observations allows us to both diagnose the presence of a magnetic field, and characterize its geometry, as discussed below.

## 3. Analysis

The high resolution, high S/N spectra allow us to derive precise fundamental parameters for the stars. Using spectrum synthesis, and by fitting model spectra directly to the observations, we derive $T_{\rm eff}$, $\log g$, $v \sin i$, microturbulence, and [Fe/H] for all the stars in our sample. The spectra can also be used to determine detailed chemical abundances (e.g. Lithium). Most of the stars have reliable distance estimates, and thus we can determine intrinsic luminosities, and from that radii and masses.

Accurate rotation periods are essential for this analysis. Thus we re-derived the rotation periods for the stars in this study. For most stars, our results support the literature photometric periods. However, for a few stars, the literature periods are inconsistent with our spectropolarimetric time series, and for those stars we use the periods we derive.

In order to derive the magnetic field strengths and geometries of the stars in this study, we use the Zeeman Doppler Imaging (ZDI; Donati et al. 2006). ZDI is a tomographic technique. The rotationally modulated variability of the line profile is inverted to reconstruct the stellar magnetic field. Stokes $V$ line profiles generated by Zeeman splitting are used. Our version of the method uses a spherical harmonic decomposition to describe the magnetic field, and the inversion process is regularized using a maximum entropy regularization in order to provide a stable unique solution. The $v \sin i$ and rotation periods derived above were used as input, as well as an inclination of the rotation axis based on $v \sin i$, radius, and period. An example magnetic map from ZDI is presented in Fig. 1.

## 4. Discussion

Even though our survey of magnetic field properties in young stars is not fully completed yet, some early trends are apparent. We observe a significant decrease of the global average magnetic field strength with age, as shown in Fig. 2. The trend is already seen at the ZAMS, albeit with a large dispersion at a given age, and considerable overlap between the mean magnetic field distributions of stars over the age range from 20 to 250 Myr.

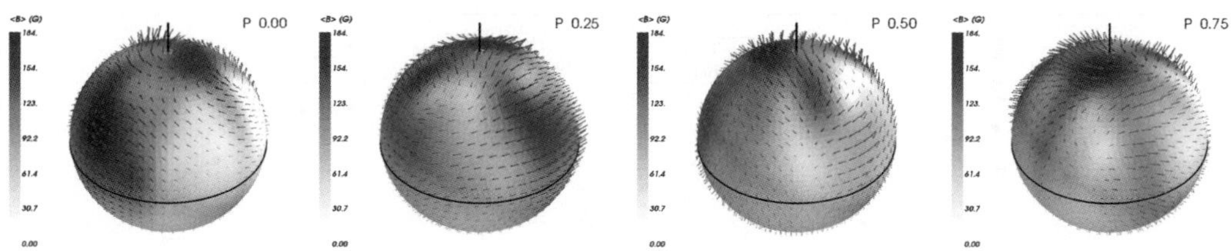

**Figure 1.** Example of a magnetic map for TYC 6349-0200-1 produced using ZDI. Arrows indicate magnetic field orientation, red arrows have a positive radial component, and blue arrows have a negative radial component.

The same trend extends all the way to the MS, as previously reported by Vidotto *et al.* (2014) using the Bcool sample (Petit *et al.*, in prep.). We find an even tighter correlation between average magnetic strength and Rossby number ($R_o$), as shown in Fig. 2. Indeed, our ZAMS sample complements and extrapolates to younger ages the $\langle B \rangle$-$R_o$ correlation seen on the MS for the Bcool sample. Only the fastest rotator of our sample, LO Peg, may show signs of magnetic saturation at very low Rossby number.

An interesting comparison sample are the classical T Tauri stars (cTTS) from the MaPP project (e.g. Donati *et al.* 2011). These are younger pre-main sequence stars that are still actively accreting and are predominantly convective. We see a clear distinction between the magnetic properties of the MaPP sample and our older stars. cTTS have much stronger magnetic fields, and their fields are much more poloidal and aligned with the stellar rotation axis (Fig. 3). This dramatic difference likely reflects a change in the internal structure of the stars, going from fully convective at the T Tauri stage to being largely radiative on the ZAMS. This structural change appears to strongly affect the type of dynamo operating in PMS stars, as suggested by Gregory *et al.* (2012). A similar difference is seen between partially and full convective M dwarfs (Morin *et al.* 2010).

The three samples are shown together in Fig. 3, where Rossby number is plotted as a function of age. It illustrates two main trends: i) the evolution of magnetic properties from the early PMS to the end of the PMS appears to be mainly driven by structural changes, with significant differences seen in magnetic strengths and topologies of PMS and ZAMS stars at a given Rossby number, while ii) the evolution of magnetic properties from the late-PMS to the ZAMS and MS strongly correlates with Rossby number, presumably reflecting the decreasing efficiency of the stellar dynamo as the stars are spun down. Beyond these clear trends, it is worth noticing the large scatter of magnetic properties seen on the ZAMS. While intrinsic variability contributes to this scatter, it could be related to a third parameter, possibly the large amount of internal differential rotation predicted at this phase of evolution (Gallet & Bouvier 2013, 2015).

## 5. Conclusion

By studying the magnetic properties of a sample of young solar-type stars, we have filled the evolutionary gap between young PMS T Tauri stars and mature main sequence stars. Comparing samples, the main finding is that magnetic properties scale primarily with internal structure during PMS evolution, and with rotation during ZAMS/early-MS evolution. The transition from a fully convective to a partly radiative interior during the PMS yields complex non-axisymmetric fields on the ZAMS, while the spin down of stars on the MS drives the decline of magnetic field strength on a longer timescale.

While we have drawn some preliminary conclusions here, the study is ongoing with a Large Program being performed at the CFHT ('The History of the Magnetic Sun', PI

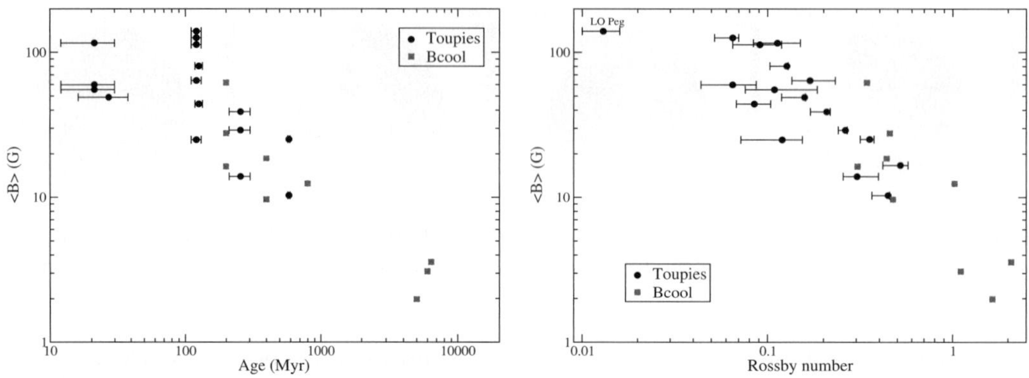

**Figure 2.** Trends in magnetic field strength with age (left) and Rossby number (right). Black points are from our sample, and red points are from the Bcool sample of Petit *et al.* (in prep.). A power law fit of the Rossby number trend produces $\langle B \rangle \propto R_o^{-1.0 \pm 0.1}$.

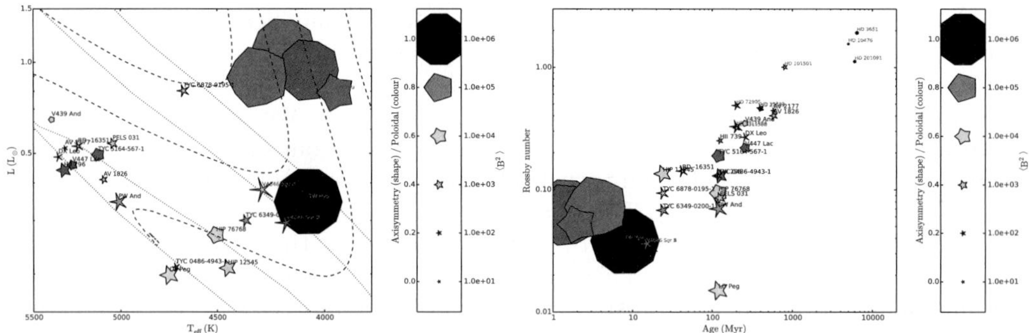

**Figure 3.** Magnetic properties of stars in our sample (black labels), stars in the Bcool sample (red labels), and stars in the MaPP sample (blue labels). Left frame: our stars and the MaPP sample in an HR diagram, with pre-main sequence evolutionary tracks. Right frame: our stars, the Bcool stars, and the MaPP stars in the age-Rossby number plane. Symbol size corresponds to magnetic field strength, symbol color corresponds to the ratio of poloidal to toroidal field, and shape corresponds to the degree of axisymmetry of the field.

P. Petit). This should help us further investigate differences in magnetic properties with rotation at a specific age, and differences with age at a specific Rossby number.

## References


Bouvier, J. 2014 *Protostars and Planets VI*, 433

Donati, J.-F. *et al.* 2006, *MNRAS*, 370, 629

Donati, J.-F. *et al.* 2011, *MNRAS*, 417, 1747

Donati, J.-F. *et al.* 1997, *MNRAS*, 291, 658

Gallet, F. & Bouvier, J. 2013 *A&A*, 556, A36

Gallet, F. & Bouvier, J. 2015 *A&A*, 577, A98

Gregory, S. G. *et al.* 2012 *ApJ*, 755, 97

Matt, S. P., MacGregor, K. B., Pinsonneault, M. H., & Greene, T. P. 2012, *ApJ*, 754, L26

Morin, J. *et al.* 2010, *MNRAS*, 407, 2269

Petit, P. *et al.* 2008 *MNRAS*, 388, 80

Vidotto, A. A. *et al.* 2014 *MNRAS*, 441, 2361

*Young Stars & Planets Near the Sun*
Proceedings IAU Symposium No. 314, 2015
J. H. Kastner, B. Stelzer, & S. A. Metchev, eds.

© International Astronomical Union 2016
doi:10.1017/S1743921315005906

# Radio Emission from Binary Stars in the AB Doradus Moving Group

## R. Azulay[1], J.C. Guirado[1,2], J.M. Marcaide[1], I. Martí-Vidal[3], and E. Ros[4,1,2]

[1]Departament d'Astronomia i Astrofísica, Universitat de València, E-46100 Burjassot, València, Spain
email: `rebecca.azulay@uv.es`
[2]Observatori Astronòmic, Universitat de València, E-46980 Paterna, València, Spain
[3]Onsala Space Observatory, Chalmers University of Technology, SE-439 92, Onsala, Sweden
[4]Max-Planck-Institut für Radioastronomie, D-53121 Bonn, Germany

**Abstract.** Precise determination of dynamical masses of pre-main-sequence stars is essential for calibrating stellar evolution models, that are widely used to derive theoretical masses of young low-mass objects. We have determined the individual masses of the pair AB Dor Ba/Bb using Australian Long Baseline Array observations and archive infrared data, as part of a larger program directed to monitor binary systems in the AB Doradus moving group. We have detected, for the first time, compact radio emission from both stars. This has allowed us to determine the orbital parameters of both the relative and absolute orbits and, consequently, their individual dynamical masses: $0.28\pm0.05\,M_\odot$ and $0.25\pm0.05\,M_\odot$. Comparisons of the dynamical masses with the prediction of pre-main-sequence (PMS) evolutionary models show that the models underpredict the dynamical masses of the binary components Ba and Bb by 10−30% and 10−40%, respectively.

**Keywords.** astrometry, binaries (including multiple): close, stars: fundamental parameters, stars: pre-main sequence

## 1. Introduction

An improvement of the calibration of PMS evolutionary tracks should necessarily come from the precise determination of dynamical masses of PMS stars. Binary stars in young, nearby moving groups are good candidates. We selected the AB Doradus moving group (AB Dor-MG) as the best-suited association to apply radio-based high-precision astrometric techniques to study binary systems. This choice is well supported by the systems mean distance to the Sun (30 pc), its reasonably well known age (50−70 Myr; Janson *et al.* 2007; Guirado *et al.* 2011), and the presence of radio emission in some of its active members (Guirado *et al.* 2006; Azulay *et al.* 2014, 2015). Following the list of AB Dor-MG members in Torres *et al.* (2008), we initiated a VLA/VLBI program to monitor binary systems known to host low-mass companions, and which are likely to present radio emission.

## 2. Long Baseline Array observations of AB Dor B

AB Dor B is a close binary which consists of two components, AB Dor Ba and AB Dor Bb, separated by ∼60 mas (Guirado *et al.* 2006; Janson *et al.* 2007). We observed this star between 2007 and 2013 with the Australian Long Baseline Array at 8.4 GHz (Azulay *et al.* 2015). With these observations, we confirmed that both components are compact and strong radio emitters. We could determine the relative position of one component

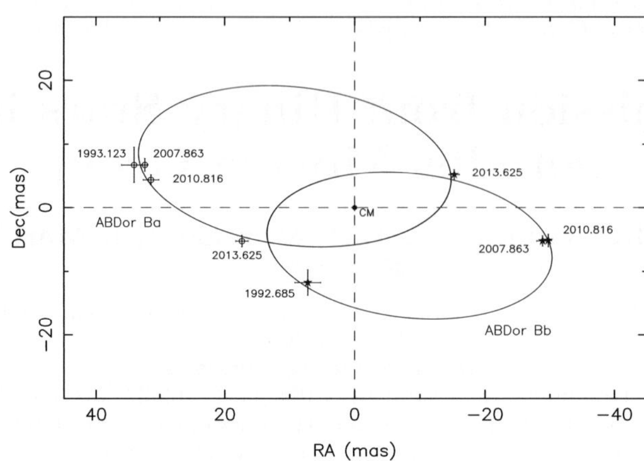


**Figure 1.** Absolute orbits of AB Dor Ba and AB Dor Bb. The positions of components Ba (circles) and Bb (star symbols) are marked along with their respective epochs. The center of mass (CM) of the system is placed at the origin.

with respect to the other and the absolute position of each component with respect to an external quasar. We also included NIR data available in the literature (Wolter *et al.* 2014, and references therein) to determine simultaneously both the relative and the individual orbits of each pair (Fig. 1). The best orbit corresponds to a period of 0.986±0.008 yr. The masses of the components are 0.28±0.05 M$_\odot$ and 0.25±0.05 M$_\odot$ for AB Dor Ba and AB Dor Bb, respectively.

## 3. Conclusions

Comparison of the dynamical masses with the prediction of PMS models show that the models underpredict the dynamical masses of the binary components Ba and Bb by 10−30% and 10−40%, respectively, underlining the known tendency of these models to underpredict the masses (Hillenbrand & White 2004; Mathieu *et al.* 2007), yet within 2-$\sigma$ of the predicted values. Some of the stellar models considered favour an age between 50 and 100 Myr for this system, meanwhile others predict older ages. Our ongoing VLBI program includes European VLBI Network 5 GHz observations of other stars belonging to the AB Dor-MG such as EK Draconis, LO Pegasus, PW Andromedae, and HD 160934. In particular, for the latter star we have determined a period of 10.33±0.06 yr and preliminary values of the individual masses of ∼0.7 M$_\odot$ for the component A and ∼0.5 M$_\odot$ for the component c (Azulay *et al.*, in preparation).

## References


Azulay, R., Guirado, J. C., Marcaide, J. M., *et al.* 2014, *A&A*, 561, A38
Azulay, R., Guirado, J. C., Marcaide, J. M., *et al.* 2015, *A&A*, 578, A16
Guirado, J. C., Martí-Vidal, I., Marcaide, J. M., *et al.* 2006, *A&A*, 446, 733
Guirado, J. C., Marcaide, J. M., Martí-Vidal, I., *et al.* 2011, *A&A*, 533, A106
Hillenbrand, L. A. & White, R. J. 2004, *ApJ*, 604, 741
Janson, M., Brandner, W., Lenzen, R., *et al.* 2007, *A&A*, 462, 615
Mathieu, R. D., Baraffe, I., Simon, M., *et al.* 2007, Protostars and Planets V, 411
Torres, C. A. O., Quast, G. R., Melo, C. H. F., *et al.* 2008, Handbook of Star Forming Regions, Volume II, 757
Wolter, U., Czesla, S., Fuhrmeister, B., *et al.* 2014, *A&A*, 570, AA95

*Young Stars & Planets Near the Sun*
Proceedings IAU Symposium No. 314, 2015
J. H. Kastner, B. Stelzer, & S. A. Metchev, eds.

© International Astronomical Union 2016
doi:10.1017/S1743921315006171

# Determining the Ages of Nearby A-Stars with Long-Baseline Interferometry

**Jeremy Jones[1], R. J. White[1], T. Boyajian[2], G. Schaefer[1], E. Baines[3], M. Ireland[4], J. Patience[5], and The CHARA Team[1]**

[1] Georgia State University, [2] Yale University, [3] Naval Research Laboratory, [4] Australian National University, [5] Arizona State University

email: jones@astro.gsu.edu

**Abstract.** We determine the age of 7 stars in the Ursa Major moving group using a novel method that models the fundamental parameters of rapidly rotating A-stars based on interferometric observations and literature photometry and compares these parameters (namely, radius, luminosity, and rotation velocity) with evolution models that account for rotation. We find these stars to be coeval, thus providing an age estimate for the moving group and validating this technique. With this technique validated, we determine the age of the rapidly rotating, directly imaged planet host star, $\kappa$ Andromedae.

**Keywords.** techniques: interferometric, stars: early-type, stars: evolution, stars: individual ($\kappa$ And), stars: rotation

## 1. The Age of the Ursa Major Moving Group

We use the Classic, CLIMB, and PAVO beam combiners on the CHARA Array to measure the interferometric visibilities of 7 stars in the Ursa Major moving group. By comparing the measured visibilities and photometry taken from the literature to modeled visibilities and photometry which takes rapid rotation into account, we determine various fundamental parameters of the stars, including radius, luminosity, and rotation speed. We compare these properties to MESA evolution models to determine ages for each of these stars and find an average age for 6 of the stars (excluding the outlier, 59 Dra) of $438 \pm 14$ Myr, in agreement with previous age estimates (e.g., King *et al.* 2003). These results are presented in Jones *et al.* (2015) and below in Fig, 1.

## 2. $\kappa$ And: A Directly Imaged Planet-Host Star

$\kappa$ Andromedae is a rapidly rotating A-star and is host to a substellar companion discovered through direct imaging by Carson *et al.* (2013). Using the technique validated with the Ursa Major moving group, we find an age of 27 Myr for the system assuming solar metallicity, implying that the companion is of planetary mass. Figure 2 shows $\kappa$ And's position on an HR-Diagram.

## References

Boyajian, T. S., et al. 2012, *ApJ*, 745, 101
Caffau, E., et al. 2011, *SoPh*, 268, 255
Carson, J., et al. 2013, *ApJ*, 763, 32
Jones, J., et al. 2015, *ApJ*, submitted
King, J. R., et al. 2003, *AJ*, 125, 1980

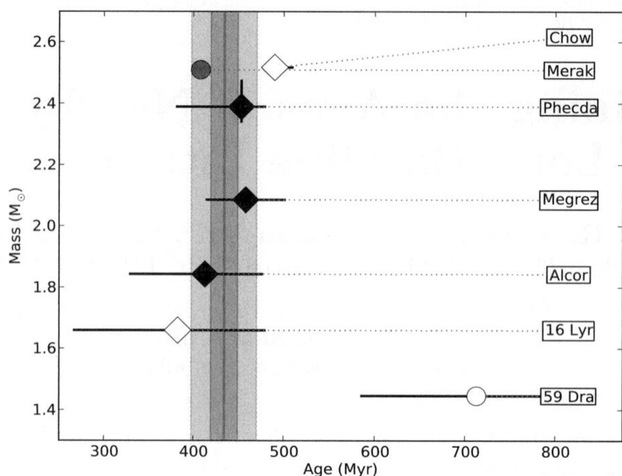

**Figure 1.** Distribution of stellar masses versus age for 7 stars in the Ursa Major moving group. The circles are slowly rotating stars ($V_e < 170$ km s$^{-1}$) and the diamonds are rapidly rotating ($V_e > 170$ km s$^{-1}$). The black points are nucleus members and the white points are stream members. The red point shows the mass and age of the nucleus member, Merak, that was previously observed by Boyajian *et al.* (2012). The dark vertical lines represent the mean ages, the broad shaded regions represent the standard deviation of ages, and the narrow shaded regions represent the the standard error in the mean. The age of the outlier, 59 Dra, was excluded when determining the mean, standard deviation, and standard error in the mean.

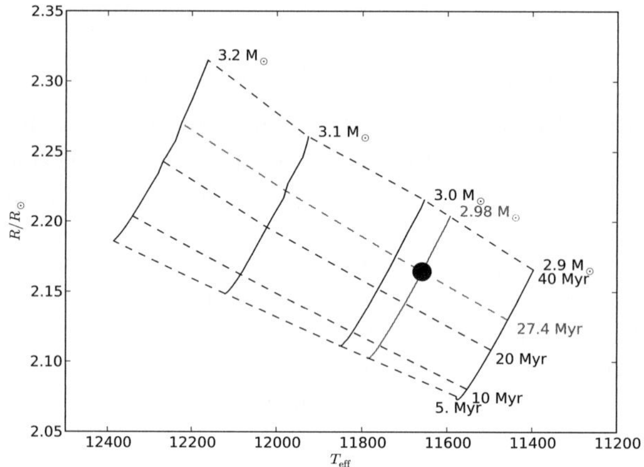

**Figure 2.** HR-Diagram showing (the black circle) average radius and temperature of the directly imaged planet host star, $\kappa$ Andromedae, compared to evolutionary mass tracks (solid lines) and isochrones (dashed lines) from the MESA evolutionary models which take stellar rotation into account. The blue mass track and isochrone are linear interpolations based on these.

## Discussion

QUESTIONER: You say you use solar metallicity for your age determination for $\kappa$ And. Which solar metallicity?

JONES: For this work, we used Z=0.02.

QUESTIONER: You should use the new solar, Z=0.0153, from Caffau *et al.* (2011).

JONES: With Z=0.0153 and the previously determined radius and temperature, I find an age for $\kappa$ And of 75 Myr. However, since the previously determined parameters are affected by the mass, this is preliminary!

*Young Stars & Planets Near the Sun*
Proceedings IAU Symposium No. 314, 2015
J. H. Kastner, B. Stelzer, & S. A. Metchev, eds.

© International Astronomical Union 2016
doi:10.1017/S1743921315006523

# Low-gravity L Dwarfs Are Likely More Variable

**Stanimir Metchev,[1,2] Aren Heinze,[2] Daniel Apai,[3] Davin Flateau,[3] Jacqueline Radigan,[4] Adam Burgasser,[5] Mark Marley,[6] Étienne Artigau,[7] Peter Plavchan,[8] and Bertrand Goldman[9]**

[1] The University of Western Ontario, Centre for Planetary Science and Exploration, Canada
email: (smetchev@uwo.ca),
[2] Stony Brook University, USA,
[3] The University of Arizona, USA,
[4] Space Telescope Science Institute, USA,
[5] University of California San Diego, Center for Astrophysics and Space Science, USA,
[6] NASA Ames Research Center, USA,
[7] Université de Montréal, Canada,
[8] Missouri State University, USA,
[9] Max-Planck-Institut für Astronomie, Germany

**Abstract.** In the "Weather on Other Worlds" Spitzer Exploration Science program, we surveyed 44 nearby L3–T8 dwarfs for spot-induced rotational variability. Among single L3–L9.5 dwarfs, we found that 80% are variable at >0.2% in the 3–5 $\mu$m wavelength range, while 36% of T0–T8 were variable at >0.4%. Taking into account viewing angle and sensitivity considerations, both of these findings are consistent with spots being present on $\sim$100% of L3–T8 dwarfs. Intriguingly, we find a tentative association (92% confidence) between low surface gravity and high-amplitude variability among L3–L5.5 dwarfs. Although we can not confirm whether lower gravity is also correlated with a higher incidence of variables, the result is promising for the characterization of directly imaged young extrasolar planets through variability.

**Keywords.** brown dwarfs – stars: low-mass – stars:rotation – stars: starspots – stars: variables: general – techniques: photometric

---

## 1. Introduction

Results from the Weather on Other Worlds (WOW) Spitzer Exploration Science program have been published in Heinze *et al.* (2013) and Metchev *et al.* (2015). The former details the data reduction methods and first discoveries, while the latter presents the sample selection, the observing methodology, and a comprehensive discussion of results. In short, the WOW sample comprised 44 targets, of which 39 are single based on ancillary high-angular resolution observations. Each target was observed for 21 consecutive hours in 12 s exposures: 14 h in IRAC channel 1 ([3.6]) and 7 h in IRAC channel 2 ([4.5]).

## 2. Variability Amplitudes of Low-gravity L3–L5.5 Dwarfs

Here we focus on the dependence of the observed variability amplitudes with surface gravity in the L3–L5.5 dwarf range. Our sample of 39 unresolved L and T dwarfs contains six individual objects that have been characterized as low- or moderately low-surface gravity dwarfs. Five of these have L3–L5.5 spectral types, and one is the $\sim$500 Myr-old T2.5 dwarf HN Peg B. The tight L3 + L5 binary SDSSp J224953.45+004404.2—not part of the unresolved sample—also has low surface gravity (Allers *et al.* 2010): yielding a total of eight low-gravity objects in the complete WOW sample.

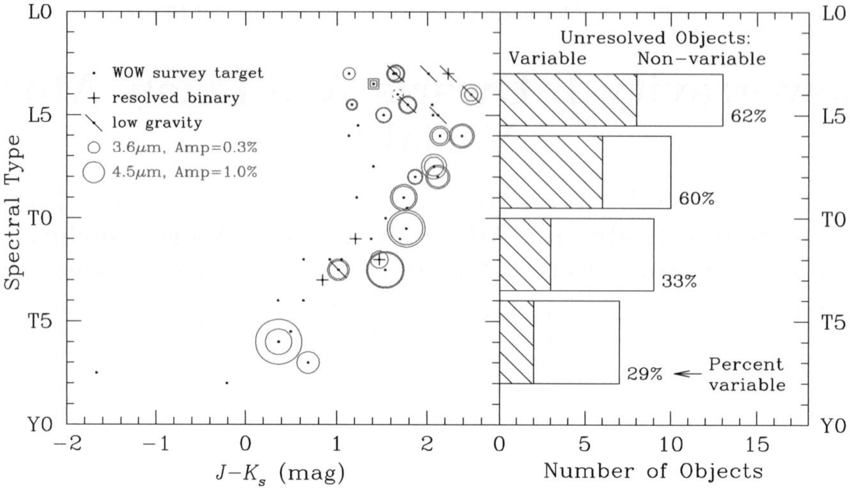


**Figure 1. Left:** Color, spectral type, and variability distribution of our 44 L3–T8 targets. Circles enclose the variable targets, with the area of the circle proportional to the variability amplitude in the IRAC [3.6] band (blue) or [4.5] band (red). The dashed blue circle encloses object 2MASS J175334518–6559559 (L4), which displays only a linear trend at [3.6], and does not have a well defined amplitude. The previously known magnetically active L3.5 dwarf 2MASSW J0036159+182110 is variable and shown with concentric squares. Known tight binaries are marked with +, and are plotted at their systemic spectral type and colour. Inclined bars denote low-gravity objects, including six L3–L5 dwarfs (one a close binary) and the T2.5 dwarf HN Peg B. **Right:** Distribution and frequency of [3.6] or [4.5] variability of the 39 objects in our unresolved sample, excluding the previously known magnetically active L3.5 variable 2MASSW J0036159+182110.

The variability fraction among the putative low-gravity L3–L5.5 dwarfs is 3/7 or 3/5, depending on whether the individual components of the non-varying binary SDSSp J224953.45+004404.2 are counted separately, or whether it is altogether excluded from the sample. Within the statistical uncertainties, this is indistinguishable from the fraction of variables among the high-gravity objects in the L3–L5.5 bin: 5/8.

However, we do detect a tentative correspondence between amplitude and surface gravity among the set of eight L3–L5.5 dwarfs that are variable. The three L3–L5.5 variables that show signatures of low gravity also have the highest [3.6]-band amplitudes in the L3–L5.5 bin (Fig. 1): 2MASS J16154255+4953211 (L4$\beta$), 2MASSW J2208136+292121 (L3$\gamma$), and 2MASS J18212815+1414010 (L4.5). The latter two objects also have the highest bin amplitudes at [4.5]. A consideration of all possible ways to choose three objects from eight shows that the three low-gravity variables would have the highest amplitudes among the eight variables in the L3–L5.5 bin in $\frac{3}{8}\frac{2}{7}\frac{1}{6} = 1.8\%$ of cases. That is, the result might appear 98.2% significant.

More generally, we would have likely considered any outcome that includes the amplitudes of the three low-gravity L3–L5.5 dwarfs among the top half in the bin. We also need to incorporate the four low-gravity L3–L5.5 dwarfs—including the individual near-equal flux components of the L3 + L5 binary SDSSp J224953.45+004404.2—that are not detected as variables. Otherwise, the exclusion of censored data could bias our conclusion. We test the significance of the result by combining all detections and non-detections in a Monte Carlo approach (Metchev *et al.* 2015). To account for the diminished sensitivity to variations from either of the components of the L3 + L5 binary, we count it as two individual objects that are half as bright. We consider as positive any outcome that includes at least three detected low-gravity L3–L5.5 variables, with [3.6] or [4.5] amplitudes all in the top half of the L3–L5.5 bin. We find that this scenario arises at random in 8% of our simulations. That is, the association between low surface gravity and enhanced variability amplitude is 92% significant.

We note that the three high-amplitude L3–L5.5 variables are also redder compared to the lower-amplitude ones (Fig. 1). That is, the higher variability amplitudes may be related to higher dust content in the atmospheres. Nonetheless, low surface gravities in ultra-cool dwarfs correspond strongly with redder photospheric colours. Hence, while low gravity by itself may not be the reason for the higher variability amplitudes, it is still likely correlated with them.

## 3. Conclusion

While we cannot conclude that low surface gravity leads to higher incidence of detectable variability, we find that low gravity and/or dustiness may be correlated with higher 3–5 $\mu$m amplitudes among variable L3–L5.5 dwarfs. This is promising for measuring the rotation periods and cloud compositions of directly imaged extrasolar giant planets through variability.

## References

Allers, K. N., Liu, M. C., Dupuy, T. J., & Cushing, M. C. 2010, *ApJ*, 715, 561
Heinze, A. N., *et al.* 2013, *ApJ*, 767, 173
Metchev, S. M. *et al.* 2015, *ApJ*, 799, 154

*Young Stars & Planets Near the Sun*
*Proceedings IAU Symposium No. 314, 2015*
*J. H. Kastner, B. Stelzer, & S. A. Metchev, eds.*

© International Astronomical Union 2016
doi:10.1017/S174392131500602X

# Rotation Periods of Nearby, Mid-to-late M Dwarfs from the MEarth Project

**Elisabeth R. Newton,**[1]* **Jonathan Irwin,**[1] **David Charbonneau,**[1]
**Zachary K. Berta-Thomspon,**[2] **Andrew A. West**[3]

[1] Harvard-Smithsonian Center for Astrophysics
*email: enewton@cfa.harvard.edu
[2] Massachusetts Institute of Technology
[3] Boston University

**Abstract.** Field stars provide important constraints for the late stages of stars' angular momentum evolution. We measured rotation periods ranging from 0.1 to 150 days for approximately 450 mid-to-late M dwarfs using photometry from the MEarth transiting planet survey. We use parallaxes, proper motions, and radial velocities to calculate galactic kinematics for these solar neighborhood M dwarfs. The velocity dispersions increase towards longer rotation periods, indicating that there is a relationship between rotation and age for these stars.

**Keywords.** stars: low-mass, stars: rotation, stars: kinematics and dynamics

## 1. Introduction

Rotation is one of the few directly measurable stellar properties and encodes the history of a star's angular momentum evolution. Late-time angular momentum loss is governed by stellar winds, which depend on the stars' magnetic field topologies. Stars ages, stellar winds, and magnetic properties are important for the detection and characterization of their planetary systems.

Field stars provide important constraints for the late stages of angular momentum evolution. These data are particularly important for fully-convective stars, which continue to undergo substantial rotational evolution at field ages. Irwin *et al.* (2011) contributed many of the currently-available measurements for fully-convective stars, with rotation periods for 41 M dwarfs from the MEarth Project (Berta *et al.*, 2012; Irwin *et al.*, 2015). Newton *et al.* (submitted) extended this analysis to the full sample of M dwarfs observed by MEarth. Our team has gathered supplementary observations of these stars: Dittmann *et al.* (2014) measured parallaxes for 1507 of the MEarth targets using MEarth photometry and Newton *et al.* (2014) obtained low-resolution near-infrared spectra of 447 MEarth targets, measuring their absolute radial velocities and estimating their metallicities. These data enable further studies of the stars with rotation periods from MEarth.

## 2. Data and analysis

The MEarth Project is an all-sky survey searching for planets transiting 3000 nearby, mid-to-late M dwarfs. MEarth-North has been operational since September, 2008 and MEarth-South was commissioned in January, 2014. Each observatory consists of eight 40cm telescopes on German Equatorial Mounts, which use red-optical filters.

We searched for stellar variability in all MEarth targets observed as of April 2015, simultaneously fitting for systematics and a sinusoidal rotational modulation (Irwin *et al.*, 2011). We compared the sinusoidal model to a null hypothesis, testing rotation periods

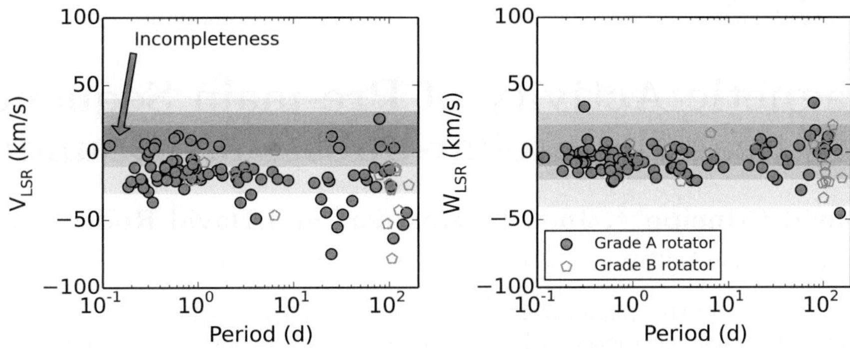


**Figure 1.** Individual components of space velocity as a function of measured photometric rotation period for stars with $M < 0.25 M_\odot$. Gray shaded regions indicate the velocities with higher incompleteness due to our proper motion limits; contours are $5 - 20\%$.

from 0.1 to 1000 days. We selected the best-fitting period and visually inspected each light curve. We qualitatively assessed the candidate period, and identified the approximately 450 stars with clear modulations as rotators. Those with unambiguous variability were classified Grade A, and those in which we were less certain were classified as Grade B.

## 3. Rotation and galactic kinematics

The rotation periods $(P)$ we detected range from 0.1 days to 150 days. We highlight the existence of slow rotators with high amplitudes of variability. These slowly-rotating stars have enough spots to produce peak-to-peak amplitudes of a percent or more.

We calculated $UVW$ velocities for 660 stars in our sample. The substructure seen in the kinematics of higher-mass stars (e.g. Nordstrom *et al.*, 2004) is seen amongst the M dwarfs, including the dynamically-created Hyades supercluster. Following the methods of Bensby *et al.* (2003), our rotators have high probabilities of belonging to the thin disk, which differs significantly from the distribution for all M dwarfs in our sample.

There is clear evidence for a rotation-age relation in all three velocity components (Fig. 1). The velocity dispersions increase with rotation period, as expected for a dynamically-heated population, and we see evidence of asymmetric drift in $V$. We use $\sigma_W$ and the Aumer & Binney (2009) age-velocity relationship to estimate the ages for different populations of rotators. We estimate that M dwarfs with $M < 0.25 M_\odot$ and $P < 10$ days are $< 2$ Gyr old and those with $70 < P < 150$ days are $4 - 8$ Gyr.

## References

Aumer, M. & Binney, J. J. 2009, *MNRAS*, 397, 1286
Berta, Z. K., Irwin, J., Charbonneau, D., Burke, C. J., & Falco, E. E. 2012, *AJ*, 144, 145
Bensby, T., Feltzing, S., & Lundström, I. 2003, *A&A*, 410, 527
Dittmann, J. A., Irwin, J. M., Charbonneau, D., & Berta-Thompson, Z. K. 2014, *ApJ*, 784, 156
Irwin, J. M., Berta-Thompson, Z. K., & Charbonneau, D., *et al.* 2015, in *18th Cambridge Workshop on Cool Stars, Stellar Systems, and the Sun*, 18, 767
Irwin, J., Berta, Z. K., & Burke, C. J., *et al.* 2011, *ApJ*, 727, 56
Newton, E. R., Charbonneau, D., Irwin, J., *et al.* 2014 *AJ*, 147, 20
Newton, E. R., Irwin, J., Charbonneau, D., *et al. submitted to ApJ*
Nordström, B., Mayor, M., & Andersen, J., *et al.* 2004, *A&A*, 418, 989

*Young Stars & Planets Near the Sun*
*Proceedings IAU Symposium No. 314, 2015*
*J. H. Kastner, B. Stelzer, & S. A. Metchev, eds.*

© International Astronomical Union 2016
doi:10.1017/S1743921315006146

# Magnetic Activity of Pre-main Sequence Stars near the Stellar-Substellar Boundary

## David Principe[1,2], Joel. H. Kastner[3] and David Rodriguez[4]

[1]Núcleo de Astronomía, Facultad de Ingeniería, Universidad Diego Portales, Santiago, Chile
email: **daveprincipe1@gmail.com**
[2]Millennium Nucleus Protoplanetary Disks, Chile
[3]Center for Imaging Science, School of Physics & Astronomy, and Laboratory for
Multiwavelength Astrophysics, Rochester Institute of Technology, Rochester, NY 14623, USA
[4]Departamento de Astronomía, Universidad de Chile, Casilla 36-D, Santiago, Chile

**Abstract.** X-ray observations of pre-main sequence (pre-MS) stars of M-type probe coronal emission and offer a means to investigate magnetic activity at the stellar-substellar boundary. Recent observations of main sequence (MS) stars at this boundary display a decrease in fractional X-ray luminosity ($L_X/L_{bol}$) by almost two orders of magnitude for spectral types M7 and later. We investigate magnetic activity and search for a decrease in X-ray emission in the pre-MS progenitors of these MS stars. We present XMM-Newton X-ray observations and preliminary results for ∼10 nearby (30-70 pc), very low mass pre-MS stars in the relatively unexplored age range of 10-30 Myr. We compare the fractional X-ray luminosities of these 10-30 Myr old stars to younger (1-3 Myr) pre-MS brown dwarfs and find no dependence on spectral type or age suggesting that X-ray activity declines at an age later than ∼30 Myr in these very low-mass stars.

**Keywords.** Stars: formation, Stars: magnetic field, Stars: late-type, X-rays: stars.

## 1. Introduction

The early evolution of magnetic activity in very low mass pre-MS stars –stars of mid-M-type, which lie near the H-burning limit of 0.08 $M_\odot$– is very poorly understood. Yet understanding their pre-MS evolution is crucial for determining the emerging differences between very low-mass MS stars and brown dwarfs. X-ray emission offers a means to indirectly probe the effects of internal and surface magnetic activity in both pre-MS and MS stars alike (Vidotto *et al.* 2014). Pre-MS and MS M-type stars are magnetically active and thus can be bright X-ray sources, as indicated by their high values of ($L_X/L_{bol}$ ∼ $10^{-3}$). However, observations of nearby late M-type MS stars suggest that stars of ∼M7 and later appear to be under luminous in X-rays (e.g., $L_X/L_{bol}$ ∼ $10^{-5}$; Berger *et al.* 2010). The narrow range of spectral types where these M stars become X-ray under luminous is roughly the same spectral type where a transition to predominantly neutral atmospheres occurs (Mohanty *et al.* 2002). Berger (2006) concluded that the decrease in X-ray activity (as well as Hα) toward late M-types is related to changes in magnetic field configuration or the decreasing ionization fractions in the atmospheres of these stars.

By determining the age at which this dramatic decrease in X-ray activity occurs for M-type stars, we can gain insight into the early pre-MS stellar evolution of such stars which lie at the low-mass-star/brown dwarf (H-burning) boundary. A recent survey combining GALEX, 2MASS, WISE and catalog proper motions have revealed a population of nearby late-M-type stars in the 10-30 Myr age range (Rodriguez *et al.* 2013) where X-ray activity of such stars has remained, until now, essentially unexplored.

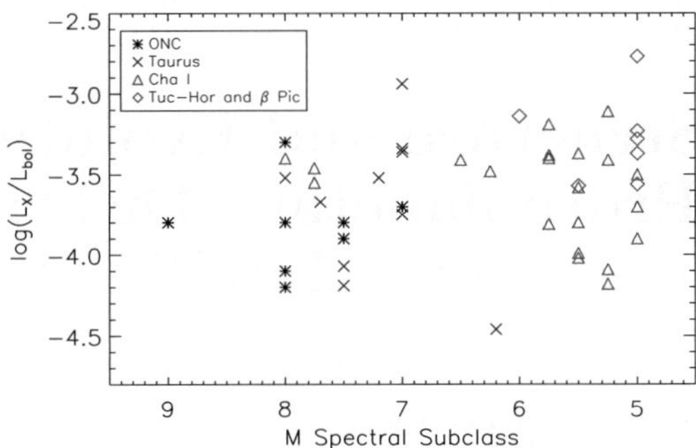

**Figure 1.** Fractional X-ray luminosity as a function of M spectral subclass for ~1-3 Myr old pre-MS stars in the Orion Nebula Cluster (asterisk), Taurus (cross), Cha I (triangle) and preliminary results of ~10-30 Myr old Tuc-Hor and $\beta$ Pic moving group members (diamonds). Figure originally from Stelzer & Micela (2007) and modified to include preliminary results.

## 2. Data and Preliminary Results

We performed XMM-Newton EPIC X-ray observations of 8 nearby ($<$70 pc) ~M5 members of the Tuc-Hor and $\beta$ Pic moving groups (ages ~ 30 Myr and 12 Myr, respectively; Rodriguez *et al.* 2013 and ref. therein). Stellar evolution models (D'Antona & Mazzitelli 1997) suggest pre-MS stars of this age and spectral type will evolve to become MS ~M7 (i.e., they may be progenitors of the under luminous MS M7 stars). Standard one and two temperature thermal plasma models were fit to the data to determine spectral parameters such as plasma temperature and $L_X$. Bolometric luminosities for each of our sources was estimated using their J band flux and the intrinsic colors of 5-30 Myr stars from Pecaut & Mamajek (2013). The fractional X-ray luminosity for each source is shown in Fig. 1 and compared to pre-MS stars of similar spectral type in younger (e.g 1-3 Myr) star-forming regions.

## 3. Conclusions

We find no trend of decreasing fractional X-ray luminosity with age in these 10-30 Myr ~M5 stars (Fig. 1). If MS stars of ~M7 and later are under luminous in X-rays, these preliminary results suggest that either X-ray activity decreases at ages later than ~30 Myr or that our sample of pre-MS stars is unusually magnetically active. The latter scenario may be more likely due to the fact that the sample from which we chose our sources required a GALEX UV detection (i.e., only M stars that were UV bright). UV emission is also an indicator of magnetic activity and thus our sample might be biased towards magnetically active (i.e., X-ray bright) pre-MS stars. More X-ray observations of both UV bright and UV faint mid-to-late M-type pre-MS stars are required to explore the age at which X-ray activity may diminish. A more detailed analysis of these data will be presented in Principe *et al.* (2015), in prep.

## References

Berger, E. 2006, *ApJ*, 648, 629

Berger, E., Basri, G., Fleming, T. A., Giampapa, M. S., Gizis, J. E., Liebert, J., Martin, E., Phan-Bao, N., & Rutledge, R. E. 2010, *ApJ*, 709, 332

D'Antona, F. & Mazzitelli, I. 1997, *Memorie della Societá Astronomia Italiana*, 68, 807

Mohanty, S., Basri, G., Shu, F., Allard, F., & Chabrier, G. 2002, *ApJ*, 571, 469

Pecaut, M. & Mamajek, E. 2013, *ApJS*, 208, 9

Rodriguez, D., Zuckerman, D., & Kastner, J. H. 2013, *ApJ*, 774, 101

Stelzer, B. & Micela, G. 2007, *A&A*, 474, 129

Vidotto *et al.* 2014, *MNRAS*, 441, 2361

*Young Stars & Planets Near the Sun*
*Proceedings IAU Symposium No. 314, 2015*
*J. H. Kastner, B. Stelzer, & S. A. Metchev, eds.*

© International Astronomical Union 2016
doi:10.1017/S1743921315006249

# The Structure and Evolution of Protoplanetary Disks: an Infrared and Submillimeter View

## Lucas A. Cieza

Núcleo de Astronomía, Universidad Diego Portales, Chile
email: lucas.cieza@mail.udp.cl

**Abstract.** Circumstellar disks are the sites of planet formation, and the very high incidence of extrasolar planets implies that most of them actually form planetary systems. Studying the structure and evolution of protoplanetary disks can thus place important constraints on the conditions, timescales, and mechanisms associated with the planet formation process. In this review, we discuss observational results from infrared and submillimeter wavelength studies. We review disk lifetimes, transition objects, disk demographics, and highlight a few remarkable results from ALMA Early Science observations. We finish with a brief discussion of ALMA's potential to transform the field in near future.

**Keywords.** protoplanetary disks, infrared: planetary systems, submillimeter: planetary systems

## 1. Introduction

Protoplanetary disks are complex systems that evolve through various physical mechanisms, including accretion onto the star (Hartmann *et al.* 1998), grain growth and dust settling (Dominik, C., & Dullemond, 2008), dynamical interactions (Artymowicz & Lubow, 1994), photoevaporation (Alexander *et al.* 2006), and planet formation itself (Lissauer 1993; Boss *et al.* 2000). However, the relative importance and timescales of these processes are still not fully understood. Describing each of these processes is beyond the scope of this review. Instead, we focus on IR and submillimeter† observational results, while providing a global view of disk evolution and their connection to planet formation theory. In §2 we discuss results from infrared observations, while in §3 we present submillimeter results previous to the commissioning of the Atacama Large Millimeter Array (ALMA). In §4 we discuss some of the highlights of ALMA Early Science observations. In §5 we speculate on connections between disk demographics and extrasolar planet statistics. In §6 we finish with a discussion on the prospects for studies with the full ALMA array and present our conclusions. Complementary discussions on planet formation, chemistry, disk evolution theory, and observational studies in other wavelength regimes can be found in other contributions to this conference proceedings.

## 2. Infrared constraints

### 2.1. *Disk lifetime*

The lifetime of protoplanetary disks is perhaps the strongest astrophysical constraint on planet formation theory, as it provides valuable information on the time available to complete the process. For rocky planets, their formation could continue beyond the

---

† Submillimeter in this context refers to the wavelength regime extending from ∼0.3 mm to a few mm.

dispersal of the gas. In the case of giant planets, the dissipation of the primordial (gas rich) disk sets a hard limit for their formation timescale. Since very small amounts of hot dust (approximately equivalent to an asteroid mass at 1000 K) are needed to produce an optically thick near-IR (NIR) excess above the stellar photosphere, NIR observations are particularly useful to trace the presence of an inner accretion disk ($r \lesssim 0.1$ AU). In fact, there is a very close correspondence between NIR excesses and accretion signatures (e.g. Hartigan *et al.* 1995), the only exception being transition disks with completely depleted inner holes.

From NIR observations of clusters at different ages, we know that $\sim80\%$ of stars with ages $\sim1$ Myr have an inner accretion disk, and that the inner-disk fraction drops close to $\sim0$ % by 10 Myr (Mamajek, 2009). From these studies we can derive a mean inner-disk lifetime of 2-3 Myr with a large dispersion: some pre-main-sequence (PMS) stars accrete for less $< 1$ Myr, other accrete for up to 10 Myr. Mid-infrared (MIR) observations with *Spitzer* later showed that most targets lacking NIR excesses also lack MIR excesses (Cieza *et al.* 2007; Wahhaj *et al.* 2010), again, with the exception of transition disks with clean inner holes. This indicates that, in general, once accretion stops and the inner disk dissipates, the entire disk is dispersed completely rather quickly, in agreement with the predictions of photoevaporation models (e.g., Alexander *et al.* 2014).

In any given cluster, disk fractions tend to be lower around higher-mass stars than lower-mass objects (Carpenter *et al.* 2006), indicating that disk dissipation proceeds faster around higher mass stars. This could be due to the combination of higher accretion and/or higher photoevaporation rates. Disk lifetime is also a function of multiplicity. In young clusters, disk frequency is close to 100% for single stars and much lower for medium-separation (5-50 AU) binaries (Kraus *et al.* 2012; Cieza *et al.* 2009). This is easy to understand, as mid-separation binaries truncate each other's outer disks, drastically decreasing the amount of material available for accretion. Spectroscopic binaries (separation $\lesssim 1$ AU) and wide binaries (separation $> 100$ AU) seem to have little effect on disk lifetimes.


### 2.2. *Disk structure and evolution*

IR observations are also useful to derive disk structures. Most protoplanetary disks ($\sim80\%$) have "full disks" extending inward to the dust sublimation radius and are optically thick at all IR wavelengths. As a result, they all have very similar spectral energy distributions (SEDs). The other 20%, the transition disks, have a wide range of structures and SEDs. Protoplanetary disks can be thought of as a series of optically thick annuli at different temperatures. Each annulus dominates the emission at a different wavelength; the closer to the star and the hotter the annulus, the shorter the corresponding wavelength. Therefore, the presence of an inner hole in the disk translates to a reduced level of NIR excess. If the hole is completely empty, no detectable excess will be seen above the stellar photosphere. If some detectable dust remains inside the cavity, an optically thin excess will still be present, but the NIR SED will stay below the typical value of an optically thick inner-disk. Similarly, if the disk contains a wide enough gap that is optically thin in the IR, its SED will show a dip at the wavelength corresponding to the temperature of the "missing" dust annulus. Furthermore, most young protoplanetary disks are flared and thus intercept and reprocess significantly more starlight than a flat disk. As disks evolve and dust grows and settles to the mid-plane, the outer disk becomes flatter with time, which translates to reduced levels of mid- and far-IR emission with respect to a fully flared disk. Finally, as the primordial disk dissipates, it becomes optically thin at all IR wavelengths, resulting in reduced levels of IR excess at all wavelengths.

In practice, most young stellar clusters and star-forming regions contain two categories of disks: normal "full" disks that are optically thick and transition disks. The latter group includes objects with inner holes and gaps (sometimes called classical transition disks, cold disks, or pre-transition disks) and objects that are physically very flat or optically thin in the IR (some times called anemic, weak-excess or homologously depleted disks)†. The relative number of these types of objects seems to be a function of age, with more transition disks in older star-forming regions and clusters (Currie *et al.* 2009).

## 3. Pre-ALMA submillimeter results

### 3.1. *Disk mass*

While most protoplanetary disks remain optically thick in the IR, they become optically thin at longer wavelengths. Submillimeter wavelength observations are hence sensitive to fundamental disk properties such as total gas and dust masses and grain size distributions, and are highly complementary to IR data. The main constituent of protoplanetary disks, $H_2$, is a very poor emitter under the relevant temperatures and densities. Circumstellar dust is much easier to observe from continuum observations. Estimating the mass of a disk, and hence its ability to form different types of planets, typically entails making very strong assumptions about the dust opacity and the gas to dust mass ratio (Williams & Cieza, 2011). Submillimeter fluxes are routinely used to estimate the masses of protoplanetary disks using:

$$M_{dust} = \frac{F_\nu d^2}{\kappa_\nu B_\nu (T_{dust})} \tag{3.1}$$

where $d$ is the distance to the target, $T$ is the dust temperature and $\kappa_\nu$ is the dust opacity. Following Beckwith *et al.* (1990), and making standard (although uncertain) assumptions about the disk temperature ($T_{dust}$ = 20 K), the dust opacity ($\kappa_\nu$ = 10[$\nu$/1200 GHz] cm$^2$g$^{-1}$), and the gas to dust mass ratio (100), Equation 3.1 becomes:

$$M_{disk}(gas + dust) = 1.7 \times 10^{-4} \left( \frac{F_\nu (1.3mm)}{mJy} \right) \times \left( \frac{d}{140pc} \right)^2 M_\odot \tag{3.2}$$

The gas to dust mass ratio of 100 that is usually adopted is appropriate for the interstellar medium, but highly questionable for protoplanetary disks that are known to undergo significant grain growth and photoevaporation. In fact, understanding the evolution of the gas (and the gas to dust mass ratio) in protoplanetary disks remains one of the main challenges in the fields of disk evolution and planet formation. The second most common molecule after $H_2$ is CO, and its isotopologues, $^{13}CO$ and $C^{18}O$, provide a viable approach to estimate the gas content in disks (Williams & Best, 2014).

From continuum submillimeter surveys of nearby star-forming regions such as Taurus and Ophiuchus, the following benchmark properties have been derived: $M_{disk} \propto M_\star$ and $M_{disk} \sim 0.5\%$ $M_\star$ (Andrews & Williams 2005, 2007; Andrews *et al.* 2013). However, the dispersion is large ($\pm$ 0.7 dex) and the submillimeter luminosities of protoplanetary disks in the same regions, and with almost identical IR SEDs, can differ by up to two orders of magnitude (e.g., Cieza *et al.* 2008; 2010), implying drastically different disk

† For a discussion on disk nomenclature, see the *Diskionary* by Evans *et al.* (2009).

masses and/or grain-size distributions. Understanding disk evolution requires both IR and submillimeter observations. Unfortunately, millimeter surveys of disks in molecular clouds and young stellar clusters clearly lag far behind their IR counterparts. While deep and complete IR censuses of disks exist for tens of regions spanning a wide range of ages and IR disk fractions, Taurus is the only region in which the entire disk population has been observed at submillimeter wavelengths with enough depth to detect the majority of the IR-detected disks (Andrews *et al.* 2013).

### 3.2. *Resolved observations*

Using interferometers and imaging synthesis techniques, nearby protoplanetary disks can be spatially resolved, providing direct information on the surface density profile of the disk, which is critical for planet formation theory. Simultaneous modeling of the SED and submillimeter images is a powerful technique to constrain the basic properties of disks. Thus far, resolved submillimeter studies have been limited to continuum observations of bright sources ($F_{mm}$ >50 mJy) and/or gas observations of optically thick lines. Therefore, they are very biased towards massive disks around relatively massive stars and provide little direct information on the gas content of disks. In particular, most imaging studies have focused on transition objects (see Williams & Cieza, 2011 for a review). These early resolved observations indicated that the surface density profiles in many of the bright protoplanetary disks are consistent with the amount of material needed for the formation of the planets in the Solar System (Andrews *et al.* 2010). With the commissioning of ALMA, we should be able to 1) extend the imaging surveys to less massive, more typical, protoplanetary disks and 2) image the disks in optically thin gas tracers such as $^{13}$CO and $C^{18}O$ in order to trace the surface density profile of the gas more directly.

## 4. ALMA Early Science results

Since the start of Early Science observations in 2011, ALMA is shedding new light on every stage of disk evolution, from the deeply embedded Class I stage† to the dissipation of the primordial disk and the transition to the debris disk phase. In the following, we highlight some of remarkable ALMA Early Science results.

### 4.1. *Rings in Class I circumstellar disks*

Long-baseline ALMA observations at 0.025" to 0.075" (3.5 – 10 AU) resolution of the Class I circumstellar disk around HL Tau revealed a set of impressive concentric gaps. These gaps show evidence for grain growth, are slightly eccentric, and many of them appear to be in resonance with each other (Partnership *et al.* 2015). All these features are highly suggestive of advanced planet formation through core accretion, although other explanations have merit, including the growth of pebbles at the snow lines of different species (Zhang *et al.* 2015). If the gaps in the HL Tau disk are in fact due to orbiting planets massive enough to carve them, it would imply that the core accretion mechanism is significantly more efficient that previously thought and that the planet formation process can be fairly advanced by the Class I stage and an age of ~1 Myr.

### 4.2. *Structures in transition objects*

High-resolution Early Science observations of transition disks have also produced some surprises such as the discovery of huge asymmetries in the outer disks of Oph IRS 48 (van der Marel *et al.* 2013), SR 21 (Perez *et al.* 2014 ), and HD 142527 (Casassus *et al.* 2013).

† The Class I stage corresponds to a disk that is still embedded in its natal envelope.

These asymmetries have been interpreted as large-scale vortices or dust traps, which could play an important role in the formation of planetesimals and planets at large radii. HD 142527 also shows a set of spiral arms (Christiaens *et al.* 2014) of unknown origin and gaseous accretion flows connecting the inner and the outer disk components of the system (Casassus *et al.* 2013). Such accretion flows have been predicted by hydrodynamical simulations of multiple planets embedded in a primordial disk (Dodson-Robinson & Salyk, 2011). All these results demonstrate that transition disks have complex structures and that the models derived from SED fitting and low-resolution images can only be considered zero-order approximations to their true structures.

### 4.3. *The primordial to debris disk transition*

As discussed in §2.2, star-forming regions include very diverse types of young stellar objects: "full" disks, disks with inner-holes and gaps, optically thin disks, and diskless stars. The evolutionary status of the optically thin disks has remained a matter of intense debate over the last few years and these objects have often been considered the final stage of the primordial disk phase: evolved, primordial (gas rich) disks where most of the primordial dust has been depleted. In an ALMA survey of 24 non-accreting PMS stars (weak-line T Tauri stars; WTTSs) with known IR excesses, Hardy *et al.* (2015) found that these systems have dust masses $\lesssim 0.3$ $M_\oplus$ and no evidence of gas, and therefore are more akin to young debris disks than to evolved primordial disks. This would indicate that they are already in a debris stage, as opposed to *transitioning* into that phase. Non-accreting PMS with IR excesses represent $\sim 20\%$ of the WTTS population (Cieza *et al.* 2007), and their incidence has been used to estimate the dissipation timescale of the primordial disks once accretion stops and Classical T Tauri stars (CTTSs) evolve into WTTSs (Wahhaj *et al.* 2010). If WTTS disks are produced by second-generation dust, it would imply that the final inside out dispersal of the primordial disk happens even faster than previously thought ($\sim 1\%$ vs $\sim 10\%$ of the disk lifetime). These ALMA results also raise questions on how to initiate the debris disk phenomenon within a few Myr of the formation of the star and why this is seen in $\sim 20\%$ of the PMS stars that have already lost their primordial disks.

## 5. Disk evolution and planet formation

It is usually assumed that "full disks" are the starting point of disk evolution and that the transition disks with holes and gaps are more "evolved" systems. However, demographic studies (Cieza *et al.* 2010; Romero *et al.* 2012) in nearby star-forming regions suggest that there are at least two different evolutionary paths that disks can follow. Some systems develop gaps and inner-holes while their disks are relatively massive and still accreting, while other objects lose most of their disk mass while keeping a perfectly "normal" IR SED for millions of years until they reach a point in which the entire disk dissipates from the inside out in a small fraction of the disk lifetime (most likely through photoevaporation). The former group includes most of the famous transition disks found in the literature. In general, these are early-type stars (early-K to A-type) have massive disks with enough material to form the Solar System and large inner-holes, again, consistent with the size of the Solar System. Despite their high profile in the literature, such systems are *not* the typical young stellar object in a nearby molecular cloud or even the typical transition disk. Instead, they represent a minority of the order of 20% of the transition disk population, and 5% of the general disk population (Cieza *et al.* 2012). It is tempting to speculate that such systems are the minority of disks that form planetary systems with one or more giant planets. On the other hand, the "typical" disk in a young (2-3 Myr) stellar cluster seems to be a low-mass accreting CTTS with a normal SED, but not enough mass left in the disk to form a giant planet (Cieza *et al.* 2015). Such systems can be considered "evolved" in the sense that they have probably already lost most of their initial disk mass and have also undergone significant grain growth and dust settling (Lada *et al.* 2006). Given the fact that rocky planets are ubiquitous in the Galaxy (Batalha *et al.* 2013), it is also tempting to speculate that these typical disks are the ones that form the more typical planetary systems, i.e., those hosting rocky planets but no giant planets.

## 6. Prospects for full-ALMA observations and conclusion

With unprecedented sensitivity and resolution, ALMA is clearly poised to revolutionize the fields of disk evolution and planet formation in the near future. It will have an impact in all stages of disk evolution. We can expect to see many detailed studies of disks in early stages, when they still remain embedded in their natal envelope. High resolution images at long wavelengths and the use of optically thin lines allow us to zoom in and peek inside the envelope emission and reveal the disk structure. Some of the questions that the full ALMA array will address include: are rings like those of HL Tau common in Class I objects? What is their origin? Are very young protoplanetary disks gravitationally unstable? Can we image collapsing clumps in the process of forming brown dwarfs or giant planets? We can also expect to see many more detailed studies of transition systems which will investigate the origin of their intriguing features (e.g., large cavities, spiral arms, asymmetries, clumps) and their connection to planet formation. However, as discused in §5, such systems are not typical and might be related to the formation of some particular types of planetary systems. If we want to learn how more common planetary systems form, we will also need to study more common protoplanetary disks. In this regard, ALMA's sensitivity will make it possible to perform complete surveys and resolve entire disk populations in nearby molecular clouds using both continuum and gas tracers. This will help us to investigate the full distribution of disk properties (gas and dust masses, sizes, surface density profiles, etc.) as a function of stellar parameters such as mass, age, and multiplicity. In particular, the observations of optically thin line tracers such as $^{13}$CO and $C^{18}O$ will allow us to place much needed constraints on the evolution of gas, which is far less established than that of the dust. Thanks to the exponential growth of extrasolar planet studies and the construction of ALMA, we are approaching an era in which we can start making direct connections between disk properties and the statistics of planetary systems we see in the Galaxy. This will help us to understand the kind of disks that form different types of planetary systems and to develop a coherent picture of the demographics of protoplanetary disks and the planets they form.

## References

Alexander, R. D., Clarke, C. J., & Pringle, J. E. 2006, *MNRAS*, 369, 229
Alexander, R., Pascucci, I., Andrews, S., Armitage, P., & Cieza, L. 2014, *Protostars and Planets VI*, 475
Andrews, S. M. & Williams, J. P. 2005, *ApJ*, 631, 1134
Andrews, S. M. & Williams, J. P. 2007, *ApJ*, 671, 1800
Andrews, S. M., Wilner, D. J., Hughes, A. M., Qi, C., & Dullemond, C. P. 2010, *ApJ*, 723, 1241
Andrews, S. M., Rosenfeld, K. A., Kraus, A. L., & Wilner, D. J. 2013, *ApJ*, 771, 129
Artymowicz, P. & Lubow, S. H. 1994, *ApJ*, 421, 651
Batalha, N. M., Rowe, J. F., Bryson, S. T., *et al.* 2013, *ApJS*, 204, 24
Beckwith, S. V. W., Sargent, A. I., Chini, R. S., & Guesten, R. 1990, *AJ*, 99, 924

Boss, A. P. 2000, *ApJL*, 536, L101

Carpenter, J. M., Mamajek, E. E., Hillenbrand, L. A., & Meyer, M. R. 2006, *ApJL*, 651, L49

Casassus, S., van der Plas, G., M, S. P., *et al.* 2013, *Nature*, 493, 191

Cieza, L., Padgett, D. L., Stapelfeldt, K. R., *et al.* 2007, *ApJ*, 667, 308

Cieza, L. A., Padgett, D. L., Allen, L. E., *et al.* 2009, *ApJL*, 696, L84

Cieza, L. A., Schreiber, M. R., Romero, G. A., *et al.* 2010, *ApJ*, 712, 925

Cieza, L. A., Schreiber, M. R., Romero, G. A., *et al.* 2012, *ApJ*, 750, 157

Cieza, L., Williams, J., Kourkchi, E., *et al.* 2015, arXiv:1504.06040

Christiaens, V., Casassus, S., Perez, S., van der Plas, G., & Ménard, F. 2014, *ApJL*, 785, L12

Currie, T., Lada, C. J., Plavchan, P., *et al.* 2009, *ApJ*, 698, 1

Dominik, C. & Dullemond, C. P. 2008, *A&A*, 491, 663

Dodson-Robinson, S. E. & Salyk, C. 2011, *ApJ*, 738, 131

Evans, N., Calvet, N., Cieza, L., *et al.* 2009, arXiv:0901.1691

Hardy, A., Caceres, C., Schreiber, M. R., *et al.* 2015, arXiv:1504.05562

Hartigan, P., Edwards, S., & Ghandour, L. 1995, *ApJ*, 452, 736

Hartmann, L., Calvet, N., Gullbring, E., & D'Alessio, P. 1998, *ApJ*, 495, 385

Kraus, A. L., Ireland, M. J., Hillenbrand, L. A., & Martinache, F. 2012, *ApJ*, 745, 19

Lada, C. J., Muench, A. A., Luhman, K. L., *et al.* 2006, *AJ*, 131, 1574

Lissauer, J. J. 1993, *ARA&A*, 31, 129

Mamajek, E. E. 2009, *American Institute of Physics Conference Series*, 1158, 3

Partnership, A., Brogan, C. L., Pérez, L. M., *et al.* 2015, *ApJL*, 808, L3

Pérez, L. M., Isella, A., Carpenter, J. M., & Chandler, C. J. 2014, *ApJL*, 783, L13

Romero, G. A., Schreiber, M. R., Cieza, L. A., *et al.* 2012, *ApJ*, 749, 79

van der Marel, N., van Dishoeck, E. F., Bruderer, S., *et al.* 2013, *Science*, 340, 1199

Wahhaj, Z., Cieza, L., Koerner, D. W., *et al.* 2010, *ApJ*, 724, 835

Williams, J. P. & Best, W. M. J. 2014, *ApJ*, 788, 59

Williams, J. P. & Cieza, L. A. 2011, *ARA&A*, 49, 67

Zhang, K., Blake, G. A., & Bergin, E. A. 2015, *ApJL*, 806, L7

*Young Stars & Planets Near the Sun*
*Proceedings IAU Symposium No. 314, 2015*
*J. H. Kastner, B. Stelzer, & S. A. Metchev, eds.*

© International Astronomical Union 2016
doi:10.1017/S1743921315006316

# Protoplanetary Disk Evolution: Singles vs. Binaries

**Sebastian Daemgen[1], Ray Jayawardhana[2], Monika G. Petr-Gotzens[3], and Elliot Meyer[1]**

[1]Department of Astronomy & Astrophysics, University of Toronto, 50 St. George Street, Toronto, ON, Canada M5H 3H4, email: **daemgen@astro.utoronto.ca**
[2]Faculty of Science, 4700 Keele Street, Toronto, ON M3J 1P3, Canada
[3]European Southern Observatory, Karl-Schwarzschildstr. 2, 85748, Garching, Germany

**Abstract.** Based on a large number of observations carried out in the last decade it appears that the fraction of stars with protoplanetary disks declines steadily between $\sim$1 Myr and $\sim$10 Myr. We do, however, know that the multiplicity fraction of star-forming regions can be as high as >50% and that multiples have reduced disk lifetimes on average. As a consequence, the observed roughly exponential disk decay can be fully attributed neither to single nor binary stars and its functional form may need revision. Observational evidence for a non-exponential decay has been provided by Kraus *et al.* (2012), who statistically correct previous disk frequency measurements for the presence of binaries and find agreement with models that feature a constantly high disk fraction up to $\sim$3 Myr, followed by a rapid ($\lesssim$2 Myr) decline.

We present results from our high angular resolution observational program to study the fraction of protoplanetary disks of single and binary stars separately. We find that disk evolution timescales of stars bound in close binaries (<100 AU) are significantly reduced compared to wider binaries. The frequencies of accretors among single stars and wide binaries appear indistinguishable, and are found to be lower than predicted from planet forming disk models governed by viscous evolution and photoevaporation.

**Keywords.** Stars: pre-main sequence; Stars: formation; circumstellar matter; binaries: general

---

## 1. Introduction

The formation of gas giant planets requires significant amounts of gas and dust to be present in the circumstellar environment of a young T Tauri star. The lifetime of protoplanetary disks is accordingly an important observable to constrain planet formation. To infer disk lifetimes, a number of previous studies have targeted young star-forming regions to measure the fraction of stars that exhibit either ongoing accretion or hot circumstellar dust or both. These fractions appear to be a strong function of the age of a star-forming region, monotonically decreasing from $\gtrsim$80% to 0% within $\sim$10 Myr (e.g., Jayawardhana et al. 2006; Fedele et al. 2010). The functional shape appears to be fit well by an exponential decay with a time constant of $\tau \approx$ 2–3 Myr (Fedele et al. 2010).

The fact that a large fraction of all young stars is bound in multiple systems (Duchêne & Kraus 2013) has a strong effect on the conclusions that can be drawn from this observation because a) the presence of stellar binary companions has a strong effect on disk evolution, and b) a large number of binary stars typically remains undetected and will "contaminate" single star measurements.

**The effect of stellar binarity on disk evolution.** Sub-mm observations show that multiple stars exhibit a much reduced total dust mass compared to undisturbed systems (Harris et al. 2012; Andrews *et al.* 2013). This can be explained by the dynamical interaction of a disk with the stellar companion. For wide binaries this leads to a truncation of the outer disk to $\sim$1/5–1/2 times the binary separation (Artymowicz & Lubow 1994). This implies shorter disk lifetimes, given that mass accretion rates in binaries are comparable to those of single stars (White & Ghez 2001; Daemgen et al. 2012). And indeed, the data of Daemgen et al. (2012, 2013, see also Kraus *et al.* 2012) imply an exponential decay of binary circumstellar disk fractions with a time constant of $\tau \approx 0.9$–$1.3\,\mathrm{Myr}$, much shorter than that found by previous studies that do not focus on binary stars.

The fact that disks in binary stars follow a different evolutionary path may help to constrain planet formation scenarios: Duchêne (2010) found that stars with close ($\rho <$ 100 AU) binary companions are less likely to be orbited by a low-mass ($\lesssim 1\,M_{\mathrm{Jup}}$) gas planet than wider binaries and single stars. One possible interpretation calls for a slow gas planet-forming process for planets $<1\,M_{\mathrm{Jup}}$ – too slow to complete in close binaries before the disk disperses. More massive planets, in contrast, may form through a rapid process that completes equally often around single stars and around even closely separated binary components where disk lifetimes are shorter.

**The difficulty of identifying single stars.** While above considerations and the fact that sample sizes are typically small paint a noisy but consistent picture of disk evolution in T Tauri *binaries*, the evolution of a disk in an undisturbed system like that of a single star cannot be easily observed. This is partly due to our lack of knowledge about the true multiplicity status of young stars: efficient high-angular resolution imaging searches for companions do not reach much below $\sim$0.1$''$ ($\sim$10 AU at the distance of the nearest star-forming regions of $\sim$100 pc), and few young stars have been subject to radial velocity (RV) monitoring over baselines $>$10 yr (equivalent to $\gtrsim$5 AU). This leaves the 5–10 AU separation range mostly unstudied for stellar companions in the best-sampled nearby regions, and typically the range of unassessed separations is even larger.

Owing to this observational effect, the fraction of undetected binary companions in any previous disk frequency survey can be $\sim$30% or larger, depending on the distance of the targeted region and the availability of suitable survey data (Kraus *et al.* 2012; Daemgen et al. 2015, *submitted*). The previously derived disk fractions in these samples accordingly represent a mixture of binary and single star disk data compromising their potential to provide physical information about disk decay.

**Disk evolution around single stars.** Applying a statistical correction for undetected binarity, Kraus *et al.* (2012) study the early evolution of single stars and find observational evidence for a disk fraction evolution that may be different from an exponential decay. In qualitative agreement with disk evolution models by Alexander & Armitage (2009), disk frequencies stay close to 100% until $\sim$2–3 Myr of stellar evolution and then rapidly decay before $\sim$6 Myr of age. Such a change would have strong implications for planet formation. For example, gas giant planets may form on much longer timescales—or start to form at a later time—than typically assumed.

While providing interesting qualitative evidence, the results reported by Kraus *et al.* (2012) are based on unresolved binaries with no information about whether circumstellar material is distributed around one or both stars. Furthermore, their study relies on the sum of a variety of disk indicators which probe different physical aspects of a disk. The values presented by Kraus *et al.* (2012) accordingly represent upper limits to the true accretion and inner dust disk frequency of single stars.

To gain a better understanding of the evolution of undisturbed disk material in the first few Myr of stellar evolution, a consistent and direct measurement of disk frequency around single stars is needed. As current instruments do not enable the identification of binary companions at all separations and mass ratios, the frequency of undetected binary companions and the disk fraction in multiples must be taken into careful consideration to infer single star values. In the following we describe our previous and ongoing efforts.

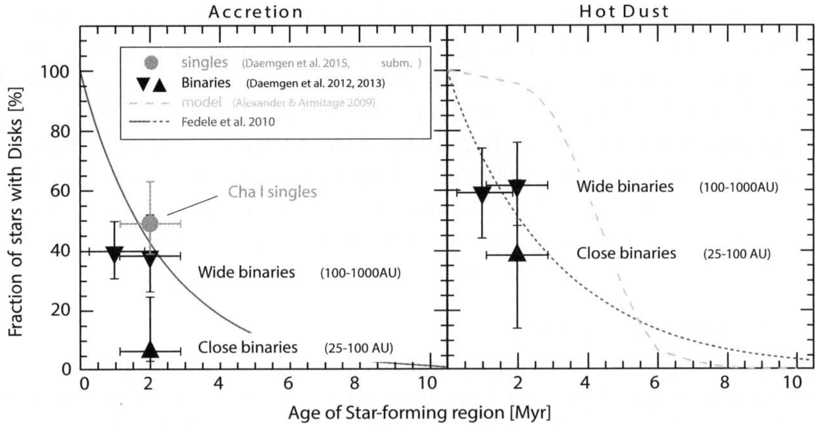


**Figure 1.** Disk frequency of single stars and components of binary stars in the Orion Nebula Cluster (1 Myr) and Chamaeleon I (2 Myr). *Left:* accretion inferred from Brackett-$\gamma$ emission. *Right:* hot circumstellar dust as inferred from $H$–$K$ and $K$–$L'$ excess (Daemgen et al. 2012, 2013, 2015). The blue continuous and dotted lines show exponential decay functions measured by Fedele et al. (2010, $\tau_{\rm accr} \sim 2.3$ Myr, $\tau_{\rm dust} \sim 3$ Myr), not corrected for undetected multiplicity. The dashed cyan line shows the disk evolution model by Alexander & Armitage (2009).

## 2. A coherent analysis of disks in single and multiple stars

We measured the frequency of disks around 52 spatially resolved multiple stars with separations between $\sim$25–1000 AU in the Orion Nebula Cluster and Chamaeleon I star-forming regions (Daemgen et al. 2012, 2013). For each stellar component, our measurements trace both the presence of ongoing accretion based on Brackett-$\gamma$ (Br$\gamma$) emission as well as hot inner dust inferred from near-infrared color excess.

We compare this binary data set to single stars in Chamaeleon I (Daemgen et al. 2015, *submitted*). In order to compile a sample with as little binary contamination as possible, we queried the high-angular resolution imaging survey by Lafrenière et al. (2008) for stars without stellar companions between $\sim$50–500 AU and brighter than $\Delta K_{\rm s} \approx 3$ mag. Stars with known radial velocity companions were removed from the sample. A total of 54 single star candidates were observed with SOFI/NTT $K$-band spectroscopy to assess the presence of emission at the wavelength of Br$\gamma$. Using a Monte Carlo simulation, we simulated the overall companion distribution of our sample and find that, even after excluding all adaptive optics (AO) and RV companions, there remains a $\sim$30% chance of multiplicity for any star in the sample. Using the results from our Chamaeleon I binary study (Daemgen et al. 2013), we correct the measured single star accretion frequency for the bias introduced by undetected binarity.

As these measurements are based on a well-defined set of suitable disk indicators (Br$\gamma$, NIR color excess) and good coverage by AO and RV companion searches, the resulting disk evolution measurements represent the most robust assessment of the relative abundance of disks around the individual components of binaries and single stars to date.

## 3. Results

Figure 1 shows our assessment of single and binary star disk fraction as a function of age. Close binaries ($\rho \leqslant 100$ AU) appear to disperse their disks on much smaller timescales than wider binaries, both for our accretion and hot inner disk measurements.

We furthermore robustly confirm that the inferred single star accretor fraction ($F = 48^{+14}_{-10}\%$; Daemgen et al. 2015, *submitted*) is about 6 times larger than that of close binaries. The single star accretor fraction appears to be slightly larger but consistent with both wide binaries and the decay curve found by previous surveys without binary correction (e.g., Fedele et al. 2010). In particular, our new measurement of the single star accretor fraction is less than $1\sigma$ larger than previous estimates of the accretion frequency in Chamaeleon I ($\sim41\%$, based on Damjanov *et al.* 2007). The data disfavor a scenario where all stars retain their disks until 2–3 Myr (Alexander & Armitage 2009).

It has recently been suggested that ages of star forming regions are much older than previously assumed (Bell *et al.* 2013). If the ages of the targeted Orion Nebula Cluster and Chamaeleon I regions turn out to be a factor of $\sim2$ larger than assumed here, then agreement between the data and the model is preserved.

## 4. Summary and Conclusions

Single stars are hard or even impossible to identify in most star-forming regions. This is because contamination with binary companions that are inaccessible to today's observational methods is on the order of $\gtrsim30\%$ for a typical star-forming region like Chamaeleon I. As close binaries ($<100\,\mathrm{AU}$) have been shown to exhibit disk evolution timescales that are shorter than those of wide binaries and single stars, typical disk frequency studies are biased by undetected binary companions. Rather than following an exponential decay, as suggested by studies without binary correction, the fraction of single stars with disks may stay constant and close to 100% until $\sim2$–3 Myr, followed by a rapid ($\lesssim3\,\mathrm{Myr}$) drop, according to disk evolution models that include viscosity, photoevaporation, and planet formation.

We conducted the first dedicated single star disk frequency measurement in Chamaeleon I. We find a significantly higher accretor fraction among single stars than around components of close binary stars ($<100\,\mathrm{AU}$) in the same region. The single star disk frequency appears to be inconsistent with the very high disk frequency at 2 Myr suggested by previous models and statistical considerations. Part of the discrepancy between the observations and models might be due to systematic age uncertainties.

More measurements of single and binary star disk frequencies are needed to infer the evolution of undisturbed disks as a function of time and to constrain the qualitative and quantitative evolution of disk material in both truncated and undisturbed disks.

## References


Alexander, R. D. & Armitage, P. J. 2009, *ApJ*, 704, 989
Andrews, S. M., Rosenfeld, K. A., Kraus, A. L., & Wilner, D. J. 2013, *Apj*, 771, 129
Artymowicz, P. & Lubow, S. H. 1994, *ApJ*, 421, 651
Bell, C. P. ., Naylor, T., Mayne, N. J., Jeffries, R. D., & Littlefair, S. P. 2013, *MNRAS*, 434, 806
Daemgen, S., Correia, S., & Petr-Gotzens, M. G. 2012, *A&A*, 540, 46
Daemgen, S., Petr-Gotzens, M. G., Correia, S., *et al.* 2013, *A&A*, 554, 43
Daemgen, S., Meyer, E., Jayawardhana, R., & Petr-Gotzens, M. G. 2015, *submitted to A&A*
Damjanov, I., Jayawardhana, R., Scholz, A., *et al.* 2007, *ApJ*, 670, 1337
Duchêne, G. 2010, *ApJ*, 709, L114
Duchêne, G. & Kraus, A. 2013, *ARA&A*, 51, 269
Fedele, D., van den Ancker, M. E., Henning, T., *et al.* 2010, *A&A*, 510, 72
Jayawardhana, R., Coffey, J., Scholz, A., *et al.* 2006, *ApJ*, 648, 1206
Harris, R. J., Andrews, S. M., Wilner, D. J., & Kraus, A. L. 2012, *ApJ*, 751, 115
Kraus, A. L., Ireland, M. J., Hillenbrand, L. A., & Martinache, F. 2012, *ApJ*, 745, 19
Lafrenière, D., Jayawardhana, R., & van Kerkwijk, M. H. 2008, *ApJ*, 689, L153
Somers, G. & Pinsonneault, M. H. 2015, *ApJ*, 807, 174
White, R. J. & Ghez, A. M. 2001, *ApJ*, 556, 265

*Young Stars & Planets Near the Sun*
Proceedings IAU Symposium No. 314, 2015
J. H. Kastner, B. Stelzer, & S. A. Metchev, eds.

© International Astronomical Union 2016
doi:10.1017/S1743921315006444

# Gas Cavities inside Dust Cavities in Disks Inferred from ALMA Observations

**Nienke van der Marel[1], Ewine F. van Dishoeck[1,2], Simon Bruderer[2], Paola Pinilla[1], Tim van Kempen[1], Laura Perez[3] and Andrea Isella[4]**

[1] Leiden Observatory, Leiden, the Netherlands
email: `nmarel@strw.leidenuniv.nl`
[2] MPE, Garching bei Munchen, Germany
[3] NRAO, Socorro, NM, USA
[4] Rice University, Houston, TX, USA

**Abstract.** Protoplanetary disks with cavities in their dust distribution, also named transitional disks, are expected to be in the middle of active evolution and possibly planet formation. In recent years, millimeter-dust rings observed by ALMA have been suggested to have their origin in dust traps, caused by pressure bumps. One of the ways to generate these is by the presence of planets, which lower the gas density along their orbit and create pressure bumps at the edge. We present spatially resolved ALMA Cycle 0 and Cycle 1 observations of CO and CO isotopologues of several famous transitional disks. Gas is found to be present inside the dust cavities, but at a reduced level compared with the gas surface density profile of the outer disk. The dust and gas emission are quantified using the physical-chemical modeling code DALI. In the majority of these disks we find clear evidence for a drop in gas density of at least a factor of 10 inside the cavity, whereas the dust density drops by at least a factor 1000. The CO isotopologue observations reveal that the gas cavities are significantly smaller than the dust cavities. These gas structures suggest clearing by one or more planetary-mass companions.

**Keywords.** planetary systems: formation, planetary systems: protoplanetary disks, techniques: interferometric, molecular data, radiative transfer

---

## 1. Introduction

Planets are formed in disks of dust and gas surrounding young stars (Williams & Cieza 2011). Of particular interest are the so-called transitional disks with cleared out dust cavities. These disks are expected to be in the middle of active disk evolution and planet formation. Transition disks are generally identified by a dip in the mid infrared part of their SED due to the lack of hot dust (Espaillat *et al.* 2014). Several mechanisms can be responsible for the appearance of a dust cavity: grain growth (Dullemond & Dominik 2005), photoevaporation (Clarke *et al.* 2001), clearing by (sub)stellar or planetary companions (Lin & Papaloizou 1979) and instabilities at the edges of dead zones (Regály et al. 2012). The latter two scenarios include trapping of the millimeter-dust in ring-like or asymmetric structures. Dust trapping can happen in a local pressure bump in the outer part of the disk due to the interaction between gas and dust (Weidenschilling 1977) and has been suggested to explain the observed structures in transition disks (Pinilla et al. 2012, van der Marel et al. 2013).

In order to distinguish between the different mechanisms, it is essential to know the distribution of dust *and* gas inside the dust cavity: in case of grain growth, both millimeter dust and gas will remain as in a full disk (Birnstiel et al. 2010); photoevaporation will clear the gas and dust simultaneously from the inside out; dead zones will only remove the dust by trapping, the gas density remains the same inside the cavity; a companion

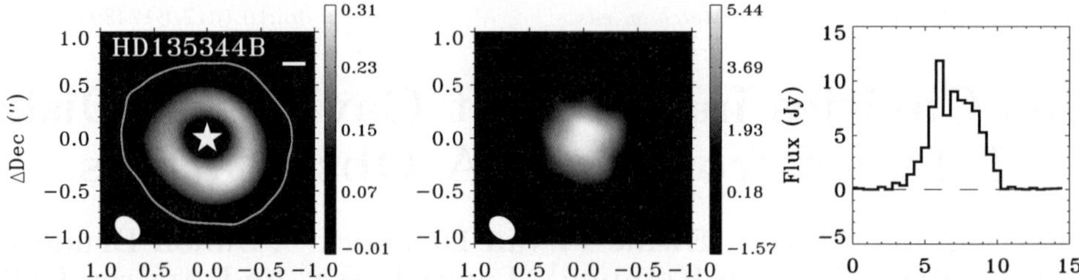

**Figure 1.** ALMA observations of the 690 GHz dust continuum and $^{12}$CO 6-5 line of HD135344B, one of six targets in the sample of the $^{12}$CO study in van der Marel et al. (2015a). In the images, the beam is indicated in the lower left corner. **Left:** Continuum image. The white bar in the upper right corner indicates the 30 AU scale and the yellow contour gives the $3\sigma$ detection limit. **Center:** zero-moment $^{12}$CO map. **Right:** $^{12}$CO spectrum integrated over the entire disk. The dashed line indicates the zero flux level.

will lower the gas density inside the dust cavity. In contrast to scattered light images at optical wavelengths, CO millimeter lines can actually quantify the gas density structure.

ALMA (Atacama Large Millimeter/submillimeter Array) allows us to zoom in and resolve the cavities in the gas and dust in transition disks, through dust continuum and CO line observations. We present a sample study of 6 transition disks that have been observed in ALMA Cycle 0 at subarcsecond resolution in continuum and $^{12}$CO. Five disks were observed in Band 9 (690 GHz or 450 $\mu$m), one disk in Band 7 (345 GHz or 850 $\mu$m).

## 2. Results

### 2.1. *Observations*

The sample consists of SR21, HD135344B, LkCa15, SR24S and RXJ1615-3255 (observed in Band 9 or 690 GHz, partially presented in Perez *et al.* 2014) and J1604-2130 (observed in Band 7 or 345 GHz, first presented in Zhang et al. 2014). One example of the disk observations in continuum and integrated $^{12}$CO emission is shown in Figure 1, the others are presented in van der Marel et al. (2015a).

This Figure shows that $^{12}$CO emission peaks inside the dust cavity, indicating that there is still gas inside the cavity. In Figure 2 the intensity cuts (cuts along the major axis for the integrated CO and continuum emission) for all six targets are presented. It is clear that all targets show gas inside the dust cavity. However, as $^{12}$CO emission is optically thick, these images do not give direct information on the gas density distribution.

### 2.2. *Modeling*

In order to quantify the gas and dust distribution, radiative transfer modeling is required. We use DALI, a full physical-chemical modeling code, to analyze the emission (Bruderer 2013). DALI solves for the dust temperatures through continuum radiative transfer, starting from a given surface density profile (Figure 3). Subsequently, DALI calculates the chemical abundances, the molecular excitation and the thermal balance of the gas, where effects such as photodissociation, freeze-out, decoupling of gas and dust temperature and UV heating are included, and finally raytraces images and spectra.

For the modeling, the dust surface density is constrained first by fitting a parameterized model with an empty dust cavity to the SED, the millimeter continuum image and the

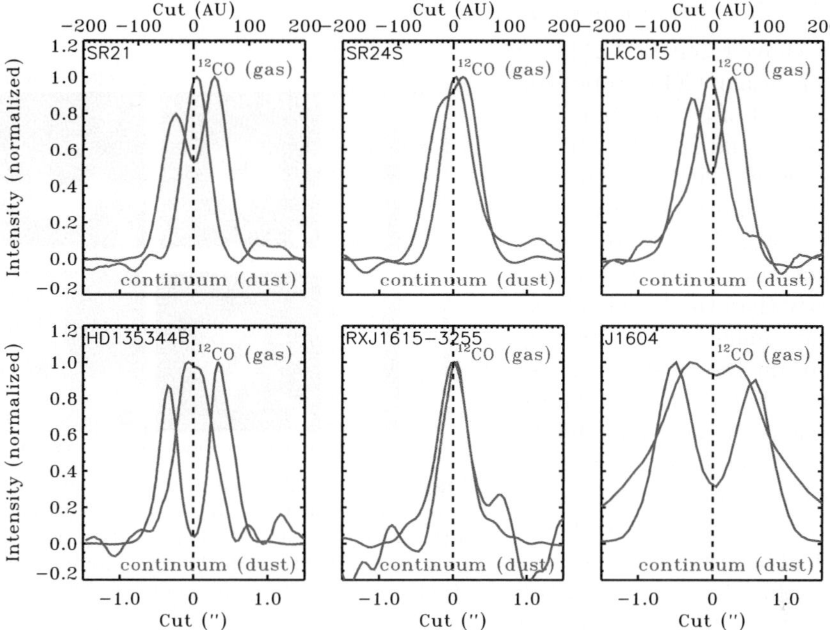



**Figure 2.** Normalized intensity cuts along the major axis of the continuum emission and the integrated $^{12}$CO map of each disk in the sample. The continuum is indicated in blue, the integrated $^{12}$CO in red. The cuts reveal that the $^{12}$CO emission is peaking inside the dust cavity.

millimeter visibilities. The latter provides the best constraints on the dust cavity size $r_{\mathrm{cav}}$. Next, the drop in dust density inside the cavity is constrained by changing $\delta_{\mathrm{dustcav}}$. For all disks it was found that $\delta_{\mathrm{dustcav}}$ is $10^{-3}$ or less, implying a dust density drop of least a factor 1000.

Using the derived dust density profile, the gas density profile is computed by taking a gas-to-dust ratio of 100 in the outer disk, and setting $\delta_{\mathrm{gas}}$ inside the cavity equal to 1, since the $^{12}$CO emission peaks inside the cavity. The resulting images are compared with the observed $^{12}$CO data, and it turns out that the model overpredicts the $^{12}$CO emission. By lowering $\delta_{\mathrm{gas}}$ by a factor 10–100 inside the cavity the emission is well reproduced for each of the six disks. The reason that optically thick $^{12}$CO decreases with $\delta_{\mathrm{gas}}$ is that the $\tau = 1$ surface shifts down, closer to the mid-plane when $\delta_{\mathrm{gas}}$ is decreased. Due to the strong vertical temperature gradient the emission thus drops slightly with $\delta_{\mathrm{gas}}$ as well (Bruderer 2013).

The different drops in density of the gas and dust inside the cavity directly points towards the companion clearing scenario for these disks, where the companion has cleared its orbit and traps the millimeter-sized dust at the edge, creating a millimeter continuum ring.

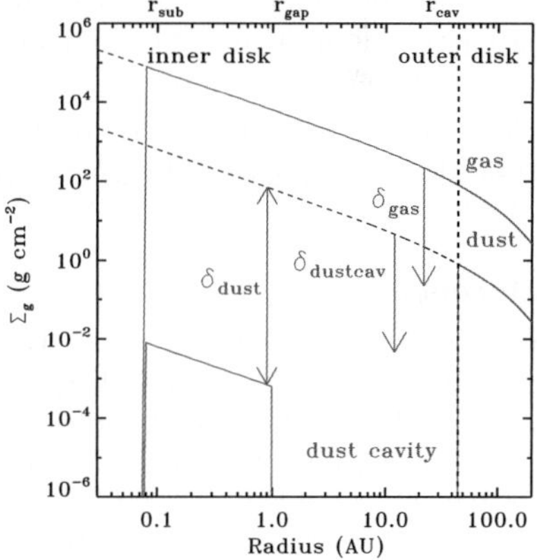



**Figure 3.** Generic surface density model of gas and dust that is used as input for DALI. Inside the cavity ($< r_{\mathrm{cav}}$), the dust density drops by a factor $\delta_{\mathrm{dustcav}}$, the gas density by a factor $\delta_{\mathrm{gas}}$. The inner disk is reproduced by an additional $\delta_{\mathrm{dust}}$. The $\delta$-values are free parameters in the model, after the outer disk density is matched to the data. Figure taken from van der Marel et al. (2015a).

142

N. van der Marel

### 2.3. *New CO isotopologue data*

Recently we have received ALMA Cycle 1 Band 7 data of CO isotopologues $^{13}$CO and C$^{18}$O of 3 transition disks, including SR21 and HD135344B (van der Marel et al. 2015b). The data of HD135344B are presented in Figure 4. CO isotopologues are less optically thick and trace the density more directly than the $^{12}$CO. The CO isotopologue images again show gas emission inside the dust cavity, but now also revealing a gas ring. The gas cavity radius is smaller than the dust cavity radius in all cases. This is consistent with the planet clearing scenario and trapping, as the dust is expected to be trapped further out than the planet orbit (Pinilla et al. 2012).

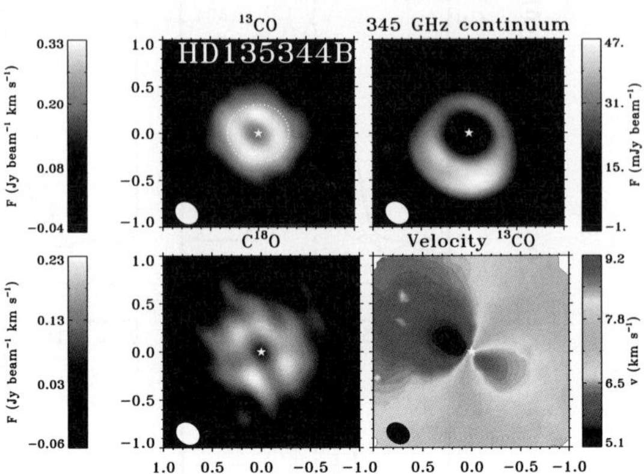

Figure 4. ALMA Cycle 1 observations of the continuum, $^{13}$CO and C$^{18}$O 3-2 lines of HD135344B. **Top left:** zero-moment $^{13}$CO map; **Top right:** Continuum map; **Bottom left:** zero-moment C$^{18}$O map; **Bottom right:** first moment $^{13}$CO map.

The previous models from the $^{12}$CO study fit well with the isotopologue data, except that the cavity radius for the gas is now seen directly and found to be smaller than that of the dust. These new data will be presented in van der Marel et al. (2015b).

## 3. Conclusions

The ALMA CO observations of the transition disks observed so far show that gas is present inside the dust cavities. Analysis of the CO emission in the six transition disks in this recent study reveals that gas cavities have smaller drops in density than the millimeter-dust cavities. In addition, CO isotopologue data reveal gas cavities to be smaller in radius than the dust cavities. These findings strengthen the conclusion that embedded planets or companions are the main clearing mechanism. Ultimately, observations should be compared directly with the output of planet-disk interaction models, rather than parametrized models, to confirm the proposed scenario.

### References


Birnstiel, T., Dullemond, C. P., & Brauer, F. 2010, *A&A*, 513, A79

Bruderer, S. 2013, *A&A*, 559, A46

Clarke, C. J., Gendrin, A., & Sotomayor, M. 2001, *MNRAS*, 328, 485

Dullemond, C. P. & Dominik, C. 2005, *A&A*, 434, 971

Espaillat, C., Muzerolle, J., Najita, J., *et al.* 2014, in *Protostars and Planets VI*, p. 497

Lin, D. N. C. & Papaloizou, J. 1979, *MNRAS*, 188, 191

Pérez, L. M., Isella, A., Carpenter, J. M., & Chandler, C. J. 2014, *ApJl*, 783, L13

Pinilla, P., Benisty, M., & Birnstiel, T. 2012, *A&A*, 545, A81

Regály, Z., Juhász, A., Sándor, Z., & Dullemond, C. P. 2012, *MNRAS*, 419, 1701

van der Marel, N., van Dishoeck E. F., Bruderer, S., *et al.* 2015, subm. to *A&A*,

van der Marel, N., van Dishoeck, E., Bruderer, S., Perez, L., & Isella, A. 2015, *A&A*, 579, 106

van der Marel, N., van Dishoeck, E. F., Bruderer, S., *et al.* 2013, *Science*, 340, 1199

Weidenschilling, S. J. 1977, *MNRAS*, 180, 57

Williams, J. P. & Cieza, L. A. 2011, *ARA&A*, 49, 67

Zhang, K., Isella, A., Carpenter, J. M., & Blake, G. A. 2014, *ApJ*, 791, 42

*Young Stars & Planets Near the Sun*
Proceedings IAU Symposium No. 314, 2015
J. H. Kastner, B. Stelzer, & S. A. Metchev, eds.

© International Astronomical Union 2016
doi:10.1017/S1743921315006237

# The Chemistry of Nearby Disks

## Karin I. Öberg

Harvard-Smithsonian Center for Astrophysics, 60 Garden St, MS 16, Cambridge, MA 02138
email: koberg@cfa.harvard.edu

**Abstract.** The gas and dust rich disks around young stars are the formation sites of planets. Observations of molecular trace species have great potential as probes of the disk structures and volatile compositions that together regulate planet formation. The disk around young star TW Hya has become a template for disk molecular studies due to a combination of proximity, a simple face-on geometry and richness in volatiles. It is unclear, however, how typical the chemistry of the TW disk is. In this proceeding, we review lessons learnt from exploring the TW Hya disk chemistry, focusing on the CO snowline, and on deuterium fractionation chemistry. We compare these results with new ALMA observations toward more distant, younger disks. We find that while all disks have some chemical structures in common, there are also substantial differences between the disks, which may be due to different initial conditions, structural or chemical evolutionary stages, or a combination of all three.

**Keywords.** astrochemistry, protoplanetary disks, ISM: molecules, radio lines: ISM

---

## 1. Introduction

Gas and dust-rich disks seem ubiquitous around young stars. The planet formation potential of these disks depends fundamentally on the disk dust and gas mass. Measuring disk masses is difficult, however. We are practically blind to the main constituent, cold $H_2$ gas. Observations of dust continuum emission is a popular proxy, assuming a constant gas-to-dust conversion ratio. Disk demographic studies relying on such measurements suggest that the material needed to assemble a planetary system similar to our Solar System is frequently present around young stars (Andrews *et al.* 2013). Even around the smallest stars, there is more than sufficient mass in grains to form several Earth-like planets. This finding is in good agreement with exo-planet statistics, including the inferred high frequency of rocky planets around M dwarfs (Dressing & Charbonneau 2015). Assuming a constant gas-to-dust ratio may be a poor approximation, however, due to a combination of dust growth into boulders that are hidden from view, decoupled dust and gas dynamics, and efficient gas photoevaporation from the disk atmosphere. CO line observations have been used as a more 'direct' tracer of the disk gas mass (Williams & Bast 2014), but suffer from their own set of constraints, including a poorly constrained chemistry and therefore $CO$-to-$H_2$ conversion factor under many disk conditions (Favre *et al.* 2013). Constraining this chemistry and therefore the utility of CO as a disk gas tracer should clearly be a priority for upcoming disk studies (Bergin *et al.* 2013). This statement can be generalized to other molecular tracers or important disk characteristics, such as temperature, density, and ionization profiles.

Understanding the chemical compositions of disks is also an end in itself. The compositions of planets depend fundamentally on the composition of the disk dust grains and disk volatiles they form from, and on how this composition varies with distance from the central star. Of especial importance for planet formation is the separation of volatiles into their gas and ice phases, which is set by the balance of adsorption ("freeze-out" of the gas phase onto solids), and desorption (back to the gas phase). We call the midplane

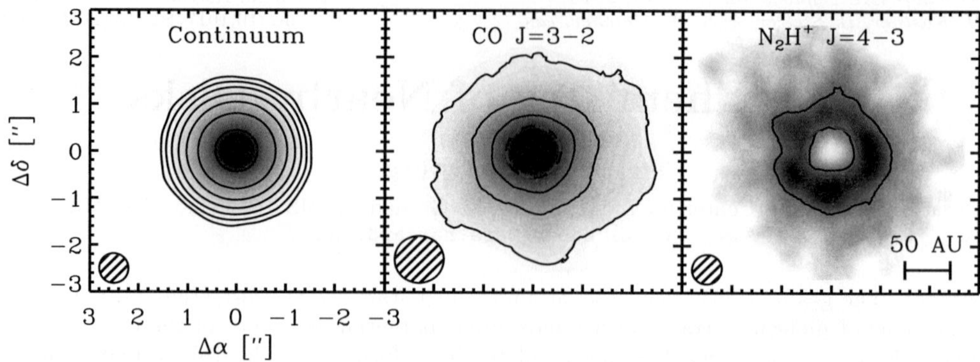

**Figure 1.** The dust continuum, CO and $N_2H^+$ emission toward TW Hya. The red dashed circle marks the inferred CO snowline location. (reprinted with permission from Qi *et al.* 2013b).

radius where freeze-out overtakes desorption for a major volatile a 'snowline'. Because of an exponential dependence of desorption rates on temperature these transitions are expected to be sharp and may thus be properly referred to as 'lines' when only considering the midplane regions. Snowlines can enhance the planet formation efficiency through a combination of increased grain surface density, rapid particle growth due to cold-head effects, pressure traps and increased grain stickiness (Ciesla & Cuzzi 2006, Johansen *et al.* 2007, Chiang & Youdin 2010, Gunlach *et al.* 2011, Ros & Johansen 2013). Snowline locations will also regulate the compositions of forming planets (Öberg *et al.* 2011) and planetesimals. In this proceeding we will only consider the CO snowline, simply because it occurs at a low temperature, i.e. far from the central star, and can thus be resolved by ALMA (§2).

A third aspect of planet formation is how Earth (and by extension, Earth-like planets and exoplanets) obtained its water. The Earth's oceans may have outgassed from the mantle, or may have been delivered by icy bodies bombarding the young Earth, or a combination of both. One approach to resolve this question is to compare the terrestrial water isotopic composition (level of deuterium enrichment) with other Solar System and interstellar water reservoirs. Using deuterium fractionation as a probe of the thermal history of terrestrial water (and other volatiles) presupposes a good understanding of the distribution and chemistry of deuterated molecules in the Solar Nebula. This chemistry is not well constrained in disks, however, and in §3 we discuss what observations of different deuterated species in the TW Hya disk and in younger disks can teach us about deuterium fractionation during planet formation.

## 2. Disk Snowlines

The locations of major snowlines depend on the volatile composition (e.g. whether most nitrogen is in $N_2$ or $NH_3$), a balance between freeze-out and thermal and non-thermal desorption rates at different disk locations (set by a combination of density, temperature and radiation fields), and disk dynamics (Öberg *et al.* 2011). Exact snowline locations are therefore difficult to predict from theory, and observational constraints are key to constrain planet formation models. The CO snowline is expected at the disk radius where the temperature drops below $\sim$20 K. Around T Tauri and Herbig Ae disks this threshold is crossed at $\sim$20–150 AU.

Extractions of CO snowline locations using CO observations is challenging because of the presence of large amounts of CO in the disk atmosphere (Qi *et al.* 2011). An alternative approach is to exploit chemical effects regulated by CO freeze-out to constrain the location of the CO snow line. Three proposed CO snow line tracers are $N_2H^+$, $H_2CO$, and $DCO^+$ (Qi *et al.* 2013, Mathews *et al.* 2013). $N_2H^+$ is expected to be a robust tracer of CO freeze-out because the presence of gas phase CO slows down $N_2H^+$ formation and speeds up $N_2H^+$ destruction. The predicted anti-correlation between gas-phase CO and $N_2H^+$ is observed in pre-stellar and protostellar environments (e.g. Caselli *et al.* 1999, Bergin *et al.* 2002, Jørgensen *et al.* 2004). Qi *et al.* (2013b) imaged $N_2H^+$ emission with the Atacama Large sub-Millimeter Array (ALMA) toward the disk around T Tauri star TW Hya and found a large ring with an inner edge at 30 AU, tracing the CO snowline (Fig. 1). The determined CO snowline location corresponds to a CO freeze-out temperature of 17 K, using a detailed disk density and temperature structure. This is a few degrees lower than expected, which may reflect a combination of model uncertainties and excess CO depletion due to chemical conversion of CO into other species in this old disk.

With ALMA similar measurements using $N_2H^+$ are possible toward samples of disks. So far only one such additional measurement exists. $N_2H^+$ observations with ALMA toward HD 163296 reveal a similar ring-like structure as previously observed toward TW Hya, but the inferred CO snowline location is substantially (10s of AU) closer to the star than expected from the inferred CO freeze-out temperature in the TW Hya disk. The HD 163296 CO snowline measurement corresponds to a CO freeze-out temperature of $\sim$25 K, 8 K higher than in the TW Hya disk (C. Qi, private communication). Such a high freeze-out temperature may be explained by dynamics (moving the snowline inwards due to drift of solids) or a water-rich ice morphology. CO is known to bind stronger to water ice compared to other CO molecules and even small amounts of water forming or freezing out with the CO ice could increase the CO freeze-out temperature by several degrees (Collings *et al.* 2004).

In summary: there are two measurements of CO snowline locations and the results are two substantially different CO freeze-out temperatures. The causes for these differences are unclear, though there are several candidates, including disk dynamics and ice chemistry. This result entails that we cannot currently use theory to predict the CO freeze-out temperature or the location of the CO snowline in a disk, including the Solar Nebula. There is a clear need for a survey of CO snowline locations toward disks to 1. constrain how CO freeze-out depends on disk structure and evolution, and 2. establish the 'typical' CO freeze-out temperature during planet formation.

## 3. Deuterium Fractionation

Deuterium fractionation is commonly used to infer (thermal) properties about volatile formation locations. Molecules become enriched in deuterium due to small differences in zero-point energies between deuterated and non-deuterated molecules, which results in preferential formation of deuterated molecules at low temperatures. Based on this understanding, the high D/H ratio in terrestrial $H_2O$ has been used to infer delivery of water from asteroids and comets, which both assembled in the colder, outer Solar System (Hartogh *et al.* 2011). The generally high D/H ratio in Solar System water has further been used to infer a cold pre-Solar origin of most Solar System $H_2O$ (Cleeves *et al.* 2014).

So far only two deuterated molecules have been detected in disks, $DCO^+$ and DCN (van Dishoeck *et al.* 2003, Qi *et al.* 2008). Figure 2 shows Submillimeter Array (SMA) observations of $DCO^+$ in the TW Hya disk. The $DCO^+$ emission presents a low SNR ring. When compared with $HCO^+$ toward TW Hya, this ring implies an increasing level of deuterium fractionation with disk radius. This is in agreement with theoretical

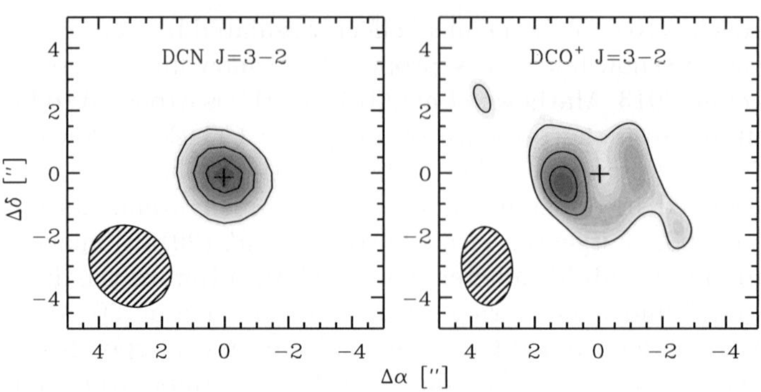

**Figure 2.** Integrated images of the TW Hya disk in (*left:*) DCN J=3-2 emission (ALMA science verification) and (*right:*) DCO$^+$ J=3-2 emission (SMA data from Qi *et al.* (2008)). The contour levels are at 50%, 75% and 90% of the peak value. The cross marks the peak of the continuum emission, which locates the position of the central star. Reprinted with permission from Öberg *et al.* (2012).

expectations, since the temperature in the disk midplane decreases with increasing disk radius. By contrast, DCN emission presents a centrally peaked feature toward the TW Hya disk (Fig. 2). This suggests that deuterium fractionation in HCN occurs at higher temperatures compared to HCO$^+$. There are clearly multiple pathways to deuterium fractionation in disks, which operate in different temperature regimes (Öberg *et al.* 2012). This already complicates the use of deuterium fractionation as a thermal history probe.

The picture is complicated further by recent observations of DCO$^+$ and DCN toward a small sample of disks with ALMA. Figure 3 shows the different DCO$^+$ and DCN 3–2 emission morphologies toward three disks around T Tauri stars AS 209, V 4046 Sgr and IM Lup. These disks were observed during 2014 and 2015 with the Atacama Large Millimeter/Submillimeter Array (ALMA; project code 2013.1.00226) with $\sim$30 antennas, spanning baselines of 15–650 m. The total on source integration time was $\sim$20 minutes. The visibility data were calibrated by ALMA staff. Each individual SPW was further phase and amplitude self-calibrated in CASA 4.2.2. In each SPW, the continuum was subtracted using line-free channels. The fully calibrated visibilities were Fourier inverted and CLEANed.

Similarly to the TW Hya disk, the DCO$^+$ emission maps always present a hole or depressions toward the warm disk center. In contrast to the previous TW Hya observations, the DCN emission maps also present central depressions toward all three disks. This may simply be an effect of the increased spatial resolution employed for these observations, however. A second similarity between TW Hya and the disk sample is that the DCN emission appears to be more compact than the DCO$^+$ emission, suggesting that cold deuterium fractionation chemistry is generally more important for DCO$^+$ production compared to DCN. These structural similarities between TW Hya and the younger disk sample in Fig. 3 are accompanied by several important differences, however.

Figure 3 shows the relative sizes of the DCO$^+$ and DCN central emission depressions vary between different disks. Toward AS 209, the radius of the central DCO$^+$ hole is smaller than the DCN hole radius. The opposite is true for the relative hole sizes of DCO$^+$ and DCN in the transition disk around V 4046 Sgr, which also presents the largest DCO$^+$ hole in the sample. IM Lup presents the most surprising morphology, a set of concentric rings in DCO$^+$ emission (Öberg *et al.* 2015, submitted to ApJ). DCN is barely detected in this source. Together with TW Hya, this small disk sample reveals that there is no simple relationship between disk radius and the deuterium pathways that results in deuterium

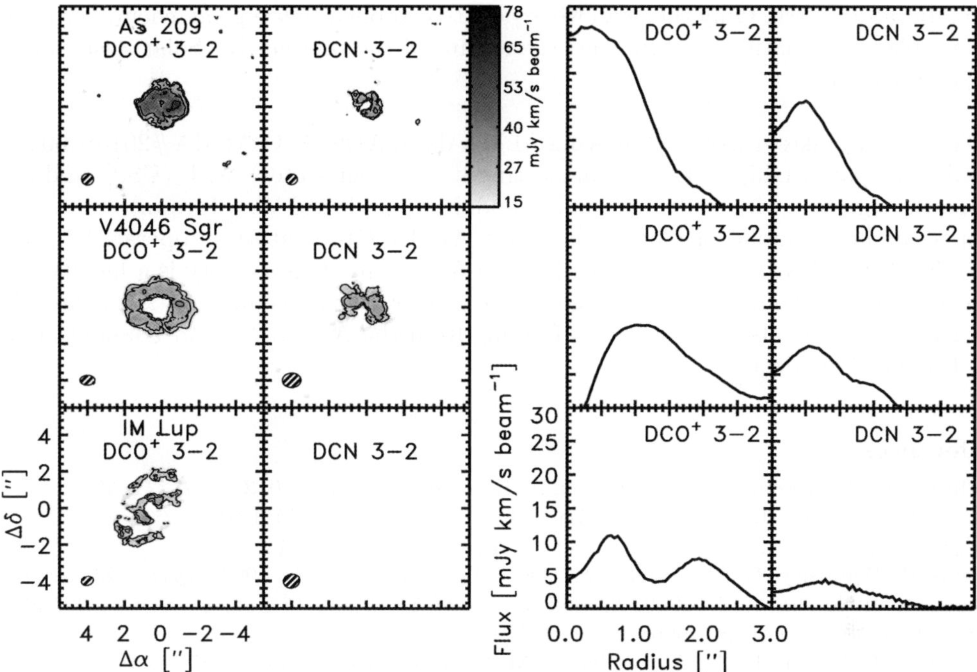


**Figure 3.** DCO$^+$ and DCN emission toward three protoplanetary disks. The left panels show the integrated emission maps of DCO$^+$ and DCN with the color scale marking the flux density in mJy km/s beam$^{-1}$ (the beam is shown in the bottom left of each panel). The right panels show the corresponding deprojected and azimuthally averaged radial profiles.

enrichments of either HCO$^+$ or HCN. In particular DCN does not always extend further in toward the central star compared to DCO$^+$. In summary: even when only considering two molecules, the deuterium enrichment pattern observed toward TW Hya does not appear to be universal in protoplanetary disks.

## 4. Conclusions

TW Hya has been and will continue to be an important starting point when studying disk chemistry (see, e.g., Kastner *et al.* 1997, Thi *et al.* 2004, Qi *et al.* 2013b). Its proximity, well-established disk density and temperature structure and rich molecular inventory makes it ideal to develop new chemical probes. The chemistry of the TW Hya disk may be far from typical, however. The disk around TW Hya is old compared to most other gas-rich circumstellar disks, and disk chemistry is time dependent. In this proceeding we compared the observed chemistry in the TW Hya disk with other, younger disks focusing on two key processes: the chemical imprint of CO freeze-out, and deuterium fractionation. In both cases we found several similarities between the TW Hya disk and the other targets, but also substantial differences. For example, the CO snowline locations in the TW Hya and HD 163296 disks do not seem to correspond to the same CO freeze-out temperature, and the more centrally peaked DCN emission compared to DCO$^+$ observed toward TW Hya is not reproduced toward all other disks. Whether these differences trace different initial conditions, present differences in disk temperature, density and radiation structures, or different chemical evolutionary stages is currently unclear. Larger surveys of disk chemistry should be able to answer some of these questions, when combining the survey results with detailed disk density, temperature and radiation structures, theory and laboratory work. Only then can we expect to establish the 'typical' chemistry present during planet formation, and further, the origins of the present-day Solar System chemical composition.

This paper makes use of the following ALMA data: ADS/JAO.ALMA#2013.1.00226.S. ALMA is a partnership of ESO (representing its member states), NSF (USA) and NINS (Japan), together with NRC (Canada) and NSC and ASIAA (Taiwan), in cooperation with the Republic of Chile. The Joint ALMA Observatory is operated by ESO, AUI/NRAO and NAOJ. The National Radio Astronomy Observatory is a facility of the National Science Foundation operated under cooperative agreement by Associated Universities, Inc. KIÖ also acknowledges funding from the Alfred P. Sloan Foundation, and the Packard Foundation.

## References

Andrews, S. M., Rosenfeld, K. A., Kraus, A. L., & Wilner, D. J. 2013, *ApJ*, 771, 129
Bergin, E. A., Alves, J., Huard, T., & Lada, C. J. 2002, *ApJL*, 570, L101
Bergin, E. A., Cleeves, L. I., Gorti, U., et al. 2013, *Nature*, 493, 644
Caselli, P., Walmsley, C. M., Tafalla, M., Dore, L., & Myers, P. C. 1999, *ApJL*, 523, L165
Chiang, E. & Youdin, A. N. 2010, *Annual Review of Earth and Planetary Sciences*, 38, 493
Ciesla, F. J. & Cuzzi, J. N. 2006, *Icarus*, 181, 178
Cleeves, L. I., Bergin, E. A., Alexander, C. M. O., et al. 2014, *Science*, 345, 1590
Collings, M. P., Anderson, M. A., Chen, R., et al. 2004, *MNRAS*, 354, 1133
Dressing, C. D. & Charbonneau, D. 2015, ArXiv e-prints
Favre, C., Cleeves, L. I., Bergin, E. A., Qi, C., & Blake, G. A. 2013, *ApJL*, 776, L38
Gundlach, B., Kilias, S., Beitz, E., & Blum, J. 2011, *Icarus*, 214, 717
Hartogh, P., Lis, D. C., Bockelée-Morvan, D., et al. 2011, *Nature*, 478, 218
Johansen, A., Oishi, J. S., Low, M.-M. M., et al. 2007, *Nature*, 448, 1022
Jørgensen, J. K. 2004, *A&A*, 424, 589
Kastner, J. H., Zuckerman, B., Weintraub, D. A., & Forveille, T. 1997, *Science*, 277, 67
Mathews, G. S., Klaassen, P. D., Juhász, A., et al. 2013, *A&A*, 557, A132
Öberg, K. I., Murray-Clay, R., & Bergin, E. A. 2011a, *ApJL*, 743, L16
Öberg, K. I., Qi, C., Wilner, D. J., & Hogerheijde, M. R. 2012, *ApJ*, 749, 162
Qi, C., D'Alessio, P., Öberg, K. I., et al. 2011, *ApJ*, 740, 84
Qi, C., Öberg, K. I., & Wilner, D. J. 2013a, *ApJ*, 765, 34
Qi, C., Öberg, K. I., Wilner, D. J., et al. 2013b, *Science*, 341, 630
Qi, C., Wilner, D. J., Aikawa, Y., Blake, G. A., & Hogerheijde, M. R. 2008, *ApJ*, 681, 1396
Ros, K. & Johansen, A. 2013, *A&A*, 552, A137
Thi, W., van Zadelhoff, G., & van Dishoeck, E. F. 2004, *ApJ*, 425, 955
van Dishoeck, E. F., Thi, W., & van Zadelhoff, G. 2003, *A&A*, 400, L1
Williams, J. P. & Best, W. M. J. 2014, *ApJ*, 788, 59

*Young Stars & Planets Near the Sun*
Proceedings IAU Symposium No. 314, 2015
J. H. Kastner, B. Stelzer, & S. A. Metchev, eds.

© International Astronomical Union 2016
doi:10.1017/S1743921315006018

# X-raying Circumstellar Material around Young Stars

## P. C. Schneider[1] and H. M. Günther[2]

[1] ESA/ESTEC
email: cschneid@cosmos.esa.int
[2] MIT
email: hgunther@mit.edu

**Abstract.** Young stars are surrounded by copious amounts of circumstellar material. Its composition, in particular its gas-to-dust ratio, is an important parameter. However, measuring this ratio is challenging, because gas mass estimates are often model dependent. X-ray absorption is sensitive to the gas along the line-of-sight while optical/near-IR extinction depends on the dust. Therefore, the absorber's gas-to-dust ratio is directly given by the ratio between X-ray and optical/near-IR extinction. We present three systems where we used X-ray and optical/near-IR data to constrain the gas-to-dust ratio of circumstellar material; from a dust rich debris disk to gaseous protoplanetary disks.

## 1. Introduction

The chemical composition of circumstellar disks is one of their most important properties. It controls many transport phenomena within the disk like accretion, angular momentum transport as well as planet formation. The interstellar medium (ISM) consists mainly of gas; only a small fraction ($\sim 1\,\%$) is in the form of dust grains. Protoplanetary disks are thought to have a gas-to-dust ratio close to the ISM value during their early evolutionary stages. Therefore, the ISM gas-to-dust ratio of 100:1 is often *assmued* when deriving the total disk mass from dust measurements. Directly deriving the total gas reservoir is more challenging as the conversion of gas emission line fluxes to total (gas) disk mass strongly depends on often uncertain model parameters.

One possibility to trace the gas is X-ray absorption. The absorption cross section is essentially independent of the chemical bonds of the absorbing material, because X-ray absorption involves the inner shell electrons while molecular bonds involve the outer (valence) electrons. Thus, we can derive the absorber's gas content from its X-ray absorption signal. Comparison with optical extinction caused by the dust along the line-of-sight enables us to measure the gas-to-dust ratio of the absorbing material.

The location of the absorber(s) along the line-of-sight often remains uncertain. Time variability of the absorption can help to overcome this ambiguity. In this proceedings, we present three cases where the location of the absorption can be determined with some certainty so that a local gas-to-dust ratio can be derived.

## 2. The debris disk around AU Mic

AU Mic is a young ($23 \pm 3$ Myrs, e.g. Mamajek *et al.* 2015) M1 dwarf at a distance of about 10 pc belonging to the $\beta$ Pic moving group (Zuckerman *et al.* 2001). AU Mic is surrounded by an edge-on debris disk, i.e., a dust rich disk. This disk, extending out to $\sim 150$ AU, has been subject to numerous studies from scattered light (e.g. Kalas *et al.* 2004, Krist *et al.* 2005), over dust emission (Liu *et al.* 2004) to absorption (Roberge *et al.*

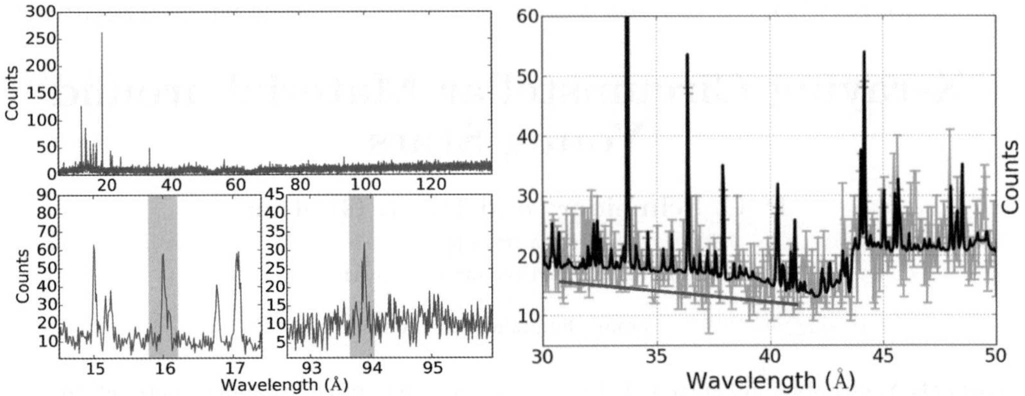

**Figure 1. Left**: *Chandra* LETGS spectrum of AU Mic. As an example, two Fe XVIII lines that were used to derive the absorbing column density are highlighted. **Right**: The carbon K-edge. The edge at 43 Å is clearly visibile, but can be entirely explained by carbon within the filter in front of the X-ray detector. Additional carbon within the AU Mic disk would lower the flux shortwards of 43 Å as indicated by the blue line.

2005) and molecular hydrogen emission (France *et al.* 2007). Given its small distance to the Earth, we can exclude substantial interstellar absorption so that any absorption signal can be safely associated with AU Mic's circumstellar disk.

Thanks to its youth and distance, AU Mic is a bright X-ray source on the sky (e.g. Linsky *et al.* 2002). We obtained a *Chandra* LETGS X-ray spectrum of AU Mic (see Fig. 1, published in Schneider & Schmitt 2012), which extends from a few Å up to $> 100$ Å. This wavelengths coverage allowed us to compare lines of the same element and ionization stage formed at largely different wavelengths. Their line ratios are given mainly by atomic physics and almost independent of the plasma temperature. Thus, these line ratios can be used to measure of the total absorption towards AU Mic. Specifically from our analysis of abundant Fe lines, we were able to limit the gas column density to $N_H < 10^{19}$ cm$^{-2}$ ($1\sigma$ conf.). Furthermore, this spectrum allowed us to directly constrain the amount of carbon in gas and small grains ($\lesssim 0.5\,\mu$m) from the depth of the carbon K-edge. Figure 1 (right) shows a zoom into the wavelengths region around the C-K edge. The edge-like structure is clearly visible. However, its depth can be entirely explained by carbon in the detector itself. We derive an upper limit on the carbon content of $N_C < 10^{18}$ cm$^{-2}$.

Comparing our upper limits with published dust masses, we find that the disk must contain more dust than gas. The measurements of Fitzgerald *et al.* 2007 suggest a dust mass around $0.01\,M_\oplus$ (France *et al.* 2007), which would correspond to a total dust column density of $7 \times 10^{-4}$ g cm$^{-2}$ or $3 \times 10^{20}$ cm$^{-2}$ for ISM abundances. That we detect less absorption implies that the dust is locked in large grains that are opaque to X-rays.

## 3. The protoplanetary disk around the CTTS AA Tau

The classical T Tauri star (CTTS) AA Tau is surrounded by a gaseous protoplanetary disk. Compared to AU Mic, AA Tau is younger ($\sim 2$ Myrs) and its disk is not viewed edge-on, but inclined by 75° with respect to the plane of the sky. Thus, the disk is not in front of the star for most of the time. The inner disk region, however, is warped and periodically eclipses the star. X-ray data showed that this inner region is gas-rich as measured by $N_H/A_V$ where $N_H$ is the equivalent hydrogen column density derived from

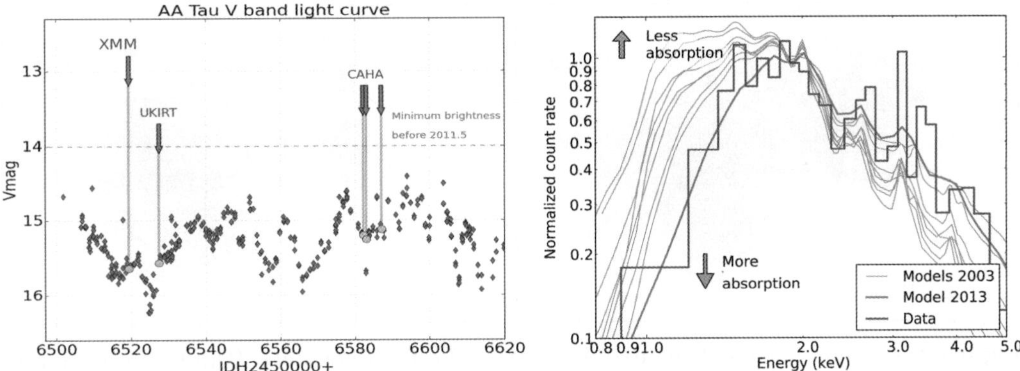

**Figure 2. Left**: Optical light curve of AA Tau around the X-ray and near-IR observations. **Right**: XMM-Newton pn spectrum of AA Tau with models normalized to the count rate at $E_{phot} = 2.0\,\mathrm{keV}$.

low-resolution X-ray spectra and $A_V$ is the dust extinction (Schmitt & Robrade 2007, Grosso *et al.* 2007).

End of 2011, the system darkened significantly in the optical ($\Delta V \approx 2$, see Fig. 2 left). Bouvier *et al.* (2013) suggest that the extra absorption is caused by disk material at larger radii ($r \gtrsim 10\,\mathrm{AU}$), which rotated into view and now obscures the line-of-sight towards the central star. That the inner disk region is obscured from view is confirmed by our high-resolution HST FUV spectrum obtained during the dim state, as the $H_2$ line width decreased compared to the bright state's data published by France *et al.* 2012. We obtained a new X-ray observation with XMM-Newton during the dim state. It reveals that the increase in absorbing column density is moderate ($N_H^{extra} = 0.5 - 1.0 \times 10^{22}\,\mathrm{cm}^{-2}$). Our contemporaneous UV to near-IR data characterize the dust extinction during the dim state. Comparing dust and gas absorption, we find a gas-to-dust ratio of $N_H/A_V = 0.8 \ldots 3.6 \times 10^{21}$, which is within a factor of two of the ISM value (Schneider *et al.*, A&A submitted).

## 4. RW Aur: Another CTTS during a dim state

RW Aur consists of two CTTSs; both of approximately solar mass. The system underwent a major dimming event towards the end of 2014 ($\Delta V = 2$ to 3 magnitudes). This event resembles a previously observed half-year long dimming, which was attributed to a tidal stream between both components (Rodriguez *et al.* 2013, Dai *et al.* 2015). However, this new dimming event might more closely resemble the AA Tau dimming with obscuration by the primary's disk itself as suggested by new infrared data (see Shenavrin *et al.* 2015).

We obtained a new *Chandra* observation of the system during a dim state that started end of 2014 (see Fig. 3). Comparing the dim state's X-ray properties with those observed previously by Skinner & Güdel (2014) during the bright state, we find an increase in absorbing column density of $N_H \approx 2 \times 10^{22}\,\mathrm{cm}^{-2}$. Antipin *et al.* (2015) find that the optical extinction appears rather gray, i.e., that the absorption is caused by grains with sizes greater than $1\,\mu\mathrm{m}$. Therefore, the dust mass might differ from the value suggested by assuming $\Delta V = A_V$ and an ISM-like grain population. As an estimate, we assume absorption by $1\,\mu\mathrm{m}$ sized grains and opacities from Draine (2003). With these assumptions, the gas-to-dust ratio of the extra absorber is below the ISM-value when comparing

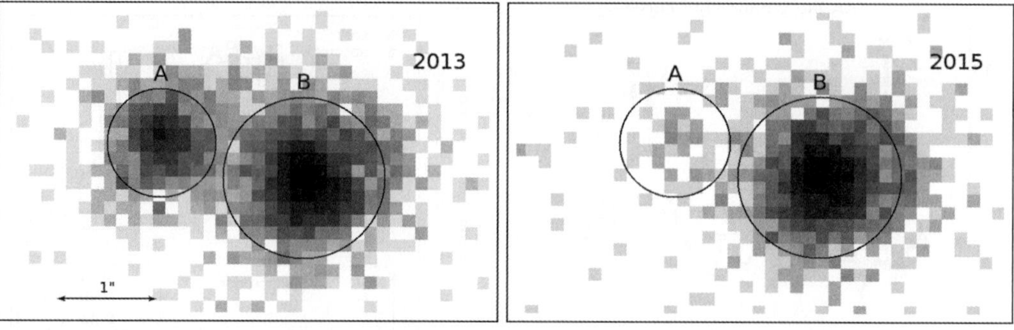



**Figure 3.** *Chandra* images of the RW Aur system. **Left**: Bright state, **Right**: dim state with the tidal stream in front of the A component.

the observed 2 mag of dust extinction with the increase in X-ray absorption (Schneider *et al.*, in prep.).

## 5. Conclusions

The gas-to-dust ratio of circumstellar material is probed by the ratio between X-ray absorption ($N_H$) and optical/NIR extinction ($A_V$). We present three systems where we applied this method. The debris disk system AU Mic contains more grains than gas, the protoplanetary disk around the CTTS AA Tau has a gas-to-dust ratio compatible with the ISM at radii of a few AU, and the absorber in the young binary system RW Aur has low gas-to-dust ratio and large grains.


## References

Antipin, S., Belinski, A., Cherepashchuk, A., Cherjasov, D., Dodin, A., Gorbunov, I., Lamzin, S., Kornilov, M., Kornilov, V., Potanin, S., Safonov, B., Senik, V., Shatsky, N., & Voziakova, O. 2015 *IBVS*, 6126, 1

Bouvier, J., Grankin, K., Ellerbroek, L. E., Bouy, H., & Barrado, D., 2013, *A&A*, 557, A77

Dai, F., Facchini, S., Clarke, C. J., & Haworth, T. J. 2015 *MNRAS*, 449, 1996

Draine, B. T. 2003, *ARAA*, 41, 241

Fitzgerald, M. P., Kalas, P. G., Duchêne, G., Pinte, C. & Graham, J. R., 2007 *ApJ*, 670, 536F

France, K., Roberge, A., Lupu, R. E., Redfield, S., & Feldman, P. D., 2007 *ApJ*, 668, 1174

France, K., Burgh, E. B., Herczeg, G. J., Schindhelm, E. Yang, H., Abgrall, H., Roueff, E., Brown, A., Brown, J. M., & Linsky, J. L., 2012 *ApJ*, 744, 22F

Grosso, N., Bouvier, J., Montmerle, T., Fernández, M., Grankin, K., & Zapatero Osorio, M. R., 2007 *A&A*, 475, 607

Kalas, P., Liu, M. C., & Matthews, B. C. 2004, *Science*, 303, 1990

Krist, J. E., Ardila, D. R., Golimowski, D. A., Clampin, M., *et al.* 2005, *AJ*, 129, 1008

Linsky, J. L., Ayres, T. R., Brown, A., & Osten, R. A., 2002 *Astron. Nachr.*, 323, 321

Liu, M. C., Matthews, B. C., Williams, J. P., & Kalas, P. G., 2004 *ApJ*, 608, 526

Mamajek, E. E. & Bell, C. P. M. 2015, *MNRAS*, 445, 2169

Roberge, A., Weinberger, A. J., Redfield, S., & Feldman, P. D., 2005 *ApJL*, 626, 105

Rodriguez, J. E., Pepper, J., Stassun, K. G., Siverd, R. J., Cargile, P., Beatty, T. G., & Gaudi, B. S, 2013 *AJ*, 146, 112

Shenavrin, V. I., Petrov, P. P., & Grankin, K. N., 2015 *IBVS*, 6143, 1S

Schmitt, J. H. M. M. & Robrade, J., 2007 *A&A*, 462, 41

Schneider, P. C. & Schmitt, J. H. M. M., 2012 *A&A*, 516, 8

Skinner, S. L. & Güdel, M. 2014 *ApJ*, 788, 101

Zuckerman, B., Song , I., Bessell, M. S., & Webb, R. A. 2001, *ApJL*, 562, L87

*Young Stars & Planets Near the Sun*
*Proceedings IAU Symposium No. 314, 2015*
*J. H. Kastner, B. Stelzer, & S. A. Metchev, eds.*

© International Astronomical Union 2016
doi:10.1017/S1743921315006420

# Photoevaporation and Disk Dispersal

## Uma Gorti

SETI Institute/NASA Ames Research Center
email: uma.gorti-1@nasa.gov

**Abstract.** Protoplanetary disks are depleted of their mass on short timescales by viscous accretion, which removes both gas and solids, and by photoevaporation which removes mainly gas. Photoevaporation may facilitate planetesimal formation by lowering the gas/dust mass ratio in disks. Disk dispersal sets constraints on planet formation timescales, and by controlling the availability of gas determines the type of planets that form in the disk. Photoevaporative wind mass loss rates are theoretically estimated to range from $\sim 10^{-10}$ to $10^{-8}$ $M_\odot$ yr$^{-1}$, and disk lifetimes are typically $\sim$ few Myr.

**Keywords.** Accretion, Planetary systems: Protoplanetary Disks, Formation

## 1. Disk Dispersal

Several lines of evidence point to dispersal mechanisms that operate in protoplanetary disks and remove most of their mass. In the solar system, gas is heavily depleted with only $\sim$10% of the initial disk mass remaining—the total mass of solids ($\sim 10^{-4}$ $M_\odot$) in planets and planetary objects is high relative to the total gas ($\sim 10^{-3}$ $M_\odot$), almost all of which is contained in the giant planets. Protoplanetary disks in star-forming regions show evidence for dispersal of dust; Haisch *et al.* (2001) first determined that the frequency of disk-bearing stars in clusters declines with cluster age on $\sim 3 - 5$ Myr timescales. Concurrent exoplanetary studies have further revealed the ubiquity of planet formation in disks around low and intermediate-mass stars, and the close correspondence between the minimum mass required to form our solar system and measured dust disk masses suggests that most of the dust in disks is depleted very quickly into planets. Gas disk dispersal is less well constrained, but is believed to also occur rapidly after star and disk formation.

Disks are initially composed of interstellar material with gas dominating their masses, and their primordial evolutionary phase ends with gas dispersal. Studies of a few nearby young disks by Zuckerman *et al.* (1995) found that CO emission was only detected from objects younger than 10 Myr, setting the earliest constraints on gas disk lifetimes. Later surveys by the Spitzer and Herschel space telescopes set upper limits on the mass of gas present in more evolved disks, $< 0.1 M_J$ within $\sim 40$ AU at ages $\sim 5-30$ Myr and $< 1 M_J$ gas within $\sim 100$ AU at ages $\sim 4-6$ Myr, respectively (Pascucci *et al.* 2006, Dent *et al.* 2013). Observations are thus consistent with gas dispersal times that are at least as long as, or similar to, dust disk dispersal times. The prevalence of gaseous envelopes around exoplanets and the low orbital inclinations of close-in Kepler planets both suggest that gas typically survives past planet formation epochs; in fact, migration in a gas disk may have shaped the architecture of many exoplanetary systems. In the light of all the above constraints, gas disk lifetimes are likely to be $\lesssim 10$ Myr.

Although several disk dispersal theories have been previously considered (e.g., see review by Hollenbach *et al.* 2000), viscous accretion and photoevaporation by stellar high energy photons are believed to be primarily responsible for removal of disk material. Viscosity depletes both gas and dust via loss of angular momentum; radial drift, if efficient, may also remove substantial amounts of larger dust and solids as the disk evolves. If all disks undergo planet formation, then most of the solids may be assembled into planetary objects. A large fraction of the gas, especially in the outer disk, is probably removed by photoevaporation whereby stellar heating of gas drives thermal winds from the disk surface.

## 2. Photoevaporation theory

Young stars generate intense high energy radiation fields at ultraviolet (FUV, $6 - 13.6$eV; EUV, $13.6 - 100$eV) and X-ray wavelengths. These photons irradiate the disk surface and can heat gas to escape temperatures when the combined thermal pressure and angular momentum support exceed gravitational pressure. Thermal winds are launched when the gas at a given radius ($r_{AU}$) is heated up to a critical temperature, $T_{crit} > (1 - 1.8) \times 10^4 \ r_{AU}^{-1}$ K for ionized/atomic flows. Typically, flows are subsonic and escape temperatures are lower than $T_{crit}$.

The wind mass loss rate, and hence the importance of photoevaporation in removing disk gas, depend on the density and temperature in the thermal wind; these in turn depend on the heating agent. The rate of change of surface density $\dot{\Sigma}_{pe} \propto \rho_b \sqrt{T_{gas}}$, where $\rho_b$ and $T_{gas}$ are the density and temperature in the flow respectively. EUV and soft X-ray ($\sim 0.1 - 0.3$ keV) photons can heat the gas to high temperatures ($\gtrsim 10^4$ K) but are easily absorbed due to their low attenuation columns and the densities of such flows are typically low. However, the physics is relatively simple as the temperature is either nearly constant in the case of EUV or can be parametrized in terms of the ionization parameter for optically thin soft X-rays. FUV and hard X-rays ($\gtrsim 1$keV) heat the gas to lower temperatures ($<$ few 1000 K) but have significantly higher penetration depths and hence higher flow densities. In this case, the determination of gas temperature is complicated by the chemistry of trace coolant species and the radiative transfer of cooling line emission that is often optically thick.

Mass loss rates estimated from theoretical models are low for EUV-driven photoevaporation ($\sim 10^{-10} \ M_\odot$ yr$^{-1}$), but $\sim 10 - 100$ times higher in the FUV and soft X-ray case. Although the mass loss rate derived differs depending on model assumptions and stellar/disk parameters, for fields typical of T Tauri stars Gorti & Hollenbach (2009) find that photoevaporation is mainly driven by FUV photons. This is partly due to the fact that accretion produces shocks on the stellar surface that generate a significant FUV flux in addition to the chromospheric/coronal flux; FUV fields are typically high around young, low and intermediate-mass stars. High EUV and X-ray fluxes can, however, also produce higher mass loss rates and could in principle dominate under certain conditions.

Mass loss rates due to photoevaporation are thus uncertain by $\sim 1 - 2$ orders of magnitude, mainly because of difficulties in assessing the amount of flux irradiating the disk surface. EUV luminosities are unknown as they are easily absorbed and nearly impossible to measure directly; this is also true for soft X-rays. However, gas irradiated by EUV and soft X-rays will be partly or fully ionized ($x_e \sim 0.1 - 1$) and Pascucci et al. (2014) recently placed constraints on the ionization levels of a few disks based on the expected free-free emission. They found that the implied EUV photon luminosity is low, even if all of the observed cm excess is attributed to free-free emission from ionized gas. EUV-driven photoevaporative mass loss rates in their disk sample are hence too low to explain inferred disk lifetimes. The role of soft X-rays is still inconclusive, as the flux of free-free emission depends on the gas temperature. The absence of a clear correlation between X-ray luminosity and free-free emission flux and the low cm excess from a known soft X-ray source, TW Hya, together suggest that soft X-rays are unimportant in driving

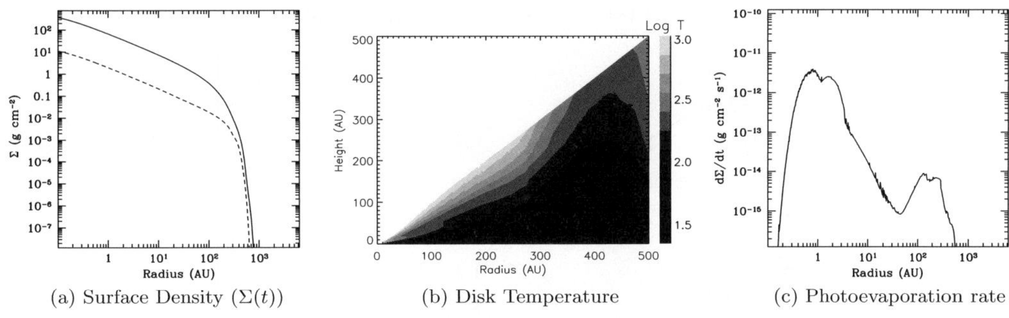

(a) Surface Density ($\Sigma(t)$)   (b) Disk Temperature   (c) Photoevaporation rate

**Figure 1.** The first panel shows the instantaneous gas surface density (solid line) of a model disk at an epoch 1 Myr after the beginning of the simulation. The dashed line is the total solid surface density which is distributed into a number of dust size bins, this distribution is calculated using the gas surface density at that epoch. At each time $t$, a thermochemical disk structure model is calculated from this $\Sigma(r,t)$ and the dust opacities, the corresponding temperature structure is shown in panel (b). From the temperature and density structure (not shown), we evaluate the height at which the photoevaporation rate ($\dot{\Sigma}_{pe}$) peaks for each radius as shown in the next panel (c). This rate is then used in the surface density evolution equation to advance $\Sigma$ to the next time step.

mass loss, but it remains to be seen if the Pascucci *et al.* (2014) results hold for a larger disk sample.

Hereafter, we assume that FUV photons (and hard X-rays, although see Gorti & Hollenbach 2009) are more relevant for photoevaporation and disk dispersal.

## 3. FUV-driven photoevaporation

FUV heating critically depends on the evolving abundance of very small grains (VSGs) and polycyclic aromatic hydrocarbons (PAHs) in the disk. Gas is indirectly heated by FUV photons via collisions with energetic electrons produced by photoelectric ejection from small dust. As the gas surface density decreases and dust grains collide, coagulate and fragment in the disk, the abundance of small grains is expected to evolve. If there are fewer small grains, the heating rate and resulting gas temperature would be lower, but on the other hand, the FUV opacity would also be reduced. Higher penetration of FUV photons would then result in higher densities in the flow. Since the photoevaporation rate $\dot{\Sigma}_{pe}$ depends on both density and temperature, the effect of a reduced or enhanced small grain abundance on the mass loss rate is not immediately apparent.

Motivated by the above, Gorti *et al.* (2015) recently investigated the effects of dust evolution on photoevaporation. Using a multi fluid (gas+solids) EUV, FUV and X-ray photoevaporation model combined with a dust grain evolution framework based on coagulation/fragmentation equilibrium, they solved for the evolution of gas and solid surface densities with time. Figure 1 shows the calculation of photoevaporation rate for one such model at a given instant of time in these simulations. A 1+1D disk structure model is used to calculate the disk structure at every epoch using the instantaneous surface density distribution as determined by a 1-D viscous evolution and disk photoevaporation model. See Gorti *et al.* (2015) for more details.

Although the dust grain size distribution significantly changes with time, photoevaporation timescales for these models are comparable to models with no dust evolution. Disk lifetimes change by less than a factor of 2, which is not significant given the uncertainties in other model input parameters (e.g., stellar spectrum, disk viscosity). Figure 2 shows the decrease in gas mass with time for models with and without dust

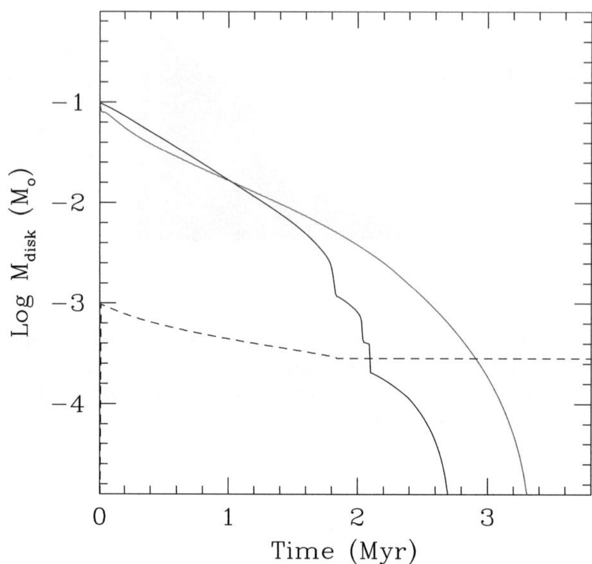

**Figure 2.** The disk gas (solid) mass and dust mass (dashed line) is shown in black for the dust evolution model and in red for a model without dust evolution. The dispersal time scales for gas in both models are similar. In the model without dust evolution, gas and dust are assumed to be always well-mixed and dust is carried along with the gas. Note that in the model with dust evolution, most of the dust is left behind after photoevaporation as very little dust is collisionally coupled to the flow.

evolution. As discussed in Gorti *et al.* (2015), the relative insensitivity of the photoevaporation rate to small grain abundance is due to the opposing effects of heating and opacity on $\dot{\Sigma}_{pe}$.

An interesting outcome of combining dust evolution with photoevaporation is that the gas/dust mass ratio in the disk changes as the disk evolves and disperses. As shown in Figure 2, most of the dust in the disk remains after the gas is removed. This result is due to two effects: (i) gas densities in the photoevaporative flow are too low for all but the smallest dust grains to be collisionally coupled with the gas, and (ii) mass loss due to photoevaporation is comparable to mass loss due to accretion (see Gorti *et al.* 2015). Low gas/dust ratios are a requirement for many planetesimal formation mechanisms (see review by Johansen *et al.* 2015), and preferential removal of gas by photoevaporation may thus aid planet formation. Although these disk evolution models do not consider planet formation and solids are restricted to sizes $< 1$ cm, the remnant dust may potentially constitute the mass reservoir out of which planetesimals and eventually planets form.

## 4. Disks around nearby young stars

Disk dispersal theory is qualitatively well understood; the main uncertainties pertain to the poorly quantified mass loss rates. If disk mass loss rates are high ($\sim 10^{-9}$ to $10^{-8}$ $M_{\odot}$ yr$^{-1}$), then photoevaporation has the potential to trigger the formation of planetesimals by lowering the gas/dust ratio and plays a pivotal role in planet formation. If mass loss rates are low ($\sim 10^{-11}$ to $10^{-10}$ $M_{\odot}$ yr$^{-1}$), then planet formation is not affected by photoevaporation; however, the dynamics and orbital architecture of planetary systems will still be influenced by the presence (or absence) of gas at these late stages.

A more accurate determination of mass loss rates is hindered by our ignorance of key inputs: the level of viscosity in the disk, stellar high energy spectra and their variation with accretion rate, disk chemistry and associated micro-physics, dust size evolution and dynamics and finally, the effect of planet formation on disk evolution.

Studies of nearby young stars provide the best opportunities to resolve some of these outstanding issues. Better determination of the stellar high energy spectra, including any possible contribution of accretion to EUV and X-ray fluxes and the variability of the high energy flux with accretion rate through high resolution, monitoring campaigns would help in better informing the theoretical models. Detailed multi-wavelength disk observations with high spatial resolution can characterize dust emission and probe particles from sub-micron to cm sizes. Gas line emission studies can help validate chemical models and shed light on heating/cooling processes. Spatial and velocity resolved observations can detect blue-shifts and asymmetries in line profiles due to photoevaporative winds and perhaps directly measure disk mass loss rates.

TW Hydra is an example of such a nearby, but perhaps atypical, system with abundant observational data and has been successfully used to develop detailed models of its gas disk structure and evolutionary status (e.g. Gorti *et al.* 2011). Similar detailed observations for other nearby disks from the X-ray to radio wavelengths will help us understand how disks evolve, form planets and eventually disperse.

## Acknowledgements

The author would like to acknowledge support from the National Science Foundation through award AST1313003 and from the NASA Astrobiology program (NNX15AG18A).

## References

Dent, W. R. F., Thi, W. F., Kamp, I., *et al.* 2013, *PASP*, 125, 477
Gorti, U. & Hollenbach, D. 2009, *ApJ*, 690, 1539
Gorti, U., Hollenbach, D., Najita, J., & Pascucci, I. 2011, *ApJ*, 735, 90
Gorti, U., Hollenbach, D., & Dullemond, C. P. 2015, *ApJ*, 804, 29
Haisch, K. E., Jr., Lada, E. A., & Lada, C. J. 2001, *ApJ*, 553, L153
Hollenbach, D. J., Yorke, H. W., & Johnstone, D. 2000, Protostars and Planets IV, 401
Johansen, A., Jacquet, E., Cuzzi, J. N., Morbidelli, A., & Gounelle, M. 2015, arXiv:1505.02941
Pascucci, I., Gorti, U., Hollenbach, D., *et al.* 2006, *ApJ*, 651, 1177
Pascucci, I., Ricci, L., Gorti, U., *et al.* 2014, *ApJ*, 795, 1
Zuckerman, B., Forveille, T., & Kastner, J. H. 1995, *Nature*, 373, 494

## Discussion

QUESTION: How would X-ray flares affect dispersal?

AUTHOR: Heating and cooling timescales are at least of the order 100 years, therefore variability on smaller timescales would get averaged out. However, if X-rays and UV were to vary on longer timescales, for example due to accretion, then this variability would impact dispersal times. For FUV, this is taken into account by including an accretion-generated field, but we do not have information to include an accretion generated component for X-rays.

QUESTION(VAN DER MAREL): What about transition disks?

AUTHOR: Transition disks are difficult to explain as photoevaporating disks in our models because the calculated disk lifetimes are at least a few Myrs, and the viscous timescales at gap opening are short, thus we would only expect to see a few % of disks transitioning because of photoevaporation. However, other groups find different results, and at least a few of the non-accreting transition disks may have been created by photoevaporation.

*Young Stars & Planets Near the Sun*
Proceedings IAU Symposium No. 314, 2015
J. H. Kastner, B. Stelzer, & S. A. Metchev, eds.

© International Astronomical Union 2016
doi:10.1017/S1743921315006158

# The Debris Disk Fraction for M-dwarfs in Nearby Young Moving Groups

**Alex Binks**

Keele University
email: a.s.binks@keele.ac.uk

**Abstract.** I present the first substantial work to measure the fraction of debris disks for M-dwarfs in nearby moving groups (MGs). Utilising the *AllWISE* IR catalog, 17 out of 151 MG members are found with an IR photometric excess indicative of disk structure. The M-dwarf debris disk fraction is $\lesssim 6$ per cent in MGs younger than $40\,\mathrm{Myr}$, and none are found in the groups older than $40\,\mathrm{Myr}$. Simulations show, however, that debris disks around M-dwarfs are not present above a $WISE\ W1-W4$ colour of $\sim 2.5$, making calculating the absolute disk fractions difficult. The debris disk dissipation timescale appears to be faster than for higher-mass stars, and mechanisms such as enhanced stellar wind drag and/or photoevaporation could account for the more rapid decline of disks observed amongst M-dwarfs.

**Keywords.** stars: late-type, stars: low-mass, stars: pre-main sequence, protoplanetary disks

## 1. Introduction

Observations of debris disks surrounding low-mass stars are rare. Whilst much work has focused on the disk frequencies for Solar-type stars, little is known of M-dwarf debris disk fractions, because most are too faint to be studied in open clusters, young field M-dwarfs are very rare and until recently few were confirmed as members of nearby, young moving groups (MGs). The $WISE$ infrared (IR) photometric catalogue has identified hundreds of IR excesses in nearby Solar-type objects (Patel, Metchev & Heinze 2014), however, very few are found in young, nearby M-dwarfs (Avenhaus, Schmid & Meyer 2012). Debris disk fractions for young M-dwarfs have been reported to be both smaller than FGK-types (Lestrade *et al.* 2009) and larger (Forbrich *et al.* 2008), however, these are limited by small number statistics.

The frequency of observed debris disks appears to be dependent on both age and spectral-type. Generally speaking, debris disks are more common around higher-mass, young stars (Wyatt 2008). The top-left panel in Fig. 1 shows that from 10 to $100\,\mathrm{Myr}$ there is roughly an overall decline from $\sim 40$ to 10 per cent in FGK stars. Although some work has investigated the disk fractions in MGs (Simon *et al.* 2012; Schneider *et al.* 2012), few have been able to provide large M-dwarf samples. Motivated by the lack of a robust M-dwarf disk fraction and the recent identification of hundreds of new M-dwarf MG members (e.g., Malo *et al.* 2013; Gagné *et al.* 2014), I use WISE photometry to attempt to measure the fraction of M-dwarfs with debris disks in MGs.

## 2. Identifying and characterising disk excess

All 151 candidates chosen for analysis have signal-to-noise ratios $> 5.0$ in the $WISE$ $W4$ band and have radial velocities to within $5\,\mathrm{km\,s}^{-1}$ of their proposed MG. The scatter in $W1-W4$ colour amongst M-dwarf field objects is $\sim 0.3$ therefore a cut of $W1-W4 >$ 1.0 is chosen to select objects indicative of IR excess (c.f., Schneider *et al.* 2012), and

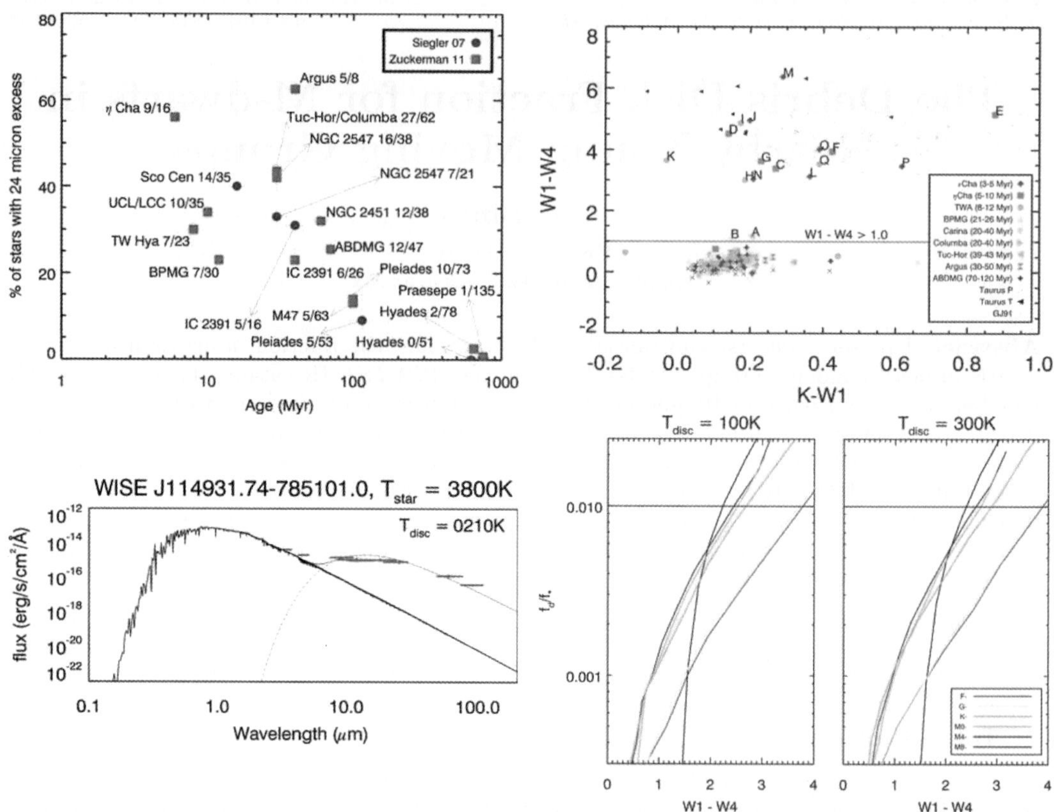

**Figure 1.** Top left: the fraction of debris disks as a function of age in groups $\lesssim 650\,\mathrm{Myr}$. Top right: $K - W1$ vs $W1 - W4$ colour-colour diagram for all M-dwarfs in this sample. Bottom left: an example SED fit. Bottom right: *WISE* sensitivity limits for debris disks.

the top right panel of Fig. 1 shows the 17 objects that qualify. Spectral class indices $(\alpha = d\log \lambda F_\lambda / d\log \lambda)$ are calculated following the procedure in Riaz & Gizis (2012) and all objects with $W1 - W4 > 1.0$ are either class II or III (see Table 1).

Spectral energy diagram (SED) fits are made to stars with IR excess and the *IRAS, IRAC, MIPS* and *AKARI* catalogues were searched for additional data at longer wavelengths. Photospheric fits are generated from either the BT-Settl models ($T_{\mathrm{eff}} > 2700\,\mathrm{K}$) or the AMES-DUSTY models (for $T_{\mathrm{eff}} \leqslant 2700\,\mathrm{K}$) and single temperature black-body fits are made to represent a disk. The best-fit minimized $\chi^2$, however for 9 objects there are insufficient data at wavelengths $> 25\,\mu\mathrm{m}$ and the fit is degenerate. Disk flux fractions ($f_{\mathrm{d}}/f_{\mathrm{bol}}$) were calculated by dividing the disk flux by the stellar flux. Objects with $f_{\mathrm{d}}/f_{\mathrm{bol}} < 0.01$ are considered as debris disks (c.f., Lagrange *et al.* 2000), objects with $0.01 \leqslant f_{\mathrm{d}}/f_{\mathrm{bol}} < 0.1$ are classed as either primordial or transitional and $f_{\mathrm{d}}/f_{\mathrm{bol}} > 0.1$ is primordial. Column 6 in Table 1 provides the disk-type measured from SED fits. The bottom-left panel of Fig. 1 shows an SED fit for an object with available IR data up to $100\,\mu\mathrm{m}$.

Each object was searched for any additional material in the literature that may point to the type of disk excess. Classification based on literary sources only assumes primacy if data has been used which wasn't available for this work. Any parameters derived in this work are compared to results in the literature if they are available. Ultimately, the final designation of a disk type is subjective in some cases, particularly if there is a lack

**Table 1.** Final designation of disk-type based on photometric excess and SED modelling.

| Name | Group | $W1-W4$ | Excess | $\alpha$ | $f_\mathrm{d}/f_\mathrm{bol}$ | Literature | Final |
|---|---|---|---|---|---|---|---|
| J010335.75−551556.6 | Tuc-Hor | 1.17 | TTT | II | N | N/A | D |
| J044356.87+372302.7 | BPMG | 1.10 | TTT | II | N/A | N/A | D |
| J084130.24−785306.3 | $\eta$ Cha | 3.39 | PPP | II | 0.04, P/T | P | P |
| J084227.02−785747.7 | $\eta$ Cha | 4.53 | TTT | III | 0.05, P/T | P | P |
| J084318.52−790518.0 | $\eta$ Cha | 5.15 | PPP | II | 0.54, P | P | P |
| J084409.09−783345.6 | $\eta$ Cha | 3.94 | PPP | II | 0.10, P/T | P | P |
| J084416.33−785907.8 | $\eta$ Cha | 3.61 | TPP | II | 0.03, P/T | P | P |
| J101209.04−312445.3 | TWA | 3.01 | TPP | III | 0.02, P/T | N/A | T |
| J111027.80−373152.0 | TWA | 4.87 | TTT | II | 0.09, P/T | T | T |
| J111835.64−793554.8 | $\epsilon$ Cha | 4.95 | TTT | III | N/A* | T | T |
| J113218.24−301952.0 | TWA | 3.64 | TPT | III | N/A* | D | T |
| J114326.57−780445.5 | $\epsilon$ Cha | 3.12 | PPP | III | N/A* | T | T |
| J114931.74−785101.0 | $\epsilon$ Cha | 6.38 | PPP | II | 0.48, P | P/D | P |
| J115504.71−791911.0 | $\epsilon$ Cha | 3.03 | TTP | III | N/A* | T | T |
| J120055.08−782029.5 | $\epsilon$ Cha | 4.01 | PPP | II | N/A* | P | P |
| J120144.32−781926.7 | $\epsilon$ Cha | 3.46 | PPP | II | N/A* | P | P |
| J120733.42−393254.2 | TWA | 3.53 | PPP | II | N/A* | T | T |

*Notes:* D = debris disk, P = primordial disk, T = transitional disk, N = no evidence for a disk, N/A = no information available (*although there are no data available for SED fits beyond 25 $\mu$m the $W1-W4$ colour is beyond the limit for a debris disk and these can only be classed as either 'P' or 'T'.

of mid-/far-IR data to constrain SED fits, however, if a sufficient amount of IR data is available then $f_\mathrm{d}/f_\mathrm{bol}$ is used as the primary indicator. In addition 7 objects are found with $W1-W4$ *less than* 1.0 but having literary sources claiming they are debris disks. SED fitting was appropriate for two of these objects, and both have $f_\mathrm{d}/f_\mathrm{bol} < 0.01$.

## 3. Sensitivity limits using *WISE*

Given that the objects in this work are relatively faint M-dwarfs, there will be a limiting $W1-W4$ for which objects can be observed with $f_\mathrm{d}/f_\mathrm{bol} \leqslant 0.01$. To explore the relationship between $f_\mathrm{d}/f_\mathrm{bol}$ and $W1-W4$, a simulation was made in which the emitting area of the disk was altered to give $f_\mathrm{d}/f_\mathrm{bol} = 0.01$, and this gave a corresponding $W1-W4$.

The results, displayed in the bottom-right panel of Fig. 1, show that the limiting $W1-W4$ colour corresponding to disk flux fractions $< 0.01$ is $\approx 2.5$ for early/mid M-dwarfs, $\approx 2.2$ for late M-dwarfs, and $\approx 3.5$ for Solar-type stars. However, if $W1-W4$ is too small it becomes indistinguishable from field stars. Therefore the simulations provide a 'window' of $1.0 < W1-W4 < 2.5$ that can indicate debris disks between the error threshold at the bottom, to where $f_\mathrm{d}/f_\mathrm{bol} = 0.01$ at the top. In Table 1, objects labelled with an asterisk in column 6 cannot be considered debris disks.

## 4. Conclusions

Only 2 of the objects with $W1-W4 > 1.0$ have $W1-W4$ values small enough to have $f_\mathrm{d}/f_\mathrm{bol} < 0.01$. Seven objects with $W1-W4 < 1.0$ were classed as debris disks based on literary sources and $f_\mathrm{d}/f_\mathrm{bol}$ (where possible). Only four MGs were found to have any debris disks; $\eta$ Cha, TWA, BPMG and Tuc-Hor. The total fraction of MG members with debris disks identified in this work is 9/151 (6 per cent), however, it can only be used as a lower limit as it is quite possible for debris disks to avoid detection with a very small $W1-W4$ excess. If the objects identified as transitional disks actually turned out to be debris disks, then these numbers change to 16/151 (11 per cent). Only five objects have a measurable disk flux fraction which is $< 0.01$. If only these objects were used, then a lower limit of $5/151$ (3 per cent) would be set.

Within these narrow detection limits several stars were identified with a modest excess; too small to be a primordial disk, and in some cases there is longer wavelength data that supports the classification as a debris disk. Without knowing how many undetected debris disks have been unaccounted for it is difficult to provide a robust M-dwarf debris disk fraction. The fraction of debris disks is small, smaller than claimed for A–K stars at similar ages, but it is hard to compare the two because of a lack of comparable sensitivity to $f_d/f_{bol}$ for stars for different masses. Higher-mass stars contribute less relative flux at IR wavelengths and are more easily detectable from a field star population of similar spectral-type.

For objects younger than 40 Myr, debris disks were detected in 9 out of 113 (8 per cent) and no debris disks were detected in the 38 objects older than 40 Myr. This is a significant difference and the confidence to which one can claim a detection rate of $< 6$ per cent in M-dwarfs older than 40 Myr is 90 per cent. Despite the sensitivity problems it is clear that the M-dwarf disks are evolving and that the evolutionary timescale for debris disk dissipation appears to be faster than for higher mass stars.

## References


Forbrich, J., Lada, C. J., Muench, A. A., & Teixeira, P. S. 2008, *ApJ*, 687, 1107

Gagné, J., Lafrenière, D., Doyon, R., Malo, L., & Artigau, É. 2014, *ApJ*, 783, 121

Lagrange, A.-M., Backman, D. E., & Artymowicz, P. 2000, *Protostars and Planets IV*, 0, 639

Lestrade, J. F., Wyatt, M. C., Bertoldi, F., Menten, K. M., & Labaigt, G. 2014, *A&A*, 506, 1455

Malo, L., Doyon, R., Lafrenière, D., Artigau, É., Gagné, J., Baron, F., & Riedel, A. 2013, *ApJ*, 762, 88

Patel, R. I., Metchev, S. A., & Heinze, A. 2014, *ApJS*, 212, 10

Riaz, B. & Gizis, J. E. 2012, *A&A*, 548, A54

Schneider, A., Melis, C., & Song, I. 2012, *ApJ*, 754, 39

Simon, M., Schlieder, J. E., Constantin, A.-M., & Silverstein, M. 2012, *ApJ*, 751, 114

Wyatt, M. C., 2008 *ARAA*, 46, 339


## Discussion

LADA: I'd like to point out that we used *Spitzer* to observe some M-dwarf debris disks in NGC 2547 and actually the disk fractions turned out to be higher in M-dwarfs than for the Solar-type stars!

AUTHOR: Yes, I would say that these results are not claiming debris disks don't exist but that they are not *observed*. Perhaps a deeper infrared survey of these objects would yield more disks.

ZUCKERMAN: Perhaps *WISE* wasn't the ideal survey for you.

AUTHOR: Maybe not, and like the simulations show, *WISE* could only pick out disks in a very narrow colour band. But for now this is all we have to work on.

*Young Stars & Planets Near the Sun*
Proceedings IAU Symposium No. 314, 2015
J. H. Kastner, B. Stelzer, & S. A. Metchev, eds.

© International Astronomical Union 2016
doi:10.1017/S1743921315006201

# Two-temperature Debris Disks: Signposts for Directly Imaged Planets?

## Grant M. Kennedy and Mark C. Wyatt

Institute of Astronomy, University of Cambridge, Madingley Road, Cambridge CB3 0HA, UK
email: gkennedy@ast.cam.ac.uk

**Abstract.** This work considers debris disks whose spectra can be modelled by dust emission at two different temperatures. These disks are typically assumed to be a sign of multiple belts, but only a few cases have been confirmed via high resolution observations. We derive the properties of a sample of two-temperature disks, and explore whether this emission can arise from dust in a single narrow belt. While some two-temperature disks arise from single belts, it is probable that most have multiple spatial components. These disks are plausibly similar to the outer Solar System's configuration of Asteroid and Edgeworth-Kuiper belts separated by giant planets. Alternatively, the inner component could arise from inward scattering of material from the outer belt, again due to intervening planets. For either scenario, the ratio of warm/cool component temperatures is indicative of the scale of outer planetary systems, which typically span a factor of about ten in radius.

**Keywords.** circumstellar matter, planet-disk interactions

---

## 1. Introduction

Debris disks are a sign of successful planetesimal formation. The radial structure of most planetesimal belts is unknown; they may lie in multiple rings analogous to the Asteroid and Edgeworth-Kuiper belts, but may also be significantly extended in a way similar to gaseous protoplanetary disks (e.g. Kalas *et al.* 2005; Su *et al.* 2009). Because they are generally detected by excess emission above the photospheric level at infra-red (IR) wavelengths, and with high resolution imaging detections being relatively rare, discerning radial structure is in general difficult. The major difficulty is that the equilibrium temperature of a dust grain depends on both distance from the star and the size and optical properties of that dust grain. Thus, the radius of an unresolved debris disk cannot be unambiguously determined from the temperature of the observed emission, as the temperature is degenerate with the sizes of grains in the disk.

This contribution summarises some aspects of our study (Kennedy & Wyatt 2014) of debris disks whose spectral energy distributions (SEDs) require dust emission at two temperatures, and are perhaps indicative of dust that resides at a range of stellocentric distances. These systems have been of particular interest due to the popular interpretation that the two temperatures arise from two distinct dust belts (e.g. Chen *et al.* 2009; Morales *et al.* 2011). One question is then whether the intervening region between the two belts contains planets, and if so, whether dynamical clearing by these planets is the reason for two-belt structure. Circumstantial evidence for such a picture is given by systems with planets that reside between two dust components, such as HR 8799 and HD 95086 (Rameau *et al.* 2013; Moór *et al.* 2013).

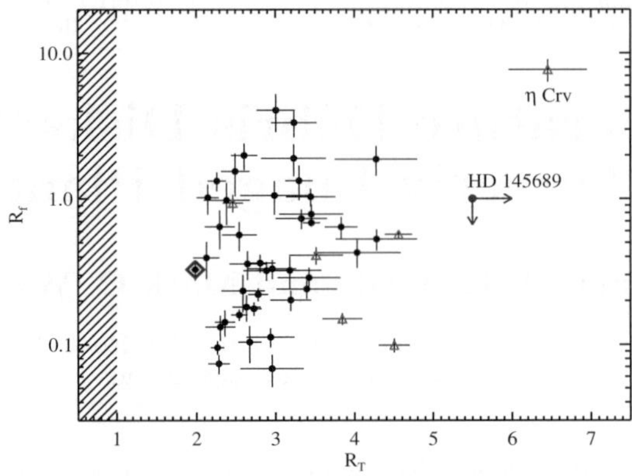


**Figure 1.** Two-temperature disk sample. Blue triangles note disks known two have multiple disk components from imaging and/or interferometry and the green diamond encloses HD 181327.

## 2. Two-temperature disk properties

One way to present two-temperature disks is the ratio of warm/cool component temperatures $R_T$ and warm/cool component disk/star (fractional) luminosities $R_f$, as shown in Fig. 1. Most two-temperature disks have fairly similar temperature ratios of 2-4, but with a range of fractional luminosity ratios. Warm components are detected with $R_f$ down to about 0.1, and those fainter than this level become difficult to detect. Biases in the sample mean that the frequency of the two-temperature phenomenon is hard to estimate, but it appears fairly common, at the tens of percent level. A clear outlier is $\eta$ Crv, with the largest temperature ratio, and one of several systems known from detailed observations to have two physically distinct dust belts.

The young F6 star HD 181327 (the dot enclosed by a green diamond in Fig. 1) is known to be concentrated in a narrow single belt. Lebreton *et al.* (2012) show that while the spectrum cannot be modelled with a single blackbody, allowing the composition of a single belt to vary can result in a good fit to the disk spectrum. Therefore, not all two-temperature disks necessarily originate from two distinct components.

## 3. Two temperatures = two belts?

Fig. 2 shows the temperature and fractional luminosity ratios against stellar temperature for our sample. In addition, the results from fitting two temperatures to model emission spectra calculated for single narrow dust belts are shown. These model grids were calculated for two different compositions and a range of dust size distributions, and the space covered by the red and blue lines shows where single narrow belts may masquerade as two belt systems due to significant dust emission at a range of temperatures.

Assuming that the minimum grain size is set by radiation pressure, two-temperature disks around A-type stars probably arise from multiple belts. A few two-temperature disks have been confirmed to have multiple belts by high resolution observations, and these comprise both A-type and Sun-like stars. For Sun-like stars, single belt models, particularly those with relatively flat size distributions, can produce two temperature disks, and this model is not conclusively ruled out because not all disks are resolved and/or detected at far-IR/mm wavelengths. Where imaging exists however, this model is disfavoured because the resolved disk size disagrees with that predicted. In addition, the flatter size distributions are steeper than those expected from collisional models and inferred from detailed modelling of well characterised systems. Therefore, in general, the assumption that two temperature disks have multiple belts should not be made without

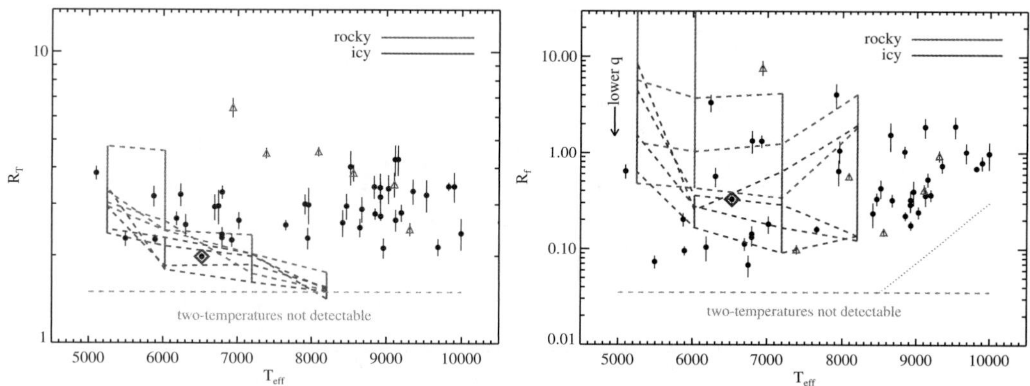

**Figure 2.** Two-temperature disk sample, with lines showing parameter space covered by single narrow belt models with "rocky" and "icy" compositions. Each vertex represents a model that resulted in a two-temperature disk.

considering the properties of those disks and their host stars, but it is likely that the bulk of two-temperature disks arise from multiple belts.

## 4. Evolution of multiple belts

A possible constraint on the origin of two-temperature disks could come from the expected collisional evolution. For example, if two-temperature disks arise from a single belt and the material compositions do not change, no significant evolution of $\mathcal{R}_f$ or $\mathcal{R}_T$ would be expected over time because the observed emission always comes from material in the same location. However, this similarity may also be expected if the warm belts are made of material delivered from the outer belt, perhaps scattered by planets (e.g. Bonsor & Wyatt 2012), in which case the brightness of the inner belt is reasonably connected to that of the outer one. On the other hand, if two temperatures arise from two independent belts (i.e. as in the Solar system), the two belts are expected to collisionally evolve at different rates. The collision rate in the disk depends strongly on orbital radius, and the brightness of a belt will start to decay when the largest objects start to collide, which will take longer for the outer belt. Therefore, for two belts that have the same initial brightness, the inner one will start to decay first, and the outer belt will follow later, and all other things being equal, in the long term the brightness difference between two belts is set by the difference in their radii.

Fig. 3 shows the evolution of $\mathcal{R}_f$ with time for our sample. No significant evolution is seen, which is suggestive of a comet delivery scenario. However, this lack of evolution does not strongly rule out the hypothesis that the two temperatures correspond to two distinct belts that are decaying due to collisions. The reason being that the warm belts are at sufficiently large radial distances that their brightness is not at odds with models of collisional evolution.

## 5. Planetary system structure

Our main conclusion is that most two-temperature debris disks comprise two disk components. Consideration of collisional models shows that these components could be two independent belts undergoing normal collisional evolution, analogous to the Solar System's Asteroid and Edgeworth-Kuiper belts. In other systems the inner components may also be linked to the outer belts via inward scattering of material by intervening

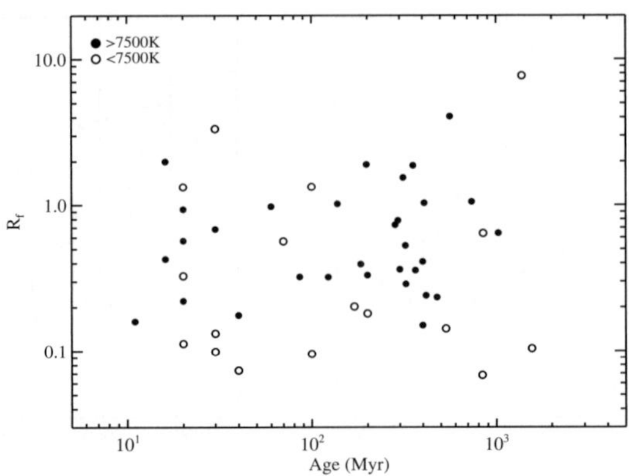


**Figure 3.** Dependence of $\mathcal{R}_f$ on stellar age. No trends are visible, but the $\sim$10 au radii of the warm components and their relatively slow collisional evolution means that the collisional two-belt scenario is not ruled out.

planets (e.g. Bonsor & Wyatt 2012). For either scenario, the ratio of warm/cool component temperatures is indicative of the scale of outer planetary systems, which typically span a factor of about ten in radius.

## Acknowledgements

This work was supported by the European Union through ERC grant number 279973.

## References


Bonsor, A. & Wyatt, M. C. 2012, *MNRAS*, 420, 2990

Chen, C. H., Sheehan, P., Watson, D. M., Manoj, P., & Najita, J. R. 2009, *ApJ*, 701, 1367

Kalas, P., Graham, J. R., & Clampin, M. 2005, *Nature*, 435, 1067

Kennedy, G. M. & Wyatt, M. C. 2014, *MNRAS*, 444, 3164

Lebreton, J., Augereau, J.-C., Thi, W.-F., Roberge, A., Donaldson, J., Schneider, G., Maddison, S. T., Ménard, F., Riviere-Marichalar, P., Mathews, G. S., Kamp, I., Pinte, C., Dent, W. R. F., Barrado, D., Duchêne, G., Gonzalez, J.-F., Grady, C. A., Meeus, G., Pantin, E., Williams, J. P., & Woitke, P. 2012, *A & A*, 539, A17

Moór, A., Ábrahám, P., Kóspál, Á., Szabó, G. M., Apai, D., Balog, Z., Csengeri, T., Grady, C., Henning, T., Juhász, A., Kiss, C., Pascucci, I., Szulágyi, J., & Vavrek, R. 2013, *ApJ*, 775, L51

Morales, F. Y., Rieke, G. H., Werner, M. W., Bryden, G., Stapelfeldt, K. R., & Su, K. Y. L. 2011, *ApJ*, 730, L29

Rameau, J., Chauvin, G., Lagrange, A.-M., Meshkat, T., Boccaletti, A., Quanz, S. P., Currie, T., Mawet, D., Girard, J. H., Bonnefoy, M., & Kenworthy, M. 2013, *ApJ*, 779, L26

Su, K. Y. L., Rieke, G. H., Stapelfeldt, K. R., Malhotra, R., Bryden, G., Smith, P. S., Misselt, K. A., Moro-Martin, A., & Williams, J. P. 2009, *ApJ*, 705, 314


## Discussion

SONG: How well can you constrain the width of the belts, for example could these systems actually just be single belts that are very wide?

KENNEDY: It's very hard to tell just from the SED. However, in all cases where we've been able to obtain high-resolution images the inner belt is distinct from the outer one.

*Young Stars & Planets Near the Sun*
*Proceedings IAU Symposium No. 314, 2015*
*J. H. Kastner, B. Stelzer, & S. A. Metchev, eds.*

© International Astronomical Union 2016
doi:10.1017/S1743921315005931

# First Results from the Disk Eclipse Search with KELT (DESK) Survey

## Joseph E. Rodriguez[1], Joshua Pepper[2,1] and Keivan G. Stassun[1,3]

[1]Department of Physics and Astronomy, Vanderbilt University, 6301 Stevenson Center,
Nashville, TN 37235, USA
email: `rodriguez.jr.joey@gmail.com`
[2]Department of Physics, Lehigh University, 16 Memorial Drive East, Bethlehem, PA 18015,
USA
email: `joshua.pepper@Lehigh.EDU`
[3]Department of Physics, Fisk University, 1000 17th Avenue North, Nashville, TN 37208, USA
email: `keivan.stassun@vanderbilt.edu`

**Abstract.** Using time-series photometry from the Kilodegree Extremely Little Telescope (KELT)
exoplanet survey, we are looking for eclipses of stars by their protoplanetary disks, specifically in
young stellar associations. To date, we have discovered two previously unknown, large dimming
events around the young stars RW Aurigae and V409 Tau. We attribute the dimming of RW Aurigae to an occultation by its tidally disrupted disk, with the disruption perhaps resulting from
a recent flyby of its binary companion. Even with the dynamical environment of RW Aurigae,
the distorted disk material remains very compact and presumably capable of forming planets.
This system also shows that strong binary interactions with disks can also influence planet and
core composition by stirring up and mixing materials during planet formation. We interpret the
dimming of V409 Tau to be due to a feature, possibly a warp or perturbation, lying at least 10
AU from the host star in its nearly edge-on circumstellar disk.

**Keywords.** Circumstellar Matter, Individual Stars: RW Aur, Individual Stars: AA Tau, Individual Stars: V409 Tau, Pre-main Sequence, Stars: Variables: T Tauri

---

## 1. Introduction

Young Stellar Objects (YSOs) are typically surrounded by protoplanetary circumstellar disks where the gas and dust eventually evolve over time through a combination of
mechanisms including accretion onto the star, dispersion by stellar winds and radiation,
and coalescence into planets. Other features of the stellar system, such as stellar companions, magnetic fields, or radiative jets, can influence the size, mass, and composition of
these disks. The evolution of protoplanetary disks and the transition of dust and gas into
planets is not well understood and directly affects our understanding of the processes
that created the thousands of planetary systems thus far discovered. Characterization of
structure in protoplanetary disks could help explain the differences between planetary
systems.

One means to better constrain the size, mass and composition of these disks arises by
observing a star being eclipsed by its circumstellar disk. Thus far, only a few of these
events have been detected and analyzed in the literature. One well-known example of this
phenomenon is $\epsilon$ Aurigae, a bright system that periodically dims every 26.1 years by $\sim$1
magnitude for a duration of almost two years (Carroll *et al.* 1991). The eclipse has been
attributed to an eclipsing binary where the companion star has a small circumstellar
disk (Kloppenborg *et al.* 2010). Another example of a large dimming event caused by
a disk eclipse occurred in 2007 around the pre-main sequence star 2MASS J14074792-

3945427. This star dimmed by $\sim$4 magnitude for $>$50 days, with the dimming believed to be caused by an occultation by a circumplanetary disk that is similar to Saturn's ring system but much larger (Mamajek *et al.* 2012). The advent of wide-field time domain surveys provides an excellent tool to search for rare eclipse events, depending on the coverage, cadence, and baseline of the survey. Using time-series photometric data from the Kilodegree Extremely Little Telescope (KELT) exoplanet survey, which has coverage of a large portion of the sky, we are searching for disk eclipsing events, specifically in young stellar associations. Our survey has already yielded two discoveries of previously unknown large dimming events toward the T Tauri stars RW Aur and V409 Tau.

## 2. KELT

The Kilodegree Extremely Little Telescope (KELT) project is a survey of bright stars ($V = 8$-$10$), with the goal to detect transiting exoplanets. The survey uses two telescopes, KELT-South (Sutherland, South Africa) and KELT-North (Sonita, Arizona), each with a $26° \times 26°$ field of view that observes with a $\sim$15 minute cadence in a broad $R$-band filter (Pepper *et al.* 2007, 2012). The KELT data are reduced using a heavily modified version of the ISIS software package, described in §2 of Siverd *et al.* 2012†. RW Aur is located in KELT-North Field 04, which is centered on $\alpha = $ 5hr 54m 14.466s, $\delta = +31°$ 44' 37". KELT-North observed this field from October 10, 2006 to September 23, 2012, obtaining 8,001 images. V409 Tau and AA Tau are both located in KELT-North Field 03, which is centered on $\alpha = $ 3h 58m 12s, $\delta = 59°$ 32' 24". KELT-North observed this field for 7 seasons from UT 2006 October 26 to UT 2013 January 9, obtaining $\sim$9100 images.

## 3. Occultation of the T Tauri Star RW Aurigae A by Its Tidally Disrupted Disk

Results for RW Aur previously appeared in Rodriguez *et al.* 2013 and are briefly summarized here. After being photometrically monitored for over 100 years (Beck & Simon 2001), with no large coherent dimming event observed during this period, the RW Aurigae system dimmed by $\sim$2 magnitudes for $\sim$180 days (Figure 1). The dimming was observed by both the KELT-North telescope and the American Association of Variable Star Observers (AAVSO). Previous millimeter observations by Cabrit *et al.* 2006 suggest that the disk around the primary component, RW Aur A, had been tidally disrupted by a recent fly-by of RW Aur B. The material that was perturbed in this interaction appears to have coalesced into a large tidal arm wrapped around RW Aur A. We attributed the large dimming observed in 2010 to an occultation of RW Aur A by a portion of the tidally disrupted material. This hypothesis has been supported by hydrodynamical modeling of the RW Aurigae fly-by (Dai *et al.* 2015). This system illustrates how binary interactions can be crucial in sculpting circumstellar disks. Because of the occultation, we furthermore were able to show that the distorted tidal arm has retained a remarkably coherent structure.

## 4. V409 Tau as Another AA Tau: Photometric Observations of Stellar Occultations by the Circumstellar Disk

Results for V409 Tau previously appeared in Rodriguez *et al.* 2015 and are briefly summarized here. Using the KELT-North data, we detected two large dimming events of the relatively unknown young star, V409 Tau (Fig. 2). This system dimmed by 1.4 mag in early 2009 and then again in early 2012. We estimate the total duration of each

† Much of the reduction software is publicly available: http://verdis.phy.vanderbilt.edu

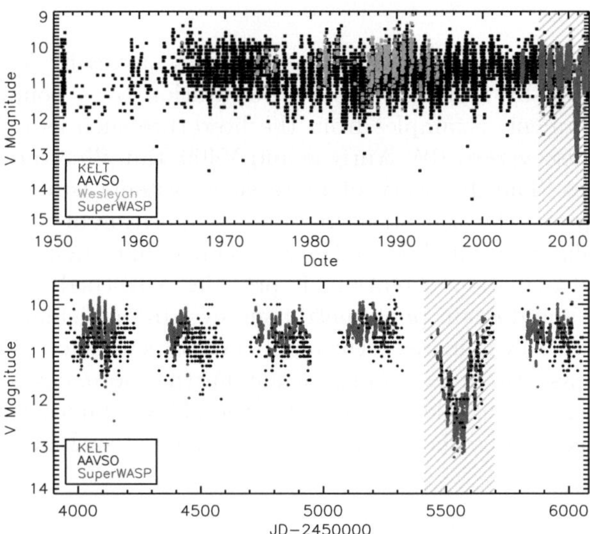

**Figure 1.** (Top) AAVSO (Black), KELT-North (Blue), SuperWASP (Red) and the Wesleyan Van Vleck (Green) light curves of RW Aur from 1950 to 2012. The KELT and SuperWASP light curves do not resolve the A and B components. The shaded region in the upper plot corresponds to the six KELT seasons, which are shown in the bottom plot. (Figure reproduced from Rodriguez *et al.* 2013)

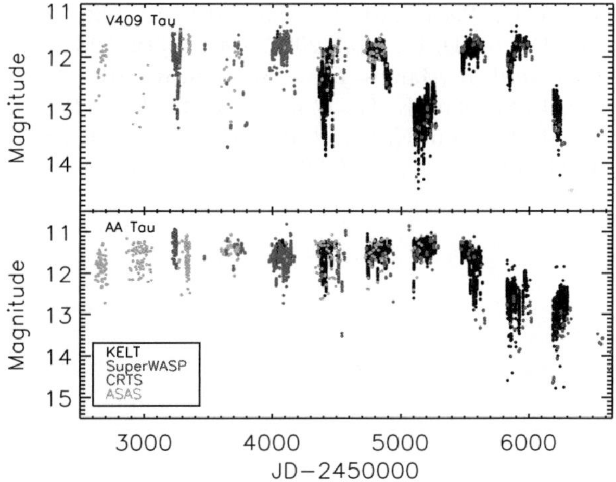

**Figure 2.** KELT-North (Black), SuperWASP (Blue), CRTS (Red) and the ASAS (Green) light curves of V409 Tau (Top) and AA Tau (Bottom) from 2004 to 2013. (Figure reproduced from Rodriguez *et al.* 2015)

event to be greater than 600 days. We also observed the well-studied dimming of AA Tau in 2011 (Bouvier *et al.* 2013). AA Tau dimmed by ∼1.5 mag in 2011 and has not recovered. Bouvier *et al.* 2013 argued that the large dimming is caused by a feature in the close to edge-on circumstellar disk occulting the host star. Our spectral energy distribution (SED) analysis indicates that V409 Tau, like AA Tau, has a close to edge-on circumstellar disk. Using our observations of AA Tau as a direct comparison, we argue that these dimming events of V409 Tau are also the result of one or more features in the edge-on circumstellar disk, at a semi-major axis of >10 AU, occulting the host star (Rodriguez *et al.* 2015).

## 5. Conclusions

Using the KELT survey, we are performing an all sky survey for YSOs presenting large dimming events, typically attributed to an occultation by a component within their circumstellar environment. Examples from the literature such as $\epsilon$ Aurigae and J1407, as well as our own discoveries, RW Aurigae and V409 Tau, illustrate the scientific value of large disk eclipses. The discovery of more such systems will permit a much wider range of studies to probe the size, structure, and composition of circumstellar disks. Our work is not only enhancing our knowledge of circumstellar environments and the early stages of stellar and planetary evolution, it provides a testbed and framework for the next generation of time domain photometric surveys. In particular, the Large Synoptic Survey Telescope (LSST) will increase the number of stars with long-baseline photometric observations by at least two orders of magnitude. Furthermore, newly discovered systems with disk eclipses will be excellent targets for the NASA James Web Space Telescope observatory, which can probe the detailed environments of such systems.

## References

Beck, T. L. & Simon, M. 2001, *AJ*, 122, 413
Bouvier, J., Grankin, K., Ellerbroek, L. E., Bouy, H., & Barrado, D. 2013, *A&A*, 557, A77
Cabrit, S., Pety, J., Pesenti, N., & Dougados, C. 2006, *A&A*, 452, 897
Carroll, S. M., Guinan, E. F., McCook, G. P., & Donahue, R. A. 1991, *ApJ*, 367, 278
Dai, F., Facchini, S., Clarke, C. J., & Haworth, T. J. 2015, *MNRAS*, 449, 1996
Kloppenborg, B., Stencel, R., Monnier, J. D., *et al.* 2010, *Nature*, 464, 870
Mamajek, E. E., Quillen, A. C., Pecaut, M. J., *et al.* 2012, *AJ*, 143, 72
Pepper, J., Pogge, R. W., DePoy, D. L., *et al.* 2007, *PASP*, 119, 923
Pepper, J., Kuhn, R. B., Siverd, R., James, D., & Stassun, K. 2012, *PASP*, 124, 230
Rodriguez, J. E., Pepper, J., Stassun, K. G., *et al.* 2013, *AJ*, 146, 112
Rodriguez, J. E., Pepper, J., Stassun, K. G., *et al.* 2015, arXiv:1505.05805
Siverd, R. J., Beatty, T. G., Pepper, J., *et al.* 2012, *ApJ*, 761, 123

*Young Stars & Planets Near the Sun*
*Proceedings IAU Symposium No. 314, 2015*
*J. H. Kastner, B. Stelzer, & S. A. Metchev, eds.*

© International Astronomical Union 2016
doi:10.1017/S1743921315006195

# Modeling of a Giant Exoring System Around the Substellar Companion J1407b

## Matthew A. Kenworthy[1] and Eric E. Mamajek[2]

[1]Leiden Observatory, Leiden University, Niels Bohrweg 2, 2333 RA, Leiden, The Netherlands
email: `kenworthy@strw.leidenuniv.nl`
[2]University of Rochester
email: `emamajek@pas.rochester.edu`

**Abstract.** We model the complex light curve of 1SWASP J140747.93-394542.6, a $\sim$16 Myr old star in the Sco-Cen OB association, with a giant ring system that fills a significant fraction of the Hill sphere of an unseen secondary companion, J1407b. The best ring model has 37 rings and extends out to a radius of 0.6 AU (90 million km), and the rings have an estimated total mass on the order of $1 M_\oplus$. The ring system has one clearly defined gap at 0.4 AU (61 million km), which we hypothesize is being cleared out by a $< 0.8 M_\oplus$ exosatellite orbiting around the secondary companion.

**Keywords.** techniques: photometric, planetary systems: formation, planets: rings

## 1. Introduction

Extended month- to year-long eclipses indicate the presence of long lived dark disks around secondary companions, including $\epsilon$ Aurigae (Guinan & Dewarf 2002; Kloppenborg *et al.* 2010), EE Cep (Mikolajewski & Graczyk 1999; Graczyk *et al.* 2003; Mikolajewski *et al.* 2005), a precessing circumbinary disk around KH 15D (Hamilton *et al.* 2005; Winn *et al.* 2006) and three systems recently discovered in the OGLE database – OGLE-LMC-ECL-17782 (Graczyk *et al.* 2011), OGLE-LMC-ECL-11893 (Dong *et al.* 2014) and OGLE-BLG182.1.162852 (Rattenbury *et al.* 2014).

1SWASP J140747.93-394542.6 (hereafter J1407) is a pre-main sequence $\sim$ 16 Myr old star, $0.9 M_\odot$, $V = 12.3$ mag K5 star at 133 pc associated with the Sco-Cen OB Association (Mamajek *et al.* 2012; van Werkhoven *et al.* 2014; Kenworthy *et al.* 2015). The Super Wide Angle Search for Planets (SuperWASP) database (Butters *et al.* 2010) showed that the star underwent a complex series of eclipses lasting $\sim$ 56 days around May 2007 and including a dimming of $> 95\%$. Mamajek *et al.* (2012) and van Werkhoven *et al.* (2014) propose that these eclipses are caused by a large ring system orbiting an unseen substellar companion, dubbed J1407b.

Fitting is performed in a two step process - we first constrain the orientation of the ring system using the gradients measured from the light curve, and then use these parameters to generate a model of the ring transmission as a function of radius from the secondary companion.

The light curve $I(t)$ of a source with finite angular size (i.e. the stellar disk of the primary) behind a tilted ring system is not time symmetric (see Figure 1). For each ring boundary, the gradient of the light curve is dependent on both the size of the star, and angle between the local tangent of the ring edge and the direction of motion. Ring structures smaller than that of the stellar diameter are smeared out by the resultant convolution with the stellar disk.

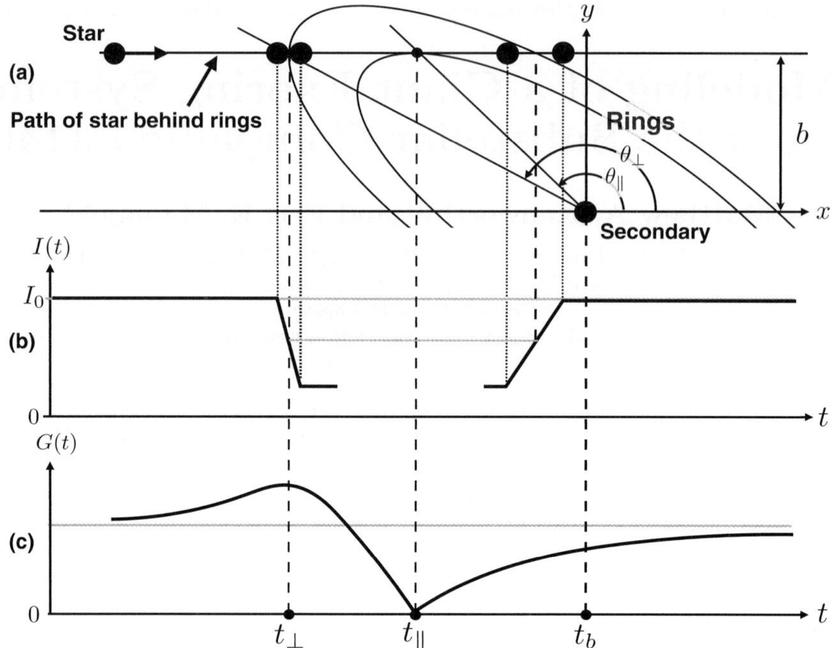


**Figure 1.** The geometry of the ring model. Panel (a) shows a ring system inclined at an angle of $i_{disk}$ and rotated from the line of relative velocity by $\phi_{disk}$. The star passes behind the ring system with impact parameter $b$ at time $t_b$. Panel (b) shows the resultant light curve $I(t)$ of the star as a function of time, demonstrating how the local ring tangent convolved with the finite sized disk of the star produces light curves with different local slopes. Panel (c) highlights the three significant epochs in the rate of change of ring radius $r$; $t_b$ at closest projected separation of the star and the secondary, $t_\perp$ where the ring tangent is perpendicular to the direction of stellar motion, and $t_\parallel$ where stellar motion is tangent to the ring. $t_\parallel$ also marks where the stellar path touches the smallest ring radius.

For a given set of ring orientation parameters $i_{disk}, \phi_{disk}, b, t_b$, we can determine the radial distances of the rings from the secondary companion (i.e. the ring radius) at any epoch, $r(t) = f(i_{disk}, \phi_{disk}, b, t_b, t)$, and also determine $dr(t)/dt$. The transmission of the disk as a function of $r$ is given by $\tau(r)$. Together with a model of the stellar disk that includes limb darkening (see van Werkhoven *et al.* 2014) and the functional form of $\tau(r)$, we can calculate the light curve of a ring system model for any epoch.

To understand our ring orientation algorithm, consider a ring system made up of alternately transparent and opaque rings whose radial width would allow complete obscuration or transmission of the stellar disk. The transmitted intensity goes from $I = I_0$ to $I = 0$ and vice versa at a rate determined by the ring velocity $v$ and the local tangent of the ring to the line of stellar motion, defined as parallel to the x-axis in our model. If the gradient of the light curve is measured close to the midpoint of the transit of a ring edge, and this quantity is plotted as a function of time, the result is the black curve in Figure 1.

The function $G(t)$ represents the maximum flux change possible between a fully transparent and fully opaque ring. For rings that have intermediate values of transmission, the resultant light gradients will lie underneath this black curve, and so the curve represents an upper bound on the light gradient for a given ring orientation. Since we do not know $\tau(r)$, we can use $G(t)$ as an upper limit and we search for ring orientations that have all measured gradients lie underneath this curve. We calculate a cost function

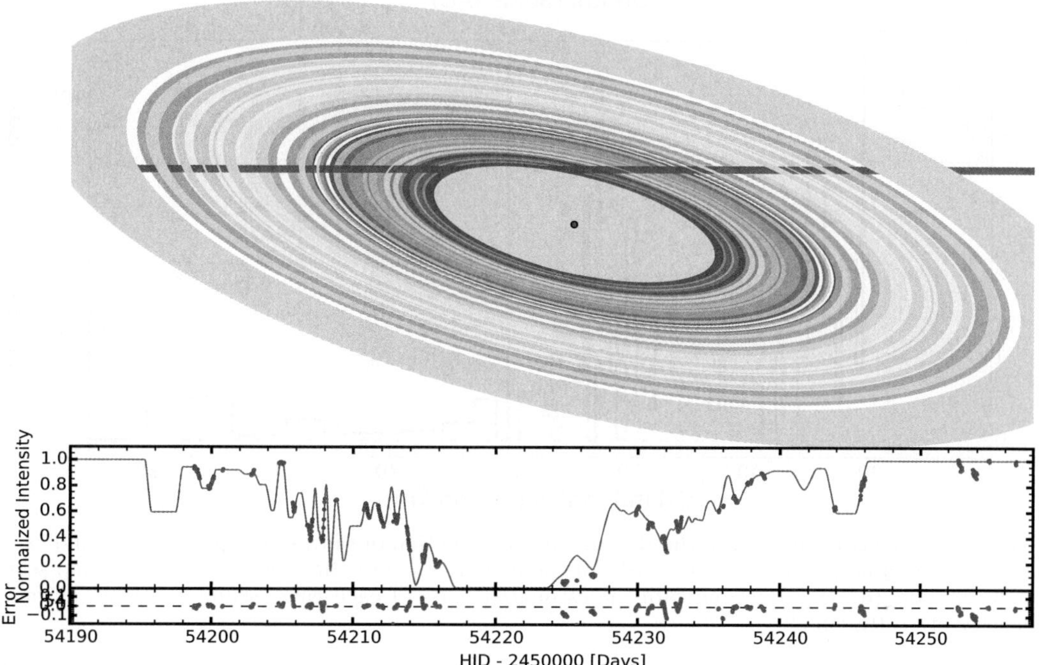



**Figure 2.** Model ring fit to J1407 data. The image of the ring system around J1407b is shown as a series of nested red rings. The intensity of the colour corresponds to the transmission of the ring. The green line shows the path and diameter of the star J1407 behind the ring system. The grey rings denote where no photometric data constrain the model fit. The lower graph shows the model transmitted intensity $I(t)$ as a function of HJD. The red points are the binned measured flux from J1407 normalised to unity outside the eclipse. Error bars in the photometry are shown as vertical red bars.

that minimises the difference between the model and measured gradients, and penalizes heavily if the measured point goes above the model point. Once the parameters for the disk are determined, we then fit for the ring transmission as a function of radius.

The number of ring edges in the light curve are estimated by counting the number of slope changes identified in the light curve and indirectly implied by the change of the light curve during daylight hours. At least 24 ring edges are required for the number of gradient changes detected in the J1407 data (van Werkhoven *et al.* 2014), but given the sparseness of the photometric coverage this number is almost certainly higher.

Figure 2 shows one possible ring solution to the J1407 photometric data, where the central eclipse is set at $t_b = 54220.65$ MJD.

There are clear gaps in all the ring model solutions explored. Gaps in the rings of Solar system giant planets are either caused directly by the gravitational clearing of a satellite or indirectly by a Lindblad resonance due to a satellite on a larger orbit. The J1407 ring system is larger than its Roche limit for the secondary companion. We take the most probable mass and period for the moderate range of eccentricities with mass $23.8 M_{Jup}$ and orbital period 13.3 yr. Gaps in the ring system are either seen directly as the photometric flux from J1407 returning to full transmission during the eclipse, or indirectly as a fit of the model to intermediate transmission photometric gradients. One ring gap with photometry is at HJD 54210, seen during the ingress of J1407 behind the ring system. The corresponding radius for this gap in the disk is seen from 59 million km to 63 million km (indicated in Figure 3). If we assume that the gap is equal to the

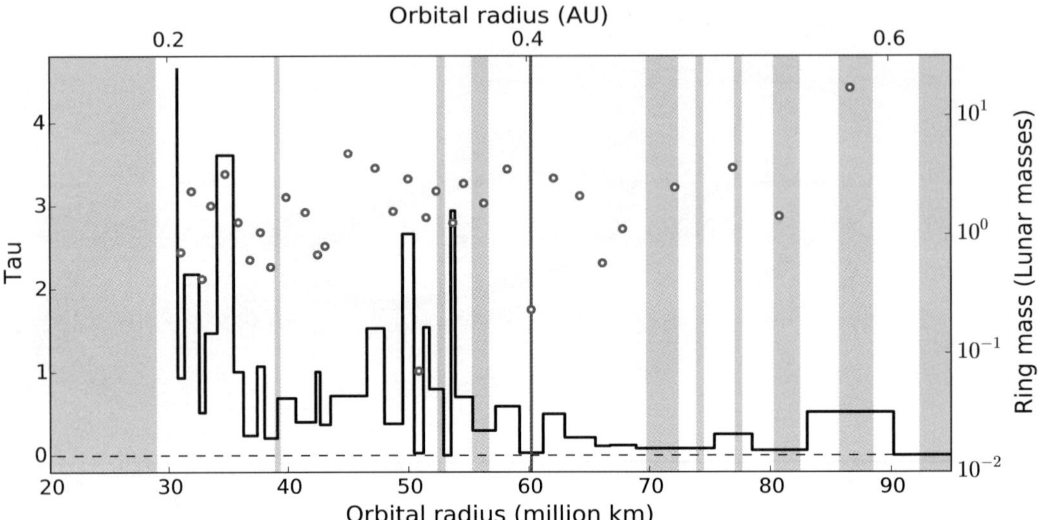


**Figure 3.** The transmission of the ring model as a function of radius. The grey regions indicate where there in no photometry to constrain the model. The blue line indicates the ring gap seen at 61 million km. Red dots indicate the estimated mass of each ring assuming a mass surface density of $\sim 50\,\mathrm{g}\,\mathrm{cm}^{-2}$.

diameter of the Hill sphere of a satellite orbiting around the secondary companion and clearing out the ring, then for the case of $23.8 M_{Jup}$ this corresponds to a satellite mass of $0.8 M_\oplus$ and an orbital period of 1.7 yr.

With simple assumptions on ring geometry and the ring plane orientation, this ring model reproduces many but not all of the nightly photometric light curves. These discrepancies imply either an error in the determined geometry of the ring plane, and/or the rings are not coplanar.

J1407 is currently being monitored both photometrically and spectroscopically for the start of the next transit. A second transit will enable a wide range of exoring science to be carried out, from transmission spectroscopy of the material, through to Doppler tomography that can resolve ring structure and stellar spot structure significantly smaller than that of the diameter of the star. The orbital period of J1407b is on the order of a decade or possibly longer.

## References


Butters *et al.* 2010, *A&A* , 520, L10
Dong, S., *et al.* 2014, *ApJ* , 788, 41
Graczyk, D., Mikołajewski, M., Tomov, T., Kolev, D., & Iliev, I. 2003, *A&A* , 403, 1089
Graczyk, D. *et al.* 2011, *Acta Astronomica*, 61, 103
Guinan, E. F. & Dewarf, L. E. 2002, in Astronomical Society of the Pacific Conference Series, Vol. 279, Exotic Stars as Challenges to Evolution, ed. C. A. Tout & W. van Hamme, 121
Hamilton, C. M., *et al.* 2005, *AJ* , 130, 1896
Kenworthy, M. A., *et al.* 2015, *MNRAS* , 446, 411
Kloppenborg, B., *et al.* 2010, *Nature* , 464, 870
Mamajek, E. E., *et al.* 2012, *AJ* , 143, 72
Mikolajewski, M., *et al.* 2005, *Ap and SS* , 296, 445
Mikolajewski, M. & Graczyk, D. 1999, *MNRAS* , 303, 521
Rattenbury, N. J., *et al.* 2014, ArXiv e-prints
van Werkhoven, T. I. M., Kenworthy, M. A., & Mamajek, E. E. 2014, *MNRAS* , 441, 2845
Winn, J. N., *et al.* 2006, *ApJ* , 644, 510

*Young Stars & Planets Near the Sun*
*Proceedings IAU Symposium No. 314, 2015*
*J. H. Kastner, B. Stelzer, & S. A. Metchev, eds.*

© International Astronomical Union 2016
doi:10.1017/S1743921315006456

# Warm Debris Disks
# with WISE and HST

## Deborah Padgett[1] and Karl Stapelfeldt[2]

[1] Code 665, NASA Goddard Space Flight Center
email: deborah.l.padgett@nasa.gov
[2] Code 667, NASA Goddard Space Flight Center
email: karl.r.stapelfeldt@nasa.gov

**Abstract.** Using 22 $\mu$m data from the Wide Field Infrared Survey Explorer (WISE), we have completed a sensitive all-sky survey for debris disks in Hipparcos and Tycho catalog stars within 120 pc. This warm excess emission traces material in the circumstellar region likely to host terrestrial planets. Several hundred previously unknown debris disk candidates were identified. We are currently performing follow-up observations to characterize the stars, companions, and circumstellar material in these systems with a variety of facilities including Keck, Herschel, and HST. Thirteen WISE debris disks have been observed to date using HST/STIS coronagraphy. Five of these disks have been detected in scattered light. One is a large and highly asymmetric edge-on disk which appears to be both warped and bifurcated.

**Keywords.** stars: circumstellar matter, stars: planetary systems, protoplanetary disks

## 1. Introduction

Circumstellar disks are the most visible signs of other solar systems. Planets and planetesimals (asteroids, comets) begin to form shortly after a star is born. The primordial gas and dust disk that gives birth to the star dissipates in 5–10 Myr due to accretion, giant planet formation, jets, and photoevaporation. Subsequently, a new "debris disk" emerges, consisting of dust grains generated by the collisions of the planetesimals and the evaporation of comets. These grains are quickly blown away by radiation pressure and stellar winds, so to exist for any significant period of time the disk must be continually replenished by the destruction of planetesimals. This is the case for the debris disk around our own Sun, the zodiacal cloud + Kuiper dust belt, formed by collisions in belts of small bodies. The presence of a debris disk therefore indicates the presence of some sort of planetary system.

Debris disks have become especially interesting to the extrasolar planet community because they can betray the presence of planets that would otherwise go unseen. Gravitational interactions between planets and dust can carve out localized gaps, create clearings (rings) within the disk, or generate warps. In our own zodiacal cloud, for instance, Earth has cleared out a region around it through resonant tidal interactions (Dermott *et al.* 1994). The A star Fomalhaut possesses a debris ring (Kalas *et al.* 2005) with 133 AU inner radius, 25 AU width, sharp inner edge, and center that is offset 15 AU from the stellar position - certain signs of a planetary mass companion sculpting the disk. A second, even more eccentric ring has been found around the solar-type star HD 202628 (Krist *et al.* 2012). Dynamical modeling of disks has matured to the point where structures can be related to the likely mass and orbital characteristics of the perturbing object (Quillen 2006, Chiang *et al.* 2009, Rodigas *et al.* 2014). In the Fomalhaut case, HST imaging Kalas *et al.* (2008) identified an unusual object orbiting near the ring inner edge (Kalas *et al.* 2013). The four planets imaged in the HR 8799 system (Marois *et al.* 2008, Marois *et al.* 2010) were subsequently shown to be orbiting in a gap between two debris belts (Su *et al.* 2009). The planet responsible for the inner disk warp in $\beta$ Pictoris was finally found by Lagrange *et al.* (2010). HD 95086 b was also imaged in a debris disk system (Rameau *et al.* 2013). It is very significant that many of the systems with directly imaged exoplanets also host bright debris disks. Mapping the structure of debris disks can thus point the way toward direct imaging of their perturbing planets.

In the 1980s IRAS provided the first evidence of debris disks with measurements of excess thermal emission around stars like Vega, Fomalhaut, Beta Pictoris and Epsilon Eridani. It was followed by ISO, Spitzer, Akari, and Herschel, which have added hundreds more potential debris disk systems to the inventory. However, as pointed observatories, ISO, Spitzer, and Herschel were able to survey only a fraction of nearby stars for infrared excess, so they cannot be expected to have identified all the brightest disks in the solar neighborhood. For solar-type stars, disks with excess emission at 22 $\mu$m occur primarily around young stars as important signposts of planetary system evolution (Rieke *et al.* 2005, Carpenter *et al.* 2009). The youth of such systems means they are excellent potential homes for bright young planets that could be directly imaged using high contrast techniques, which might also reveal disk structure unresolved at the low resolution currently available for most thermal IR imagery.

## 2. WISE Debris Disk Candidates

In 2010, the Wide-field Infrared Survey Explorer (WISE; Wright *et al.* 2010) performed an all-sky search for debris disks in the mid-infrared. With 22 $\mu$m sensitivity more than 100 times that of IRAS at 25 $\mu$m, WISE is capable of finding warm debris disks around G stars within 100 pc. With our knowledge of the WISE catalog and data products, we merged the all-sky cryogenic WISE catalog with the Hipparcos catalog and found over 566 stars with greater than $4\sigma$ mid-IR excess (from either [3.4]−[22] or [12]−[22]) within 120 pc, of which 350 were not known to have debris disks in the pre-WISE era (McDonald *et al.* 2012, Patel *et al.* 2014, etc.). Merging the Tycho-2 catalog with WISE yielded more than 450 additional debris disk candidates with proper motions consistent with distances less than 120 pc (Padgett *et al.* (2015a), in preparation). The number of $4\sigma$ excess detections decreased for the newer ALLWISE data release, as the uncertainties for saturated source photometry for [3.4] (W1) and [4.5] (W2) magnitudes increased with the addition of warm WISE data for those bands. These are high-quality detections with position matches between Hipparcos and WISE better than 0.3 arcsec, smoothly continuous photometry with 2MASS measurements, and individually vetted by visual checks of the WISE and 2MASS images. Ten percent of these sources show WISE colors consistent with 12 $\mu$m excess as well as 22 $\mu$m excess. Half of the WISE/HIP candidates have radial velocities, and known members of young associations are well represented. Some previously unstudied sources are apparently comoving with known associations but are at large spatial separations (Padgett *et al.* 2015a, in preparation).

The WISE debris disk candidate stars are in the process of being characterized at a variety of wavelengths. About 120 solar-type and A stars from this sample were observed at 70 $\mu$m and 160 $\mu$m by Herschel/PACS. Nearly all of the sources were detected at these wavelengths, although about 20% show enough of an offset between the stellar position and the long wavelength emission to raise suspicion of chance alignment between the star and a background star-forming galaxy. This level of contamination is not surprising for the Herschel sample since 98 of the targets are solar-type stars. Of the stars with convincing far-IR excess, the derived dust temperatures range from 60 - 150 K with roughly equal numbers below and above 90 K. A significant number have evidence for at least two temperature components in their spectral energy distributions (Padgett

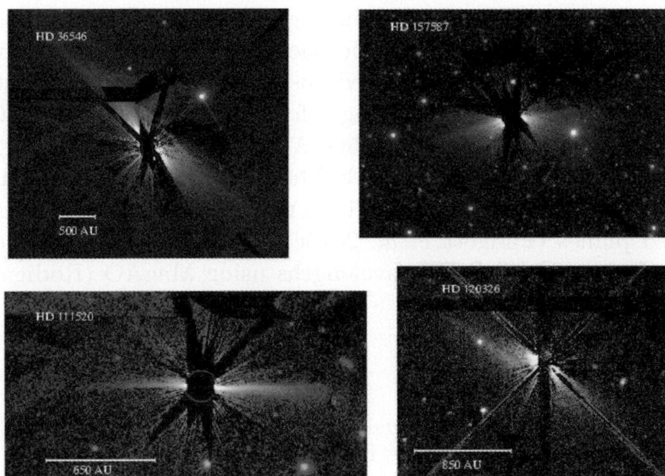

**Figure 1.** STIS coronagraph images of WISE-discovered warm debris disks.

*et al.* 2015b, in preparation). The WISE debris disk candidates are also the subject of an ongoing Keck near-IR coronographic survey for substellar companions (Hinkley *et al.* 2014).

## 3. Hubble Space Telescope Coronography of WISE Disks

Debris disks are difficult to see in scattered light, as even the brightest have total optical fluxes of only a fraction of a percent of their stars'. The first debris disk ever imaged was $\beta$ Pictoris (Smith & Terrile 1984), which was observed using a coronagraph to reduce the glare from the star. After many fruitless years of searching from the ground, other debris disks were finally resolved at millimeter, infrared, and optical wavelengths beginning in the late 1990s: HR 4796, Vega, Fomalhaut, $\epsilon$ Eridani, and HD 141569A; see Wyatt (2008) for a review. During the past decade the number of resolved debris disks has nearly quadrupled. However, only about half of these are detected in scattered light where 0.1 arcsec resolution imaging is available (see the compilation at http://circumstellardisks.org). Some large disks have also been resolved in far-infrared thermal emission by Herschel (Matthews *et al.* 2010, Lohne *et al.* 2012, Acke *et al.* 2012, Booth *et al.* 2013). Many of the brightest disks have been imaged using adaptive optics coronagraphy. Yet the disks imaged thus far represent spectral types A through M with ages ranging from a few million to several billion years: too sparse a sample from which to derive any correlations that could reveal the physics behind the evolution of these systems.

The optical depths of debris disks (which determine their scattered light brightnesses) are proportional to the strength of their infrared excess. Disks with fractional infrared luminosities $L_{dust}/L_{star} > 10^{-4}$ have been shown to be detectable in scattered light by HST. Not all such disks have been detected, however, as some may be smaller than the inner working angle of the coronagraph, have low albedos or have face-on orientations that make detection more difficult. In HST cycles 19 and 20, we observed a total of 13 WISE debris disk candidates using the STIS coronagraphic wedge. These were selected for high $L_{dust}/L_{star}$, proximity, and brightness of excess at WISE 22 $\mu$m and/or Herschel 70 $\mu$m. They were each observed for two orbits at different roll angles, using self-subtraction to mitigate the remaining scattered light from the point spread function (PSF). Six of the targets showed no scattered light beyond the PSF residuals. Two were resolved as star-galaxy pairs, demonstrating the continuing contamination issue posed by superposition of bright stars with the extragalactic background. Five have significant scattered light consistent with the presence of resolved dust structures around the star. Fig. 1 shows the roll-subtracted STIS coronagraph images for four of these objects. One spectacular example is an asymmetric edge-on system 500 AU in radius around HD 111520, an F star in Lower Centarus-Crux (Fig. 1 lower left). The disk appears warped and is five times brighter on the NW side than on the SE. The NW side is also clearly bifurcated, with two distinctly inclined planes (Padgett *et al.* 2015c, in preparation). We are in the process of characterizing these disks at NIR wavelengths using MagAO (Rodigas *et al.* 2015, in preparation).

## References


Acke, B., Min, M., Dominik, C., Vandenbussche, B., Sibthorpe, B., Waelkens, C., Olofsson, G., Degroote, P., Smolders, K., & Pantin, E., *et al.* 2012, *A&A*, 540, 125

Booth, M., Kennedy, G., Sibthorpe, B., Matthews, B. C., Wyatt, M. C., Duchene, G., Kavelaars, J. J., Rodriguez, D., Greave, J. S., Koning, A. *et al.* 2013, *MNRAS*, 428, 1263

Carpenter, J. M., Mamajek, E. E., Hillenbrand, L. A., & Meyer, M. R. 2009, *ApJ*, 705, 1646

Chiang, E., Kite, E., Graham, J. R., & Clampin, M. 2009, *ApJ*, 693, 734

Dermott, S. F., Jayaraman, S., Xu, Y. L., Gustarfson, B. A. S.., & Liou, J. C. 1994, *Nature*, 369, 719

Hinkley, S., Mawet, D., Stapelfeldt, K., Padgett, D., Morales, F., & Serabyn, E. 2014, *Proceedings of the conference Thirty years of beta Pic and debris disks studies*, A.-M. Lagrange & A. Boccaletti, eds., p.40

Kalas, P., Graham, J. R., & Clampin, M. 2005, *Nature*, 435, 1067

Kalas, P., Graham, J., Chiang, E., Fitzgerald, M., Clampin, M., Kite, E., Stapelfeldt, K., Marois, C., & Krist, J. 2008, *Science*, 322, 1345

Kalas, P., Graham, J. R., Fitzgerald, M. P., & Clampin, M. 2013, *ApJ*, 775, 56

Krist, J. E., Stapelfeldt, K. R., Bryden, G., & Plavchan, P. 2012, *AJ*, 144, 45

Lagrange, A.-M., Bonnefoy, M., Chauvin, G., Apai, D., Ehrenreich, D., Boccaletti, A., Gratadour, D., Rouan, D., Mouillet, D., Lacour, S., & Kaspar, M. 2010, *Science*, 329, 57

Lohne, T., Eiroa, C., Augereau, J.-C., Ertel, S., Marshall, J. P., Mora, A., Absil, O., Stapelfeldt, K., Thebault, P., Bayo, A., *et al.* 2012, *Astronomische Nachrichten*, 333, 441

Marois, C., Macintosh, B., Barman, T., Zuckerman, B., Song, I., Patience, J., Lafreniere, D., & Doyon, R. 2008, *Science*, 322, 1348

Marois, C., Zuckerman, B., Konopacky, Q. M., Macintosh, B., & Barman, T. 2010, *Nature*, 468, 1080

Matthews, B. C., Sibthorpe, B., Kennedy, G., Phillips, N., Churcher, L., Duchene, G., Greaves, J. S., Lestrade, J.-F., Moro-Martin, A., & Wyatt, M. C., *et al.* 2010, *A&A*, 518, 135

McDonald, I., Zijlstra, A., & Boyer, M. 2012, *MNRAS*, 427, 343

Patel, R., Metchev, S., & Heinze, A. 2014, *ApJS*, 212, 10

Quillen, A. C. 2006, *MNRAS*, 372, 14

Rameau, J., Chauvin, G., Lagrange, A.-M., Meshkat, T., Boccaletti, A., Quanz, S. P., Currie, T., Mawet, D., Girard, J. H., Bonnefoy, M., & Kenworthy, M. 2013, *ApJ*, 779, 26

Rieke, G. H., Su, K. Y. L.., Stansberry, J. A., Trilling, D., Bryden, G., Muzerolle, J., White, B., Gorlova, N., Young, E. T., Beichman, C. A., Stapelfeldt, K. R., & Hines, D. C. 2005, *ApJ*, 620, 1010

Rodigas, T. J., Malhotra, R., & Hinz, P. M. 2014, *ApJ*, 780, 65

Smith, B. & Terrile, R. 1984, *Science*, 226, 1421

Su, K. Y. L., Rieke, G. H., Stapelfeldt, K. R., Malhotra, R., Bryden, G., Smith, P. S., Misselt, K. A., Moro-Martin, A., & Williams, J. P. 2009, *ApJ*, 705, 314

Wright, E. L., Eisenhardt, P. R. M.., Mainzer, A. K., Ressler, M. E., Cutri, R. M., Jarrett, T., Kirkpatrick, J. D., Padgett, D., McMillan, R. S., Skrutskie, M. *et al.* 2010, *AJ*, 140, 1868

Wyatt, M. 2008, *ARAA*, 46, 339

*Young Stars & Planets Near the Sun*
*Proceedings IAU Symposium No. 314, 2015*
*J. H. Kastner, B. Stelzer, & S. A. Metchev, eds.*

© International Astronomical Union 2016
doi:10.1017/S1743921315006389

# Comprehensive Census and Complete Characterization of Nearby Debris Disk Stars

## Tara H. Cotten and Inseok Song

University of Georgia
email: **tara@physast.uga.edu**
email: **song@uga.edu**

**Abstract.** Debris disks are intimately linked to planetary system evolution since the rocky material surrounding the host stars is believed to be due to secondary generation from the collisions of planetesimals. With the conclusion and lack of future large scale infrared excess survey missions, it is time to summarize the history of using excess emission in the infrared as a tracer of debris and exploit all available data as well as provide a comprehensive study of the parameters of these important objects. We have compiled a catalog of infrared excess stars from peer-reviewed articles and performed an extensive search for new debris disks by cross-correlating the Tycho-2 and AllWISE catalogs. This study will conclude following the thorough examination of each debris disk star's parameters obtained through high-resolution spectroscopy at various facilities which is currently ongoing. We will maintain a webpage (www.debrisdisks.org) devoted to these infrared excess sources and provide various resources related to our catalog creation, SED fitting, and data reduction.

**Keywords.** stars: evolution - circumstellar

## 1. Introduction

The InfraRed Astronomical Satellite (*IRAS*) laid the foundation for large-scale surveys of infrared excess detections in the 1990s following the discovery of the debris disk around Vega (Aumann et. al. 1984). This type of excess emission in the infrared maps the debris in regions analogous to the Kuiper and Asteroid belts. With the completion of the most recent all-sky infrared mission (WISE; *Wide-Field Infrared Survey Explorer*, Wright *et al.* 2010) and without the advent of new dedicated surveys, the amalgamation of all infrared data today will serve as a collective legacy for the detection of debris disk stars. In addition, as technology advances to directly detect planets embedded in mature stellar systems, the result of this study will provide ideal targets for exoplanet investigations. The goal of this study is to compile the most comprehensive catalog of high-fidelity infrared excess stars through a combination of prior studies and new detections using the Tycho-2 catalog and the recent release of the WISE catalog (AllWISE).

Though there have been hundreds of debris disk stars published in the literature including those from WISE data searches (i.e. Wu *et al.* 2013, Patel *et al.* 2014), only a small fraction of those objects have comprehensive stellar and disk information necessary for uniting theories of planetary formation to observations. To thoroughly explore the relationship between disks and a planetary system, stellar information such as effective temperature, metallicity, rotational velocity, radial velocity, and age indicators (CaII H & K, Li 6708 Å) using optical spectroscopy must be obtained for the catalog of infrared excess stars provided by Cotten & Song (2015, in prep.). In particular, one such relationship is based on the trend found in which higher mass planet host stars have higher

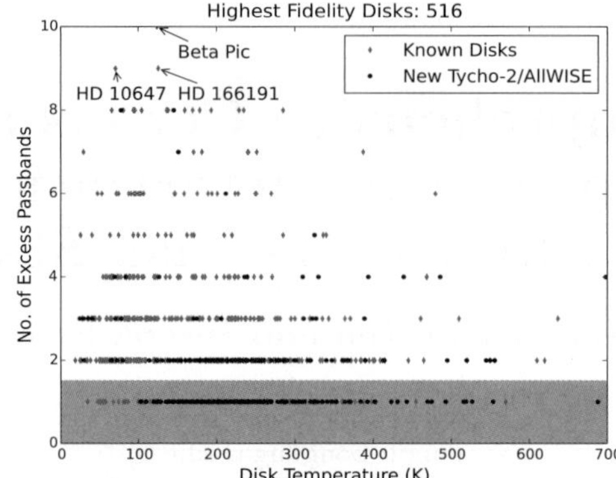


**Figure 1.** (*Left*) Number of passbands in the infrared which display excess emission. The stars that have excess confirmed by more than one wavelength make up the high fidelity infrared excess sample.

metallicities (Fischer & Valenti 2005). Greaves *et al.* (2006) examined a limited sample of debris disk stars to link the dusty disk to the metal-rich star but were unable to recognize any trend. Moreover, Trilling *et al.* (2007) and Rodriguez & Zuckerman (2012) investigated whether multiplicity would promote or hinder planetary system formation and evolution, especially since stars prefer to form in multiples (Patience *et al.* 2008 and reference therein). While many factors play a role in these systems, Trilling *et al.* (2007) and Rodriguez & Zuckerman (2012) produced contradictory results regarding whether debris is common among multiple star systems. Finally, the connection between the rotation of the host star and the surrounding dusty debris disk has only recently been examined and remains an open question. Specifically, Sierchio *et al.* (2010) and Mizusawa *et al.* (2012) outlined contrasting findings, however, the use of $v \sin i$ as an indicator for rotational velocity has an inherently large uncertainty due to the inclination angle.

The literature concludes that the connection between stellar properties and dust evolution can only be fully confirmed with larger samples and accurate stellar information. To aid in this endeavor, the creation of the complete census of infrared excess stars from the literature and a new Tycho-2 and AllWISE correlation is described in Section 2. Section 3 itemizes the ongoing observations and future plans to investigate the relationships described above with the largest sample of debris disk stars to date.

## 2. Creation of Census

The creation of the complete census of debris disk stars began with the selection of a couple of survey publications involving infrared excess (Rhee *et al.* 2007 and Rieke *et al.* 2005) and a review paper by Wyatt (2008). Citations by and of these articles amassed 193 papers with 863 unique sources of infrared excess. Given the improvements to infrared excess detections by more recent surveys such as WISE, we re-evaluated the amount of excess emission above the photosphere including all available photometry as well as inspected the AllWISE catalog images to remove sources that may be contaminated by nearby stars, extended cirrus contribution, or background infrared sources. In addition, since most of the previously known excess sources are *Hipparcos* stars, we used the accurate parallax measurements to only maintain stars within 125 pc. The selections from the literature constitutes 530 stars.

Generating the largest census would be incomplete without a full exploitation of the AllWISE catalog containing infrared photometry at 3.5, 4.6, 12, and 22$\mu$m for over 740 million sources. Therefore, we chose the Tycho-2 catalogue of the 2.5 million brightest

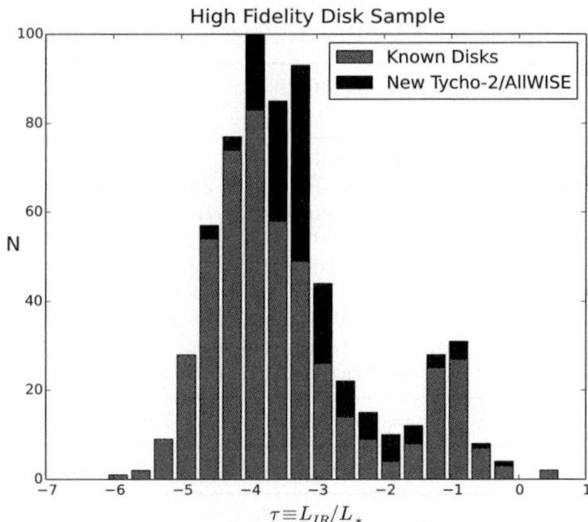



**Figure 2.** (*Right*) Histogram of the fractional dust luminosity for our high fidelity sample of infrared excess stars. The secondary peak of significantly dusty sources is comprised of T Tauri and Herbig Ae/Be objects, although recent investigation found a handful of extremely dusty debris disk stars which are currently under further investigation.

stars (Høg *et al.* 2000) as a compatible match given the reliable optical photometry and proper motions. Lacking accurate *Hipparcos* parallax measurements, we used the total proper motion ($\mu_{tot}$) as a proxy for distance and selected sources with $\mu_{tot} > 25$ mas yr$^{-1}$ corresponding to within 200 pc. These stars cross-correlated with the AllWISE catalog using a five arc second search radius returned over 513,000 stars. A number of further criteria were implemented to ensure high quality photometry for SED fitting, that we can eliminate as many giants as possible since only *Hipparcos* stars have precise distance measurements, and finally that the stars show a significant amount of excess above the photosphere. The summary of these criteria are included in Table 1 and more thorough description can be found in Cotten & Song (2015, in prep).

Ruling out any duplicates found between the literature search and this new Tycho-2/AllWISE search, assembled 963 unique infrared excess stars. Assessment of the quality of the infrared excess detection required the development of a measure portrayed in Figure 1 defined to be the number of infrared passbands displaying excess emission above the photosphere. Since we have performed extensive SED fitting using Tycho-2 *B* and *V*, *2MASS* J, H, and $K_S$, and AllWISE W1 and W2 to define the stellar photosphere, we then examined the amount of excess for all available infrared data which could include AllWISE, IRAS, MSX, Akari, *Spitzer*, and *Herschel*. As can be seen in Figure 1, there are a number of stars having excess in only one passband (usually AllWISE W4) in the grey region. These stars will be maintained in our study, but the stars outside of this region (516 unique sources) display the highest fidelity infrared excess through confirmation by an additional passband. Lastly, Figure 2 displays the fractional dust luminosity ($\tau$) for the highest fidelity sample demonstrating there are two populations of dustiness where the extremely dusty disks are mostly comprised of young T Tauri or Herbig Ae/Be stars, however, a handful of stars found through our new method show significant amounts of dust and appear to be true debris disk stars.

## 3. Future Work

Characterization of the complete catalog of infrared excess sources will enable a full analysis of the relationship between a star, disk, and potential planets and improve upon past studies by expanding the number of debris disk stars by more than three times.

**Table 1.** Summary of High Fidelity Infrared Excess Star Selection Criteria

| Selection Criteria | No. of Stars Remaining |
|---|---|
| AllWISE Catalog | $\sim 740$ million |
| Total Proper Motion for Dist. Proxy $> 25$ mas/yr | 515518 |
| 5.0 arcsec Cross-Correlation between Tycho-2 and AllWISE | 513478 |
| Well-Measure AllWISE photometry (no upper limits) | 263955 |
| HIP CMD and various color-color diagrams to exclude giants | 245874 |
| Reliable Goodness-of-Fit ($\chi^2$) from SED fitting | 243584 |
| Significance of Excess ($> 5\sigma$ in W3 *OR* W4 in each temperature division) | 4537 |
| Image Analysis and Inspection to remove contaminators | 1030 |
| New Infrared Excess Stars within 125 pc | 595 |
| Known Infrared Excess Stars within 125 pc in Literature | 530 |
| Final Combined Sample of High Fidelity Infrared Excess Stars (excluding duplicates) | 516 |

The parameters we seek to obtain include age, metallicity, surface gravity, stellar rotation, radial velocity, and multiplicity. Majority of these parameters can only be obtained through high-resolution optical spectroscopy and so we searched the spectral archives of instruments such as FEROS, ELODIE, SOPHIE, HARPS, and UVES so we do not re-observe these stars. Due to nearly half of the final sample lacking archival spectra, we have distributed sources using numerous parameters such as declination, brightness, and dustiness to collaborators across the globe as well as proposed for additional telescope time at various facilities. Our census may combine a heterogeneous collection of spectra, but we plan to derive the aforementioned stellar parameters in a uniform manner. In addition, following the publication of the complete census (Cotten & Song 2015), we will make our catalog publicly available through the webpage: http://www.debrisdisks.org. The webpage will serve not only as a catalog of photometric and spectroscopic information, it will also contain interactive plotting capabilities for the reduced spectra as well as our Python SED fitting algorithm.

## References


Aumann, H. H., Beichman, C. A., & Gillett, F. C. *et al.* 1984, *ApJ*, 278, 23

Fischer, D. A. & Valenti, J. 2005, *ApJ*, 622, 1102

Greaves, J. S., Fischer, D. A., & Wyatt, M. C. 2006, *MNRAS*, 366, 283

Høg, E., Fabricius, C., & Makarov, V. V. *et al.* 2000, *A&A*, 355, 27

Mizusawa, T. F., Rebull, L. M., & Stauffer, J. R. *et al.* 2012, *AJ*, 144, 135

Patel, R. I., Metchev, S. A., & Heinze, A. 2014, *ApJS*, 786, 10

Patience, J., Akeson, R. L., & Jensen, E. L. N. 2008 *ApJ*, 677, 616

Rhee, J. H., Song, I., Zuckerman, B., & McElwain, M. 2007, *ApJ*, 660, 1556

Rieke, G. H., Su, K. Y. L, & Stansberry, J. A. *et al.* 2005, *ApJ*, 620, 1010

Rodriguez, D. R. & Zuckerman, B. 2012, *ApJ*, 745, 147

Sierchio, J. M., Rieke, G. H., & Su, K. Y. L., *et al.* 2010, *ApJ*, 712, 1421

Trilling, D. E., Stansberry, J. A., & Stapelfeldt, K. R., *et al.* 2007, *ApJ*, 658, 1289

Wright, E. L., Eisenhardt, P. R. M., & Mainzer, A. K., *et al.* 2010, *AJ*, 140, 1868.

Wu, C.-J., Wu, H., & Lam, M.-I. *et al.* 2013, *ApJS*, 208, 29

Wyatt, M. C. 2008, *ARA&A*, 46, 339


## Discussion

METCHEV: In your plot of significance of excess histograms, how many stars are at $-5\sigma$?

COTTEN: I can't tell you the exact number, but I know it is significantly less than that the number of excess sources.

*Young Stars & Planets Near the Sun*
*Proceedings IAU Symposium No. 314, 2015*
*J. H. Kastner, B. Stelzer, & S. A. Metchev, eds.*

© International Astronomical Union 2016
doi:10.1017/S1743921315006614

# Debris Disks in Nearby Young Moving Groups in the ALMA Era

## Á. Kóspál and A. Moór

Konkoly Observatory, Research Centre for Astronomy and Earth Sciences, Hungarian
Academy of Sciences, PO Box 67, 1525 Budapest, Hungary
email: `kospal@konkoly.hu, moor@konkoly.hu`

**Abstract.** Many members of nearby young moving groups exhibit infrared excess attributed to circumstellar debris dust, formed via erosion of planetesimals. With their proximity and well-dated ages, these groups are excellent laboratories for studying the early evolution of debris dust and of planetesimal belts. ALMA can spatially resolve the disk emission, revealing the location and extent of these belts, putting constraints on planetesimal evolution models, and allowing us to study planet-disk interactions. While the main trends of dust evolution in debris disks are well-known, there is almost no information on the evolution of gas. During the transition from protoplanetary to debris state, even the origin of gas is dubious. Here we review the exciting new results ALMA provided by observing young debris disks, and discuss possible future research directions.

**Keywords.** circumstellar matter, submillimeter: planetary systems, planet–disk interactions

---

## 1. Introduction

Nearly all young stars are encircled by massive circumstellar disks that serve as reservoirs for mass accretion and provide the necessary primordial gas and dust for the formation of planetesimals and planets. According to observations, gas-rich protoplanetary disks are depleted within about 10 million years and evolve into gas-poor, tenuous, dusty debris disks (e.g, Williams & Cieza 2011). Debris disks are composed of second generation dust, where individual grains are rapidly removed mainly due to stellar radiation forces, but the dust is continuously replenished by collisional erosion or evaporation of previously formed planetesimals (Wyatt 2008).

Many young stars ($<200\,\mathrm{Myr}$) in the solar neighbourhood belong to moving groups, i.e., gravitationally unbound, loose associations, whose members have common origin and move through space together (e.g., Kastner, this volume). A significant fraction of their members exhibits excess emission at infrared (IR) wavelengths (e.g., Zuckerman *et al.* 2011). While in some cases this may come from long-lived protoplanetary disks, the majority can be attributed to circumstellar debris dust. Many well-studied debris systems, e.g., $\beta$ Pic, AU Mic, HR 4796A, belong to young moving groups. With their proximity and well-dated ages between 8 and 200 Myr, these groups offer an excellent laboratory for studying the evolution between the primordial and debris stages, as well as the early evolution of debris dust and, thereby, of planetesimal belts. This time range overlaps with the expected last phase of the formation of terrestrial planets (Raymond et al. 2014) and the initiation of collisional cascade and production of observable debris material in the outer disk regions (Kenyon & Bromley 2008). However, most debris disks around members of nearby young moving groups have remained spatially unresolved so far, providing only limited information about disk properties.

With its unprecedented sensitivity and spatial resolution, the recently commissioned Atacama Large Millimeter/submillimeter Array (ALMA) is opening a new era in the investigation of debris disks by allowing to observe both the thermal emission of cold dust and the rotational lines of different gas molecules (if present). By studying ALMA continuum data we can map the spatial distribution of mm-sized grains, which dominate the disk emission at submm/mm wavelengths. These large dust grains are less subject to dynamical removal effects than smaller grains and are therefore the best tracers of the structure of the parent planetesimal belts. By determining the location and extent of these belts, we can put contraints on planetesimal formation and evolution models and the underlying dynamical processes. It also allows to probe traces of the interaction between giant planets and the disk. Molecular line observations with ALMA allows to measure the amount of gas in debris disks and how it evolves during the transition from protoplanetary to debris state, a largely unexplored field so far. In the following we review open questions concerning the dust and gas content of young debris disks, the latest ALMA results in the field, and promising future directions.

## 2. Stirring of debris disks

Planetesimals in protoplanetary disks are on low eccentricity and low inclination orbits because of the damping effect of gas. Therefore, collisions between them occur at low relative velocities resulting in the merging of bodies, even after the amount of gas decreases. For destructive collisions with effective dust production, the motion of planetesimals needs to be stirred. In the self-stirring scenario proposed by Kenyon & Bromley (2008), Pluto-sized bodies embedded in the outer disk can initiate a collisional cascade by perturbing the orbits of neighbouring smaller planetesimals. As large planetesimals build up slower at larger disk radii, the collisional cascade is ignited in the inner disk first and then the active dust production propagates outward. As the disk evolves, planetesimals in the inner regions are ground down, leading to a decline in the local dust production. This results in an outwardly increasing dust surface density whose maximum coincides with the region where Pluto-sized bodies have just been formed (Kennedy & Wyatt 2010). This is very different from the surface density profiles observed in protoplanetary disks. The pace of the outward propagation of the stirring front depends on the disk surface density: in an initially denser, more massive disk, the outwards spread is faster (Kenyon & Bromley, 2008). A giant planet or a stellar companion can also excite the motion of planetesimals via its secular perturbations (planetary stirring, Wyatt 2005). If the perturber is located closer to the star than the planetesimal belt, then this mechanism also results in an inside-out disk stirring whose timescale could be even shorter than that of self-stirring (Mustill & Wyatt 2009).

Observational verification of specific aspects of this picture, however, are not yet conclusive. By mapping the spatial distribution of mm-sized debris dust grains, we can evaluate several predictions of stirring models. Using ALMA, MacGregor et al. (2013) and Ricci *et al.* (2015) successfully resolved the radial density surface density profiles in two young broad debris disks around AU Mic and HD 107146. In both cases a rising profile was found at least in the outer regions of the disks, consistently with the predictions of stirring models. The site of active dust production in stirring models is expected to propagate outward as a function of age. Previous observations have already suggested a weak correlation between disk radii and ages (e.g., Rhee *et al.* 2007; Eiroa *et al.* 2013). However, disk radii in these studies were estimated from their characteristic temperature and the stellar luminosity, assuming blackbody grains. Because of the well-known degeneracy between the equilibrium temperature and size of grains, these estimates are not unambigous (e.g. Krivov 2010). By measuring debris disk sizes with ALMA in well-dated systems with different ages, e.g., in different young moving groups, this predicted trend can be reliably verified. In unusually extended young debris disks, the self-stirring model would require too massive initial disks. In such cases, the self-stirring scenario is unfeasible, thus, they can be considered as prime candidates for planetary stirring. Indeed, the debris disks around HR 8799, $\beta$ Pic, and HD 95086 – systems with known outer giant planets – fall into this category (Moór et al. 2015). This principle can be used to search for additional planetary stirring candidates with ALMA. Since the potential targets are young, using advanced adaptive optics systems such as the Gemini Planet Imager or VLT/SPHERE (Macintosh *et al.* 2008, Beuzit *et al.* 2010), even the perturber(s) could be identified.

## 3. Signatures of planet-disk interaction

Debris dust grains and the planetesimals from which they are derived are the smallest constituents of a planetary system that may include more massive planets as well. As the system evolves, the planet(s) and the planetesimal disk can interact in many ways. The different types of gravitational perturbations can cause various kind of footprints, such as offsets, spirals, gaps, or clumps in the disk. Secular perturbations from a misaligned or eccentric planet can cause warps or tightly wound spirals in the disk (Matthews et al. 2014a). These structures then propagate through the disk, as the gravitational effect of the planet extends to outer and outer regions. By forcing the distant planetesimals on intersecting orbits, this may result in more frequent collisions that occur with higher relative velocities, eventually leading to the initiation of a collisional cascade. After a time, the secular perturbations from an eccentric planet make the whole disk also to be eccentric (e.g.,Kalas *et al.* 2005).

Resonant perturbations occur when dynamical frequencies, typically of the mean motions, are a simple integer ratio of each other. Resonances can lead to either stabilization or destabilization of the orbits, causing overpopulated or underpopulated regions in a debris disk (Matthews *et al.* 2014a, and references therein). As a consequence of overlapping first-order mean-motion resonances, the region close to the planet's orbit becomes chaotic, and dust particles orbiting in this zone are short-lived. Therefore, the chaotic zone is evacuated and a gap is formed. In our Solar System, the Kirkwood gaps in the asteroid belt are formed via such mean motion resonances with Jupiter. Overpopulated regions can also be formed via resonant captures, such as the Trojan asteroids (in 1:1 resonance with Jupiter) or the Plutinos (in 3:2 resonance with Neptune) in our Solar System. In a debris disk, two different migration mechanisms are proposed to lead to resonant clumps in the dust distribution. Outward migration of a planet can trap planetesimals into its external resonances, the erosion of these planetesimals then produce an enhanced dust population. Smaller grains can escape from these regions because of radiation pressure. Large particles, however, are trapped in the resonance and cause a local enhancement that can be traced in submillimeter/mm maps of the disk (Wyatt, 2003). Alternatively, in those tenuous debris disks where the timescale of Poynting-Robertson drag is shorter than the collisional timescale, dust grains produced in an outer planetesimal belt can spiral inward due to drag forces into the mean-motion resonances with an inner planet (Kuchner & Holman, 2003).

ALMA observations of Fomalhaut's disk showed that the mm-sized dust is concentrated in a very narrow, vertically thin ring, whose inner and outer edges are sharply truncated (Boley et al. 2012). This ring morphology can be best explained with the presence of two shepherding planets located at the inner and the outer edges, which are likely to be less massive than $3\,M_\oplus$. HD 107146 is encircled by a very broad debris disk. ALMA images of the radial distribution of dust revealed a decrease in the surface brightness profile at intermediate radii (Ricci et al. 2015). One possible interpretation is that this is related to a gap carved by a few Earth-mass planet in the dusty disk.

Previously, based on spatially resolved millimeter images of nearby bright debris disks around Vega, $\epsilon$ Eri, and HD 107146, several clumpy structures were reported and the presence of these inhomogeneities were linked to large grains trapped in mean motion resonances with unseen planets (Holland et al. 1998; Greaves *et al.* 1998; Corder *et al.* 2009). Most of these structures, however, turned out to be dubious, because subsequent more sensitive and higher resolution observations could not confirm their presence (e.g., Piétu *et al.* 2011; Hughes et al. 2011). Thanks to its excellent sensitivity, ALMA can detect possible resonant structures of even lower contrast in the brightness distribution of debris disks. Recent ALMA maps of $\beta$ Pic revealed a bright region in the southwestern part of the disk. This clump be seen both in the distribution of CO gas and mm-sized dust grains (Dent *et al.* 2014) and it also coincides well with an enhancement of micron-sized grains observed at resolved mid-IR images (Telesco *et al.* 2005). Both the gas and dust in $\beta$ Pic may have secondary origin. Based on current data, the clump can be attributed to frequently colliding icy planetesimals trapped in a mean motion resonance with an outwardly migrating planet. Alternatively, the clump may be a result of a recent collision between two Mars-sized icy comets (Dent et al. 2014).

The observed disk asymmetries and structures can be used to deduce the orbit and other parameters of the planet(s) responsible for them and can also provide an opportunity to constrain the dynamical history of the planetary system. In young systems, wide separation giant planets predicted in this way can be confirmed using new generation adaptive optics systems. Moreover, such observations of the perturbers will be able to help in the evaluation and improvement of methods we use in the interpretation of structures seen in the dust distribution. Indeed now we may already know two potential benchmark systems. HR 8799 and HD 95086 are two young A-type stars with outer giant planets detected by direct imaging (Marois *et al.* 2010; Rameau *et al.* 2013). Both systems harbor a warm inner dust belt and a broad cold outer disk. The giant planet(s) are located between the two dusty regions (Matthews et al. 2014b; Moór *et al.* 2013a; Su *et al.* 2009, 2015). ALMA observations of these disks are ongoing.

## 4. Gas in young debris disks

Debris disks are expected to be gas-poor systems with a gas-to-dust ratio significantly lower than in protoplanetary disks. The detection of the presumably little amounts of gas is a challenging task. Despite substantial efforts, we know only eight debris disks where gas has been detected. In the pre-ALMA era, the edge-on orientation of disks around $\beta$ Pic and HD 32297 allowed the detection of their gas content via absorption lines (Slettebak et al. 1975; Hobbs *et al.* 1985; Redfield 2007). The gaseous disks around HD 172555, HD 181296, and AU Mic were identified based on O I , C II, and fluorescent $H_2$ line emissions, respectively (Riviere-Marichalar *et al.* 2012, 2014; France et al. 2007), while CO gas has been observed in disks around 49 Ceti, HD 21997, and HD 131835 (Zuckerman et al. 1995; Moór *et al.* 2011; Moór *et al.* in prep). All of these systems are probably younger than 50 Myr, and with the exception of HD 32297 they belong to young moving groups or associations. Interestingly, apart from the M-type AU Mic, all of the host stars are A-type. The origin of gas in these disks is uncertain. In most cases the gas might be secondary, i.e., similarly to dust grains it is continuously replenished via erosion of larger bodies. Collisions between icy planetesimals, evaporation of comets or icy grains, and photon-induced desorption of solids can lead to gas release. However, because of their youth, we cannot exclude that some of these disks are hybrid in the sense that they retain their residual primordial gas, while the dust component have secondary origin.

The first observations of gaseous debris disks with ALMA provided examples of both types. By mapping the dust and gas in the disk around $\beta$ Pic, Dent *et al.* (2014) found that the spatial distribution of the two components is similar, both showing a characteristic bright clump in the southwest part of the disk. The short lifetime of CO and other dynamical arguments make it probable that the gas in this system has secondary origin, probably indicating intense collisional activity (Fernández et al. 2006; Dent *et al.* 2014). ALMA observations of HD 21997 showed that the CO gas is not co-located with the mm-sized dust and thereby the planetesimals, but there is a dust-depleted inner gas disk (Kóspál et al. 2013; Moór et al. 2013b). Moreover, the disk's total CO content is more than a thousand times larger than that of $\beta$ Pic. These cannot be explained with secondary gas models, implying that this disk instead harbors residual primordial gas and is the first example of a hybrid disk (Kóspál *et al.* 2013). HD 21997 may be a unique object or may be the tip of an iceberg of many fainter hybrid disks. In the latter case, the gas-rich phase of disk evolution could be significantly longer than previously thought, also affecting the formation of planets.

## 5. Conclusions

Before ALMA, less than a dozen debris disks were spatially well resolved at millimeter wavelengths. These studies constituted the first steps towards understanding the evolution of young debris systems. ALMA is now in the process of multiplying the sample of resolved debris disks, thereby allowing to study more general trends. The first observations have already produced remarkable results. ALMA confirmed that the rising brightness profile seen in certain debris disks is consistent with the stirring models. It also identified disks where the derived initial mass was too high for the self-stirring scenario, pointing to planetary stirring. It imaged structures that hint at so-far unknown planets shepherding the dust or clearing gaps. Finally, it detected a large amount of molecular gas, leading to the discovery of the first hybrid disk where the dust is secondary but the gas component is still primordial. In the coming years, using an increasingly larger sample, we will be able to answer several fundamental questions concerning the dust and gas content of debris disks. Dust continuum observations will be used to, e.g., verify whether the location of the dust-producing region expands with time, to identify asymmetric dust distributions and their relationship with planets. Gas line observations with ALMA are the best way to study the cold gas content of debris disks, providing a possibility to study how the gas evolves, and what mechanism produces the gaseous material. These exciting new results ahead of us will transform the research of debris disks, and – in connection with the age and environmental information provided by their membership in young moving groups – will shed light onto a period of very active planet formation.

## Acknowledgment

This work was supported by the Momentum grant of the MTA CSFK Lendület Disk Research Group. A.M. acknowledges support from the Bolyai Research Fellowship of the Hungarian Academy of Sciences.

# References

Beuzit, J.-L., Boccaletti, A., Feldt, M., *et al.* 2010, *Proceedings of the workshop "Pathways Towards Habitable Planets"*, V Coudé du Foresto, D. M. Gelino, and I. Ribas, eds., p. 231

Boley, A. C., Payne, M. J., Corder, S., *et al.* 2012, *ApJL*, 750, L21

Corder, S., Carpenter, J. M., Sargent, A. I., *et al.* 2009, *ApJL*, 690, L65

Dent, W. R. F., Wyatt, M. C., Roberge, A., *et al.* 2014, *Science*, 343, 1490

Eiroa C., *et al.* 2013, *A&A*, 555, A11

Fernández, R., Brandeker, A., & Wu, Y. 2006, *ApJ*, 643, 509

France, K., Roberge, A., Lupu, R. E., Redfield, S., & Feldman, P. D. 2007, *ApJ*, 668, 1174

Greaves, J. S., Holland, W. S., Moriarty-Schieven, G., *et al.* 1998, *ApJL*, 506, L133

Hobbs, L. M., Vidal-Madjar, A., Ferlet, R., Albert, C. E., & Gry, C. 1985, *ApJ*, 293, L29

Holland, W. S., Greaves, J. S., Zuckerman, B., *et al.* 1998, *Nature* , 392, 788

Hughes, A. M., Wilner, D. J., Andrews, S. M., *et al.* 2011, *ApJ*, 740, 38

Kalas, P., Graham, J. R., & Clampin, M. 2005, *Nature*, 435, 1067

Kenyon, S. J. & Bromley, B. C. 2008, *ApJS*, 179, 451

Kennedy G. M. & Wyatt M. C. 2010, *MNRAS*, 405, 1253

Kóspál, Á., Moór, A., Juhász, A., *et al.* 2013, *ApJ*, 776, 77

Krivov A. V. 2010, *RAA*, 10, 383

Kuchner M. J. & Holman M. J. 2003, *ApJ*, 588, 1110

MacGregor, M. A., Wilner, D. J., Rosenfeld, K. A., *et al.* 2013, *ApJL*, 762, L21

Macintosh, B. A., Graham, J. R., Palmer, D. W., *et al.* 2008, *Proceedings of the SPIE*, N. Hubin, C. E. Max, P. L Wizinowich, eds., Vol 7015, article id. 701518

Marois C., Zuckerman B., Konopacky Q. M., Macintosh B., & Barman T. 2010, *Nature*, 468, 1080

Matthews B. C., Krivov A. V., Wyatt M. C., Bryden G., & Eiroa C. 2014a, *Protostars and Planets VI*, H. Beuther, R. Klessen, C. Dullemond, Th. Henning, eds., p. 521-544.

Matthews B., Kennedy G., Sibthorpe B., Booth M., Wyatt M., Broekhoven-Fiene H., Macintosh B., & Marois C. 2014b, *ApJ*, 780, 97

Moór, A., Ábrahám, P., Juhász, A., *et al.* 2011, *ApJ*, 740, L7

Moór, A., *et al.* 2013a, *ApJ*, 775, L51

Moór, A., Juhász, A., Kóspál, Á., *et al.* 2013b, *ApJ*, 777, 25

Moór, A., Kóspál, Á., Ábrahám, P., *et al.* 2015, *MNRAS*, 447, 577

Mustill A. J. & Wyatt M. C. 2009, *MNRAS*, 399, 1403

Piétu, V., di Folco, E., Guilloteau, S., Gueth, F., & Cox, P. 2011, *A&A*, 531, L2

Rameau J., *et al.* 2013, *ApJ*, 779, L26

Raymond S. N., Kokubo E., Morbidelli A., Morishima R., & Walsh K. J. 2014, *Protostars and Planets VI*, H. Beuther, R. Klessen, C. Dullemond, Th. Henning, eds., p. 595-619.

Redfield, S. 2007, *ApJ*, 656, L97

Rhee J. H., Song I., Zuckerman B., & McElwain M. 2007, *ApJ*, 660, 1556

Ricci, L., Carpenter, J. M., Fu, B., *et al.* 2015, *Apj*, 798, 124

Riviere-Marichalar, P., Barrado, D., Augereau, J.-C., *et al.* 2012, *A&A*, 546, L8

Riviere-Marichalar, P., Barrado, D., Montesinos, B., *et al.* 2014, *A&A*, 565, A68

Slettebak, A. 1975, *ApJ*, 197, 137

Su, K. Y. L., Rieke, G. H., Stapelfeldt, K. R., *et al.* 2009, *ApJ*, 705, 314

Su, K. Y. L., Morrison, S., Malhotra, R., *et al.* 2015, *ApJ*, 799, 146

Telesco, C. M., Fisher, R. S., Wyatt, M. C., *et al.* 2005, *Nature*, 433, 133

Williams, J. P. & Cieza, L. A. 2011, *ARA&A*, 49, 67

Wyatt, M. C. 2003, *ApJ*, 598, 1321

Wyatt, M. C. 2005, *A&A*, 440, 937

Wyatt, M. C. 2008, *ARA&A*, 46, 339

Zuckerman, B., Forveille, T., & Kastner, J. H. 1995, *Nature*, 373, 494

Zuckerman B., Rhee J. H., Song I., & Bessell M. S. 2011, *ApJ*, 732, 61

*Young Stars & Planets Near the Sun*
*Proceedings IAU Symposium No. 314, 2015*
*J. H. Kastner, B. Stelzer, & S. A. Metchev, eds.*

© International Astronomical Union 2016
doi:10.1017/S1743921315006304

# Cosmic Rays and the Origin of Volatiles in Protoplanetary Disks

**Germán Chaparro-Molano[1] and Inga Kamp[2]**

[1] Universidad ECCI, Bogotá, Colombia
email: gchaparrom@ecci.edu.co
[2] Kapteyn Astronomical Institute, Rijksuniversiteit Groningen, The Netherlands
email: kamp@astro.rug.nl

**Abstract.** The origin of water and other volatiles in protoplanetary disks can be either interstellar or due to chemical processing during the protoplanetary disk phase. Depending on the strength of the ionization field present during this stage, an active chemical evolution in the protoplanetary disk midplane can lead to formation of complex volatiles on timescales shorter than the disk dissipation timescale. For this reason, we investigate the effects of cosmic rays and the usually neglected cosmic ray induced UV ionization field in time dependent chemical models of protoplanetary disks. These results are benchmarked against our current knowledge of the chemical composition of cometary ices. We conclude that water and other, more complex volatiles can be preserved in the ice mantles of dust grains. This ice mantle growth can also have a significant impact on the dust opacity and hence on the temperature profile of the disk midplane. This effect will be observable in the near future with ALMA.

**Keywords.** Astrochemistry, Comets: General, Protoplanetary Disks, (ISM:) Cosmic Rays

## 1. Introduction

Observations of comets present to date a fairly inhomogeneous dataset, with comets from different families (Jupiter family, Oort cloud) and observations taken at different orbital times before and after aphelion. Bockelee-Morvan (2010) summarize the abundance pattern emerging from those measurements. Water is the dominant species ($\sim$80%), followed by $CO$, $CO_2$, $CH_3OH$, $CH_4$, $NH_3$, and $H_2CO$ (in order of decreasing abundance). There are about two orders of magnitude spread in $CO$/water abundance and about one order of magnitude for other species. The reason can be different evolutionary tracks, formation histories, formation locations or different orbital phases at which the data was taken (Ferrín 2013). However, if no important sources of ionization are present in the disk midplane, we cannot expect primordial material to be chemically processed into the more complex species found in comets. For this reason, we propose that cosmic ray induced processes can be a significant source of ionization in comet-forming regions.

Cosmic rays can penetrate down into the midplane of a typical T Tauri protoplanetary disk (PPD) beyond $\sim$1 AU (depending on the surface density power law). This is where planetesimals are expected to form from small, bare and ice-coated dust grains. The composition of the ices that form during the protoplanetary disk stage can be strongly influenced by cosmic ray driven chemistry. This includes secondary ionization processes that can be enhanced up to an order of magnitude, in the case of the cosmic ray induced UV field (Chaparro Molano & Kamp 2012).

## 2. Model predictions for ices

We use the thermo-chemical disk modeling code ProDiMo (Woitke *et al.* 2009). The gas thermal balance and chemistry are solved consistently using a network of 90 species

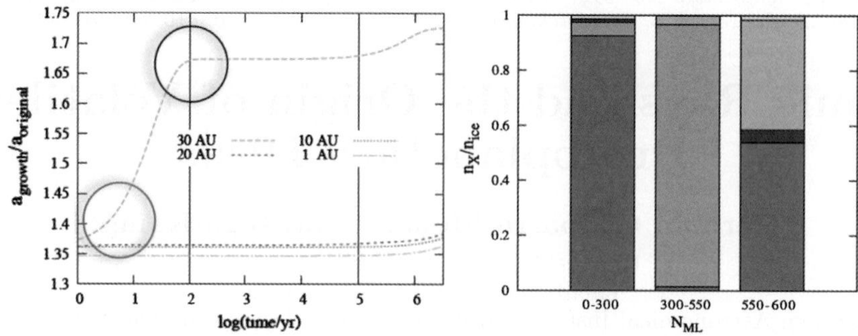

**Figure 1.** *Left*: Growth factor evolution of a 0.2 $\mu$m dust grain due to ice mantle deposition. Red: original ISM mantle. Black: PPD mantle. *Right*: Composition of ice mantles at 30 AU.

(gas and ice) and 1100 reactions (mostly UMIST, Woodall *et al.* 2007). In this first assessment, we neglect surface chemistry. After solving the disk thermal balance, we run a time-dependent chemistry solver using molecular cloud abundances as initial conditions.

Cosmic ray induced photoprocesses favor $CO_2$ ice formation at the expense of water, $CH_4$ and CO. Reactions with $He^+$ (created by cosmic rays) destroy CO, releasing oxygen to form $CO_2$ and water. After 1 Myr, cosmic ray induced UV photons ionize $CH_3$, leading to $CH_5^+$ and subsequent recombinations that form stable $CH_4$.

Acharyya *et al.* (2011) found that ice mantles can increase the grain size by $\sim$30% in cold molecular clouds. We account here for grain growth in two ways: (1) PPD>ISM size distribution (2) ice mantle growth. The first $\sim$300 ML of mostly water ice form during the first 10 Myr (ISM dust, 20 K, $10^6$ cm$^{-3}$). The remaining $\sim$600 ML build up in the PPD model (PPD grains, 20 K, $7\times10^{10}$ cm$^{-3}$). A 0.2 $\mu$m grain grows by a factor $\sim$1.8 due to ice mantle deposition (Fig. 1).

## 3. Conclusions

We thus propose a possible scenario for the formation of ice mantles on the surface of dust grains: (1) The first ice mantles form during the lifetime of molecular clouds, typically a few Myr (e.g. Tielens & Hagen 1982, Chang *et al.* 2007). In this stage, surface chemistry can already lead to the formation of more complex organic species such as methanol. (2) These ices are processed during the formation of a star+disk system, which takes less than 1 Myr (e.g. Visser *et al.* 2009). (3) During the protoplanetary disk stage, a second ice mantle forms. Evolutionary timescales in this stage can be up to a few Myrs. These ice mantles are formed in layers whose composition strongly depends on the distance to the proto-Sun and on the strength of the local ionization field. This effect also leads to a shift in the grain size distribution function, which will affect the disk midplane temperature profile.

## References

Acharyya, K., Hassel, G. E., & Herbst. E. 2011, *ApJ*, 732, 73

Bockelee-Morvan, D. 2010, in *Physics and Astrophysics of Planetary Systems, EAS Publication Series*, 41, 313

Chaparro Molano, G. & Kamp, I. 2012, *A&A*, 537, A138

Chang, Q., Cuppen, H. M., & Herbst, E. 2007, *A&A*, 469, 973

Ferrín, I. 2013, *MNRAS*, 442, 2, p.1731

Tielens, A. G. G. M. & Hagen, W. 1982, *A&A*, 114, 245

Visser, R., van Dishoeck, E. F., Doty, S. D., & Dullemond, C. P. 2009, *A&A*, 495, 881

Woitke, P., Kamp, I., & Thi, W.-F. 2009, *A&A*, 501, 383

Woodall, J., Agúndez, M., Markwick-Kemper, A. J., & Millar, T. J. 2007, *A&A*, 466, 1197

*Young Stars & Planets Near the Sun*
*Proceedings IAU Symposium No. 314, 2015*
*J. H. Kastner, B. Stelzer, & S. A. Metchev, eds.*

© International Astronomical Union 2016
doi:10.1017/S1743921315006262

# T Tauri Disk Lifetime in the Lupus Association

**P. A. B. Galli[1], C. Bertout[2], R. Teixeira[1] and C. Ducourant[3,4]**

[1]Instituto de Astronomia, Geofísica e Ciências Atmosféricas, Universidade de São Paulo, Rua do Matão, 1226 - Cidade Universitária, 05508-900, São Paulo - SP, Brazil.
email: `galli@astro.iag.usp.br`
[2]Institut d'Astrophysique, 98bis, Bd. Arago, 75014 Paris, France.
[3]Univ. Bordeaux, LAB, UMR 5804, F-33270, Floirac, France.
[4]Observatoire Aquitain des Sciences de l'Univers, CNRS-UMR 5804, 33270 Floirac, France.

**Abstract.** In a recent study, we derived individual distances for a sample of pre-main sequence stars that define the comoving association of young stars in the Lupus star-forming region. Here, we use these new distances to investigate the mass and age distributions of Lupus T Tauri stars and derive the average disk lifetime in the Lupus association based on an empirical disk model.

**Keywords.** stars: variables: T Tauri, Herbig Ae/Be - stars: formation - stars: evolution

## 1. Introduction

It has been hypothesized for a long a time (Bertout 1989) that classical T Tauri stars (CTTSs), which are actively accreting from circumstellar disks, become weak-line T Tauri stars (WTTSs), which show no spectroscopic evidence for accretion, when their disks dissipate. This implies that the timescale of disk dissipation should be only a few Myr to explain both subgroups coexisting in the Hertzsprung-Russell diagram (HRD). Bertout *et al.* (2007) obtained for the first time the average lifetime of a circumstellar disk in terms of the mass of the parent star in the Taurus-Auriga star-forming region. However, the evolution of circumstellar disks is expected to depend on environmental criteria. Here, we investigate whether the different environment of the Lupus star-forming region affects the lifetime of the circumstellar disks in this region.

## 2. Physical properties of the Lupus comoving stars

The sample of stars used in thus study is based on the comoving stars of the Lupus association with individual distances derived by Galli *et al.* (2013). Using the photometric and spectroscopic information available in the literature we compute the photospheric luminosity of 92 T Tauri stars (TTSs), and estimate their masses and ages from the Siess *et al.* (2000) evolutionary models. Fig. 1 illustrates the HRD of the Lupus association.

We investigated the mass and age distribution of the CTTSs and WTTSs in Lupus and concluded from KS-tests and Wilcoxon rank sum tests that the probability of both TTS subclasses being drawn from the same age distribution is very low. The CTTSs are on average younger than WTTSs and their distribution in the HRD (see Fig. 1) is shifted towards lower masses and late spectral types. We also confirmed in this work that the more dispersed "off-cloud" WTTSs surrounding the Lupus clouds are older than the "on-cloud" population of WTTSs located in the direction of the main star-forming clouds of the complex (Lupus 1–4).

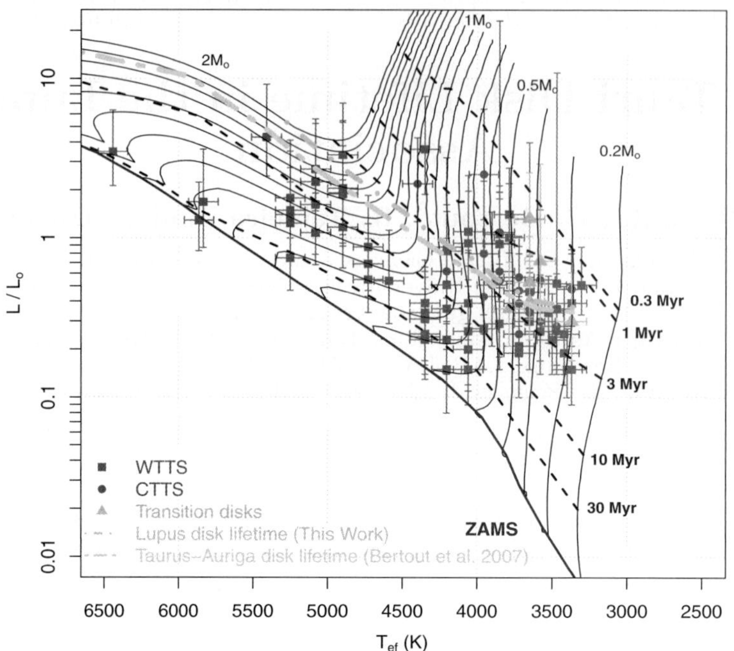


**Figure 1.** HRD of the Lupus association using the Siess *et al.* (2000) models.

## 3. The evolution from CTTS to WTTS

Following the disk model developed by Bertout *et al.* (2007), we assume that the disk mass of a TTS is given by $M_d = \alpha(M/M_\odot)^\beta$ while the mass accretion rate (in units of $M_\odot/yr$) is given by $\dot{M}_{acc} = \gamma(M_\star/M_\odot)^{2.1}$. The disk lifetime $\tau_d = M_d/\dot{M}_{acc}$ can be written as $\log \tau_d = \log(\alpha/\gamma) + (\beta - 2.1)\log(M_\star/M_\odot)$, where $\alpha$, $\beta$ and $\gamma$ are parameters to be determined from the modeling, using the stellar masses derived in this work.

The model predicts that the star is a CTTS, if $t_\star \leqslant \tau_d(M_\star)$, and a WTTS, if $t_\star > \tau_d(M_\star)$. We compare the mass distribution of the modeled CTTSs (and WTTSs) with the observed CTTSs (and WTTSs) in our sample, and run a KS-test to find the best match between these populations for a given combination of $\alpha$, $\beta$ and $\gamma$. We vary these parameters in the range of $5 \leqslant \log(\alpha/\gamma) \leqslant 8$ and $-4 \leqslant \beta \leqslant 4$ in steps of 0.01. Doing so, we find $\log(\alpha/\gamma) = 6.53 \pm 0.05$ and $\beta = 2.65 \pm 0.10$. This implies that the average disk lifetime in the Lupus association predicted by our model is $\tau_d \simeq 3 \times 10^6 \, (M_\star/M_\odot)^{0.55}$yr.

The average disk lifetime given by this model for the Taurus-Auriga association (see Bertout *et al.* 2007) is $\tau_d \simeq 4 \times 10^6 \, (M_\star/M\odot)^{0.75}$ yr. Thus, our heuristic model predicts that the two associations may have different disk lifetime mass dependencies. Although this result will need confirmation with statistically more significant samples, it gives a first indication that disk lifetimes can be different in different star-forming regions.

## References

Bertout, C. 1989, *ARA&A*, 27, 351
Bertout, C., Siess, L., & Cabrit, S. 2007, *A&A*, 473, L21
Galli, P. A. B., Bertout, C., Teixeira, R., & Ducourant, C. 2013, *A&A*, 558, A77
Siess, L., Dufour, E., & Forestini, M. 2000, *A&A*, 358, 593

*Young Stars & Planets Near the Sun*
Proceedings IAU Symposium No. 314, 2015
J. H. Kastner, B. Stelzer, & S. A. Metchev, eds.

© International Astronomical Union 2016
doi:10.1017/S1743921315005979

# Rings of $C_2H$ in the Molecular Disks Orbiting TW Hya and V4046 Sgr

J. H. Kastner[1], C. Qi[2], U. Gorti[3], P. Hily-Blant[4], K. Oberg[2],
T. Forveille[4], S. Andrews[2], and D. Wilner[2]

[1]Lab. for Multiwavelength Astrophysics, Rochester Institute of Technology; jhk@cis.rit.edu
[2]Harvard-Smithsonian Center for Astrophysics
[3]SETI Institute & NASA Ames Research Center
[4]Université Grenoble Alpes & Institut de Planétologie et d'Astrophysique de Grenoble

**Abstract.** We have used the Submillimeter Array (SMA) to image, at $\sim 1''$ resolution, $C_2H(3\text{--}2)$ emission from the molecule-rich circumstellar disks orbiting the nearby, classical T Tauri star systems TW Hya and V4046 Sgr. The SMA imaging reveals that the $C_2H$ emission exhibits a ring-like morphology within each disk; the radius of the inner hole of the $C_2H$ ring within the V4046 Sgr disk ($\sim 70$ AU) is somewhat larger than than of its counterpart within the TW Hya disk ($\sim 45$ AU). We suggest that, in each case, the $C_2H$ emission likely traces irradiation of the tenuous surface layers of the outer disks by high-energy photons from the central stars.

**Keywords.** protoplanetary disks, stars: individual (TW Hya, V4046 Sgr), stars: pre-main sequence

## 1. Introduction

Our single-dish radio telescope molecular line surveys of the evolved protoplanetary disks orbiting the nearby, actively accreting T Tauri star systems TW Hya and V4046 Sgr established that both display strong $C_2H$ line emission (Kastner *et al.* 2014). As $C_2H$ is likely the dissociation product of more complex hydrocarbons and organics, it potentially provides a diagnostic of disk irradiation by high-energy (UV and X-ray) photons generated via accretion and coronal activity at the central T Tauri stars. With this as motivation, we have used the Submillimeter Array (SMA) to image, at $\sim 1''$ resolution, $C_2H(3\text{--}2)$ emission from the TW Hya and V4046 Sgr disks.

## 2. SMA observations of $C_2H$ emission from TW Hya and V4046 Sgr

**TW Hya:** The results of SMA imaging of $C_2H$ emission from this single, nearby ($D = 54$ pc) $\sim 0.8$ $M_\odot$ T Tauri star — the namesake of the $\sim 8$ Myr-old TW Hya Association — are reported in Kastner *et al.* (2015; hereafter K15). These observations have revealed that the $C_2H$ line surface brightness exhibits a ring-like emission morphology, with an inner hole of radius $\sim 45$ AU (Fig. 1). Based on comparison of the SMA $C_2H$ imaging results for TW Hya with previous continuum and line emission imaging and spectroscopy of that disk, it appears that the $C_2H$ abundance is enhanced in warm, low-density, disk surface layers that lie beyond the CO "ice line" (as traced by $N_2H^+$ emission; Fig. 1).

**V4046 Sgr:** This close ($\sim 2.4$-day) binary system, a member of the $\sim 20$ Myr-old $\beta$ Pic Moving Group, consists of nearly equal, $\sim 0.9$ $M_\odot$ components (Rosenfeld *et al.* 2012). The circumbinary molecular disk orbiting V4046 Sgr is somewhat larger than that of TW Hya ($R \sim 350$ AU vs. $R \sim 200$ AU; Rosenfeld *et al.* 2012; Qi et al. 2013), and — in contrast to the nearly pole-on TW Hya — is viewed at intermediate inclination ($i \approx 34°$; Rosenfeld *et al.* 2012). Extended- and compact-configuration SMA observations

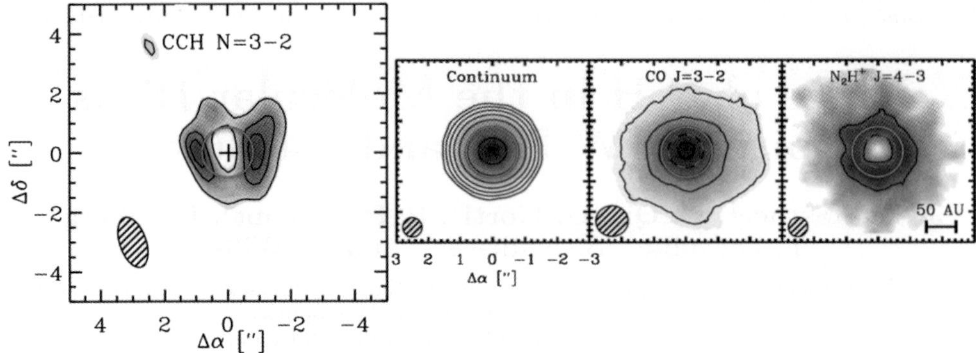

**Figure 1.** Interferometric imaging of the TW Hya disk. Left panel: SMA $C_2H(3–2)$ image, from K15. Left center, right center, and right panels: ALMA 372 GHz continuum, SMA CO(3–2), and ALMA $N_2H^+$(4–3) images, from Qi *et al.* (2013). The red circles indicate the $\sim$45 AU inner radius of the $C_2H$ ring, as determined from model fitting described in K15.

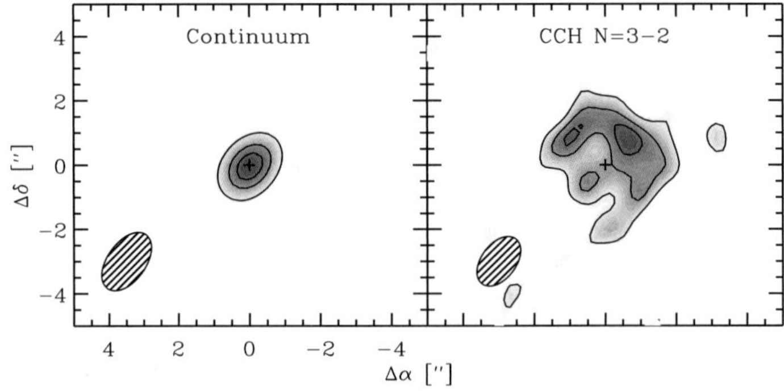

**Figure 2.** SMA $C_2H(3–2)$ image of the disk orbiting V4046 Sgr (right) alongside the 260 GHz continuum image obtained from the same (extended+compact configuration) dataset (left).

of $C_2H(3–2)$ emission from the V4046 Sgr disk were acquired in 2015 March and May, respectively; the resulting image is displayed in Fig. 2. It is evident that V4046 Sgr, like TW Hya, displays a ring-like $C_2H$ morphology, but with a more asymmetric appearance (perhaps due to its intermediate disk inclination) and somewhat larger inner $C_2H$ "hole" radius ($\sim$70 AU).

## 3. Remarks

We (K15) have proposed that the $C_2H$ ring in the TW Hya disk may trace particularly efficient photo-destruction of small grains and/or photodesorption and photodissociation of hydrocarbons (principally, $C_2H_2$) derived from grain ice mantles in the surface layers of the outer disk. Comparative modeling of the two SMA datasets will establish whether the V4046 Sgr $C_2H$ ring is in fact larger and more asymmetric than that of TW Hya and, more generally, should shed further light on the mechanism(s) responsible for production of $C_2H$ in evolved, protoplanetary disks.

## References


Kastner, J. H., Hily-Blant, P., Rodriguez, D. R., Punzi, K., & Forveille, T. 2014, *ApJ*, 793, 55
Kastner, J. H., Qi, C., Gorti, U., *et al.* 2015, *ApJ*, in press (arXiv:1504.05980; K15)
Qi, C., Öberg, K. I., Wilner, D. J., *et al.* 2013, *Science*, 341, 630
Rosenfeld, K. A., Andrews, S. M., Wilner, D. J., & Stempels, H. C. 2012, *ApJ*, 759, 119

*Young Stars & Planets Near the Sun*
Proceedings IAU Symposium No. 314, 2015
J. H. Kastner, B. Stelzer, & S. A. Metchev, eds.

© International Astronomical Union 2016
doi:10.1017/S174392131500589X

# Vertical Shear Instability in the Solar Nebula

## Min-Kai Lin and Andrew N. Youdin

Department of Astronomy and Steward Observatory,
University of Arizona, 933 North Cherry Avenue, Tucson, AZ 85721, USA
email: minkailin@email.arizona.edu, youdin@email.arizona.edu

**Abstract.** We quantify the thermodynamic requirement for the Vertical Shear Instability and evaluate its relevance to realistic protoplanetary disks as a potential route to hydrodynamic turbulence.

**Keywords.** accretion, accretion disks, hydrodynamics, instabilities, methods: analytical

## 1. Introduction

A purely hydrodynamic route to turbulence has important implications for the evolution of cold protoplanetary disks and solids within them. One such candidate is the Vertical Shear Instability (VSI) operating in disks where the orbital frequency $\Omega$ depends on the height $z$ away from the disk midplane. While astrophysical disks generally possess vertical shear, VSI also requires short thermal timescales. We determine a quantitative thermodynamic requirement for the VSI, and apply our results to realistic disk conditions. We find the VSI can operate effectively at 5—50AU in a typical protoplanetary disk, with characteristic growth times of 30 orbits.

## 2. The need for rapid cooling for the VSI

A radial variation in a disk's entropy or temperature profile leads to vertical shear, $\partial_z \Omega \neq 0$. This is a source of free energy, and may thus lead to instability (Goldreich & Schubert 1967; Urpin 2003; Barker & Latter 2015), provided the vertical lengthscale of the disturbance is much larger than its radial lengthscale. However, in stably stratified, irradiated protoplanetary disks (Chiang & Youdin 2010), the VSI also requires rapid cooling to overcome the stabilizing influence of vertical buoyancy (Nelson *et al.* 2013).

More specifically, for a vertically isothermal disk with radial temperature dependence $T \propto r^q$, the cooling or thermal relaxation timescale $t_c$ must be sufficiently small for the VSI to operate effectively:

$$t_c \Omega_\mathrm{K} < \frac{h|q|}{\gamma - 1}, \tag{1}$$

where $\Omega_\mathrm{K}$ is the Keplerian frequency, $h$ is the disk aspect ratio, and $\gamma$ is the adiabatic index (Lin & Youdin 2015). Since protoplanetary disks are typically thin, $h \ll 1$, Eq. 1 implies that $t_c \ll \Omega_\mathrm{K}$ is required for the VSI. That is, the thermal timescale must be significantly shorter than the dynamical timescale, which reflects the fact that vertical shear is weakly destabilizing, while vertical buoyancy is strongly stabilizing.

## 3. Applicability of the VSI in realistic protoplanetary disks

We solve the hydrodynamic linear stability problem for axisymmetric perturbations in a disk model of the Solar Nebula described in Chiang & Youdin (2010). We include in

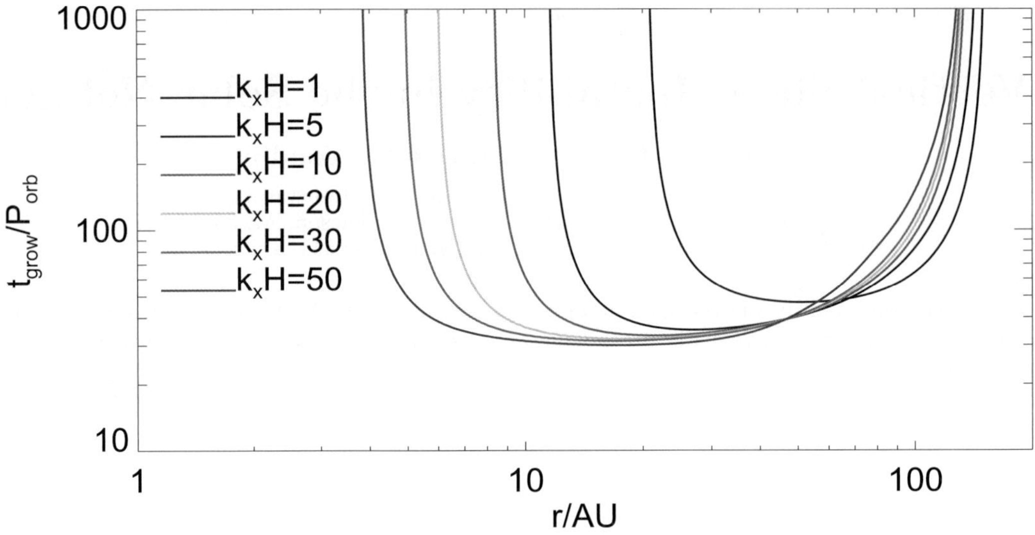



**Figure 1.** Characteristic growth times of the Vertical Shear Instability in a typical protoplanetary disk model. Here, $k_x$ is the radial wavenumber of the disturbance under consideration and $H$ is the disk scale-height.

the energy equation a realistic cooling function based on dust opacity, which depends on $z$ as well as the disturbance lengthscale. (For details, see Lin & Youdin 2015.)

In Fig. 1 we plot characteristic VSI growth times for a range of radial perturbation wavenumbers. Growth times increase rapidly at $\lesssim$ 5AU and $\gtrsim$ 100AU as the cooling times violate Eq. 1 in these regions. Thus, the VSI is most effective at intermediate radii. Notice also the growth times are strongly dependent on lengthscale toward the optically-thick inner disk, with smaller-scale disturbances being able to operate further inwards. In the optically-thin outer disk, cooling times are independent of $k_x$ and all scales are stabilized beyond approximate the same radius.

## 4. Conclusions

Vertical shear is ubiquitous in astrophysical disks. Such configurations are unstable if the cooling timescale is significantly shorter than the dynamical timescale (Eq. 1) in a thin disk. This can be satisfied for small-scale disturbances at 10s of AU in typical protoplanetary disk models with realistic cooling. We have thus shown that the VSI is dynamically important in the outer parts of protoplanetary disks. This is expected to have significant implications for mass transport and dust evolution in these regions.

## References


Barker, A. J. & Latter, H. N. 2015, *MNRAS*, 450, 21
Chiang, E. & Youdin, A. N. 2010, *Annual Review of Earth and Planetary Sciences*, 38, 493
Goldreich, P. & Schubert, G. 1967, *ApJ*, 150, 571
Lin, M.-K. & Youdin, A. N. 2015, *ApJ*, submitted, ArXiv: 1505.02163
Nelson, R. P., Gressel, O., & Umurhan, O. M. 2013, *MNRAS*, 435, 2610
Urpin, V. 2003, *A&A*, 404, 397

*Young Stars & Planets Near the Sun*
*Proceedings IAU Symposium No. 314, 2015*
*J. H. Kastner, B. Stelzer, & S. A. Metchev, eds.*

© International Astronomical Union 2016
doi:10.1017/S1743921315006043

# HD 76582's Circumstellar Debris Disk

## J. P. Marshall on behalf of the SONS consortium

School of Physics, UNSW Australia, High Street, Kensington, Sydney, NSW 2052, Australia
email: jonty.marshall@unsw.edu.au

**Abstract.** The debris disk host star HD 76582 was observed at 450 $\mu$m and 850 $\mu$m as part of
the JCMT/SCUBA-2 debris disk legacy survey 'Sub-millimetre Observations of Nearby Stars'
(SONS). The sub-millimetre data are inconsistent with a disk undergoing a steady-state collisional cascade. Combining the sub-millimetre (sub-mm) measurements with mid- and far-
infrared measurements from Spitzer and Herschel, we simultaneously model the disk's thermal
emission and radial extent in a self-consistent manner.

**Keywords.** stars:individual: HD 76582 – stars: circumstellar matter – infrared: planetary systems – sub-millimetre: planetary systems

---

The presence of a cool circumstellar dust disk around a mature, main-sequence star
is a visible signpost of a planetary system. These disks are the remnants of asteroidal
and cometary bodies broken up in mutual collisions, and are therefore known as debris
disks. Debris disks are composed of rings of icy and rocky bodies ranging in size from
micrometre-sized dust grains to kilometre-sized planetesimals. These structures are analogous to the Edgeworth-Kuiper belt in our solar system. Understanding these structures
is fundamental to obtaining a full picture of the outcomes of the planet formation process,
and the evolution of planetary systems.

Typically, debris disk host stars are identified through the detection of infrared excess
above the stellar photosphere from near-infrared to sub-mm wavelengths. Surveys to date
have identified disks around 20% of main-sequence stars. Interpretation of this emission
is often limited by the lack of resolved imaging of the disk to constrain its architecture.

Here we present observations and preliminary interpretation of the Herschel/PACS far-
infrared and JCMT/SCUBA-2 sub-mm observations of the debris disk around HD 76582,
an F0 IV star at a distance of 46 pc, taken as part of the 'SONS' survey.

We have compiled data spanning optical to far-infrared wavelengths to complement
the SONS sub-mm photometry. The complete SED is shown in Fig. 1. The shape of
the SED, with its mid-infrared rise, broad far-infrared excess and steep sub-mm decline,
precludes the adoption of a single temperature (narrow annulus) model for the disk.

The disk is seen to be extended in the Herschel far-infrared images, providing a limit
on the size of the outer system. However, the angular resolution is too poor to offer any
constraint on the disk's architecture, or the location of the inner edge of the debris disk.
At sub-mm wavelengths the disk is detected at both 450 $\mu$m and 850 $\mu$m, see Fig. 2.

The steep sub-millimetre slope of HD 76582's disk is of particular interest. A power-
law disk model requires an exponent of 7 for the grain size distribution, compared to
typical values of 3-4 expected from theory. Disks with similar slopes have been observed
at far-infrared wavelengths by the Herschel Open Time Key Programme 'DUNES'.

A simple three component power-law model with astrosilicate dust composition fitted
to the data reproduces the observations tolerably well. The model is comprised of two
10 au-wide annuli, and a broad component lying between these two rings. The narrow
annuli follow standard assumptions regarding min/max grain size, surface brightness and

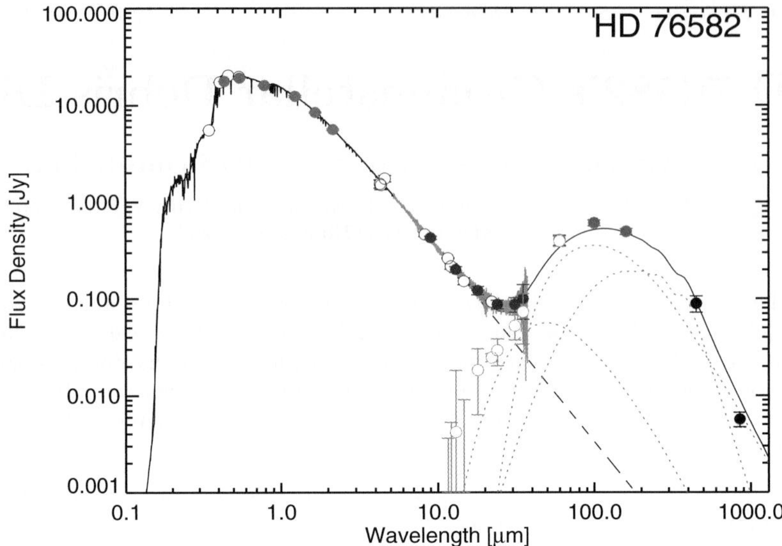

**Figure 1.** SED of HD 76582: green points denote optical/near-infrared photometry used to scale the photosphere model, blue points (and line) are the Spitzer/IRS data, red points are the Herschel/PACS data, black points are the JCMT/SCUBA-2 data, and the white points are other ancillary data. Excess measurements are shown as grey, hollow data points. The solid line is the total star+disk model, whilst the dashed and dotted lines denote the star and disk individual contributions, assuming a three-component disk model (see later).

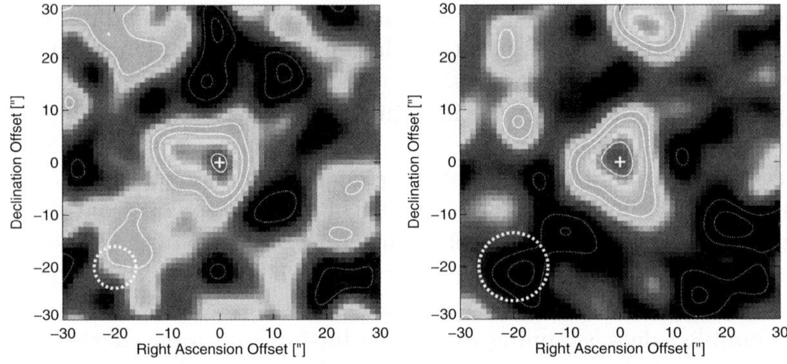

**Figure 2.** JCMT/SCUBA-2 images of HD 76582 at 450 $\mu$m (left) and 850 $\mu$m (right). The contours denote -3, -2, 2, 3, and 5-$\sigma$ steps for background r.m.s. values of 0.306 mJy/arcsec$^2$ at 450 $\mu$m and 0.005 mJy/arcsec$^2$ at 850 $\mu$m. Dashed contours are negative values. The white '+' denotes the stellar position. Image scale is 1" per pixel. The instrument beam is denoted by the dashed circle in the bottom left. Orientation is north up, east left.

size distribution, but the broad component has a minimum grain size of 50 $\mu$m and the steep grain size exponent mentioned previously.

Whilst a two component fit can replicate the data, it leaves significant residual flux in the images compared to the three component fit. We do not absolutely favour a three component fit however, due to the number of free parameters available rendering its adoption physically unmeaningful.

*Young Stars & Planets Near the Sun*
Proceedings IAU Symposium No. 314, 2015
J. H. Kastner, B. Stelzer, & S. A. Metchev, eds.

© International Astronomical Union 2016
doi:10.1017/S1743921315006122

# Sensitive Identification of Nearby Debris Disks via Precise Calibration of WISE Data

**Rahul Patel[1], Stanimir Metchev[1,2,3], Aren Heinze[1] and Joe Trollo[1]**

[1]Stony Brook University - email: `rahul.patel.1@stonybrook.edu`
[2]Univeristy of Western Ontario - email: `smetchev@uwo.edu`
[3]Centre for Planetary Science and Exoploration

**Abstract.** Using data from the WISE All-Sky Survey, we have found > 100 new infrared excess sources around main-sequence Hipparcos stars within 75 pc. Our empirical calibration of WISE photospheric colors and removal of non-trivial false-positive sources are responsible for the high confidence (>99.5%) of detections, while our corrections to saturated W1 and W2 photometry have for the first time allowed us to search for new infrared excess sources around bright field stars in WISE. The careful calibration and filtering of the WISE data have allowed us to probe excess fluxes down to roughly 8% of the photospheric emission at $22\mu m$ around saturated stars in WISE. We expect that the increased sensitivity of our survey will not only aid in understanding the evolution of debris disks, but will also benefit future studies using WISE.

**Keywords.** infrared: stars, stars: circumstellar matter, methods: data analysis, statistical

---

## 1. Introduction

Micron sized dust grains in debris disks are generated from the grinding down of asteroids and comets, whose collisions are induced by the gravitational stirring of planetary bodies. Thus, the presence of a debris disk can act as a signpost for planetary systems. The majority of disks, discovered by IRAS, ISO, and the Spitzer Space Telescope (see reviews by Zuckerman 2001 & Wyatt 2008) are cold disks ($T_D < 100$ K) analogous to the Kuiper Belt in the Solar System. Warm disks ($T_D \sim 150 - 300$ K) probe activity in the terrestrial planet zone of the star. These warm disks are typically identified by an IR excess between $10$–$30\mu m$, but are relatively rare ($\sim 4\%$ for solar type stars; Trilling *et al.* 2008).

To compensate for the low incidence of warm disks, data from the Wide Field Infrared Survey Explorer all-sky mission (WISE; Wright *et al.* 2010) can be used to identify a large number of warm disks. This is due to the mission's increased sensitivity and resolution compared to the Infrared Astronomical Satellite (IRAS). The contemporaneous coverage of WISE in the W1, W2, W3, and W4 bands (3.4, 4.6, 12, and $22\mu m$, respectively) has previously been used to identify new disks using the W3 and W4 bands. In our survey, we include bright saturated stars in WISE as well as decrease our sensitivity to large sources of false-positives to probe faint dust levels, making this work complementary to previous studies.

## 2. Identification of Bona-Fide Excesses

We used five WISE colors to search for excesses at W3 or W4 (W1-W4, W2-W4, W3-W4, W1-W3 and W2-W3). The details of our sample can be found in Patel *et al.* (2014). In short, we used main-sequence Hipparcos stars within 120 pc, that lie outside the galactic plane ($|b| > 5°$). We included stars up to 4.5 mag in W1 and 2.7 mag in W2 by applying corrections to saturated photometry (Fig. 2 in Patel *et al.* 2014). Using the larger 120 pc set of stars as a parent sample, we accurately determined the

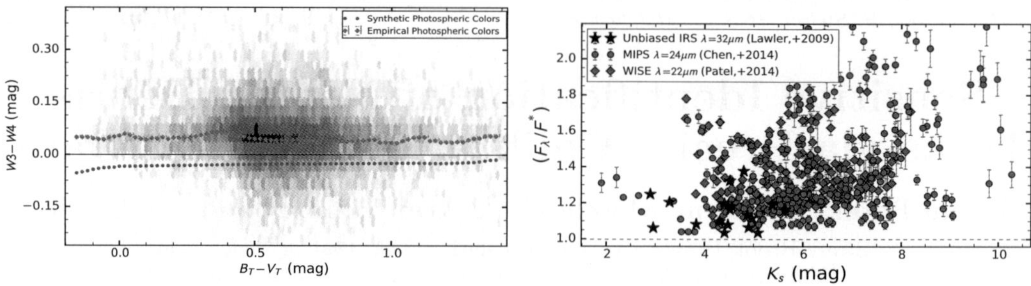

**Figure 1. Left:** A color-color diagram for Hipparcos main-sequence stars within 120 pc, showing the trend between IR and $B_T - V_T$ photospheric colors. We subtracted this dependence to eliminate the photospheric contribution, and the residuals (excess) were used to assess the 99.5% confidence level above which stars were selected as excess candidates. The synthetic colors were derived using NextGen models convolved to the WISE bandpasses. **Right:** Comparison of relative fluxes for excesses identified from Spitzer/MIPS (Chen *et al.* 2014), unbiased sample of Spitzer/IRS (Lawler *et al.* 2009) and our own WISE detections.

photospheric behavior of WISE colors. The left panel in Fig. 1 shows the empirically identified photospheric trend we used to subtract from our WISE colors. Over-plotted are synthetic photospheric colors that show a systematic offset from the empirical trend. Without accounting for the difference between the empirical and synthetic photospheric colors, the magnitude of the WISE excess can be overestimated.

Candidate debris disk stars were selected from each color based on the significance of their color excess at confidence levels > 99.5%. Along with a large array of filters to remove false-positives (see Sec. 2 in Patel *et al.* 2014), we verified our excess detections by assessing the significance of their weighted color excess (Patel *et al.* 2015). We also identified and removed contaminated stars by using higher resolution images from the unWISE survey (Lang 2014) and compared the relative centroid offsets between W3 and W4 for each star. Significant shifts indicated stars whose photometry were contaminated by nearby or background objects within the WISE beam.

## 3. Results and Discussion

We identified 237 excesses at either W3 or W4 within 75 pc of the Sun. Our survey has increased the total number of disks within 75 pc by 25%. Our study has also increased the census of excesses at $10-30\mu m$ by 35%. We show that 16–22% of A stars and 1.4–1.8% of FGK stars possess W4 excesses. The incidence of W3 excesses is much smaller; 0.8–1.0% of A stars and 0.04–0.08% of solar type stars possess a W3 excess. We detect IR excesses that are on average 30% above the photosphere at $22\mu m$, and in a few cases as little as 8% above the photosphere (right panel Fig. 1). Although WISE is not as sensitive to excesses as Spitzer/MIPS and Spitzer/IRS, this is not surprising, considering that those surveys were pointed and deeper.

### References

Chen, C., *et al.*, 2014, *ApJS*, 211, 25
Lang, D., 2014, *AJ*, 147,108
Lawler, S., *et al.* 2009, *ApJ*, 705, 89
Patel, R., *et al.* 2014, *ApJS*, 212, 10
Patel, R., *et al.* 2015, in prep
Trilling, D. E., *et al.* 2008, *ApJ*, 674, 1086
Wright, E. L., *et al.* 2010, *AJ*, 140, 1868
Wyatt, M., 2008, *ARA&A*, 46, 339
Zuckerman, B., 2001, *ARA&A*, 39, 549

*Young Stars & Planets Near the Sun*
*Proceedings IAU Symposium No. 314, 2015*
*J. H. Kastner, B. Stelzer, & S. A. Metchev, eds.*

© International Astronomical Union 2016
doi:10.1017/S1743921315006432

# HD141569A: Disk Dissipation Caught in Action

**J. Péricaud[1,2], E. Di Folco[1,2], A. Dutrey[1,2], J.-C. Augereau[3,4], V. Piétu[5] and S.Guilloteau[1,2]**

[1] Univ. Bordeaux, LAB, UMR 5804, F-33270, Floirac, France
email: `jessica.pericaud@obs.u-bordeaux1.fr`
[2] CNRS, LAB, UMR 5804, F-33270, Floirac, France
[3] Université' Grenoble Alpes, IPAG, 38000 Grenoble, France
[4] CNRS, IPAG, 38000 Grenoble, France
[5] IRAM, 300 rue de la piscine, F-38046 Saint Martin d'Hères, France

**Abstract.** Debris disks are usually thought to be gas-poor, the gas being dissipated by accretion or evaporation during the protoplanetary phase. HD141569A is a 5 Myr old star harboring a famous debris disk, with multiple rings and spiral features. I present here the first PdBI maps of the $^{12}$CO(2-1), $^{13}$CO(2-1) gas and dust emission at 1.3 mm in this disk. The analysis reveals there is still a large amount of (primordial) gas extending out to 250 AU, i.e. inside the rings observed in scattered light. HD141569A is thus a hybrid disk with a huge debris component, where dust has evolved and is produced by collisions, with a large remnant reservoir of gas.

**Keywords.** stars: circumstellar matter, protoplanetary disks, radio-lines: stars.

---

## A debris disk still containing gas

HD141569A is a $5\pm3$ Myr old star (Merín *et al.* 2004), of spectral type B9.5V/A0Ve, located $116\pm8$ pc away (van Leeuwen 2007). With a stellar mass of 2 $M_\odot$, the HD141569A system appears to be in an intermediate evolutionary stage between protoplanetary and debris disks.

A debris disk has been discovered first by *IRAS*, with an infrared excess of the same order of magnitude as $\beta$ Pictoris ($L_{disk}/L_\star = 8 \times 10^{-3}$; Sylvester *et al.* 1996). Optical images reveal that the debris disk around HD141569A is very complex, with a double-ring architecture, a large inner depletion within 125 AU, and arc and spiral features (e.g. Augereau *et al.* 1999, Biller *et al.* 2015). The dust appears to be of second generation origin, i.e. produced by collisions, as indicated by the timescale for collisions of $\sim 10^4$ years which is 100 times less than the age of the star (Boccaletti *et al.* 2003).

In addition to its impressive debris disk, CO gas has been detected around HD141569A (Zuckerman *et al.* 1995; Dent *et al.* 2005). NIR CO and other atomic lines have also been observed (Goto *et al.* 2006; Thi *et al.* 2014). The inferred total remnant mass of gas has thus been estimated in the range 80-135 $M_\oplus$ (Jonkheid *et al.* 2006).

We present here the first resolved maps of the $^{12}$CO $J$=2-1 and $^{13}$CO $J$=2-1 emission lines, which we obtained in 2014/2015 with the Plateau de Bure Interferometer array. The

**Table 1.** Best fit parameters from DiskFit gas modeling

| | Inclination (°) | Postion Angle (°) | $R_{out}$ (AU) | $R_{in}$ (AU) | $T_0$ (K) | $q$ |
|---|---|---|---|---|---|---|
| $^{12}CO$ | $54.4 \pm 0.4$ | $86.1 \pm 0.2$ | $254 \pm 3$ | $22 \pm 1$ | $44 \pm 2$ | $0.35 \pm 0.05$ |
| $^{13}CO$ | $57 \pm 2$ | $88 \pm 1$ | $253 \pm 15$ | $21 \pm 7$ | $16 \pm 4$ | $0.2 \pm 0.3$ |

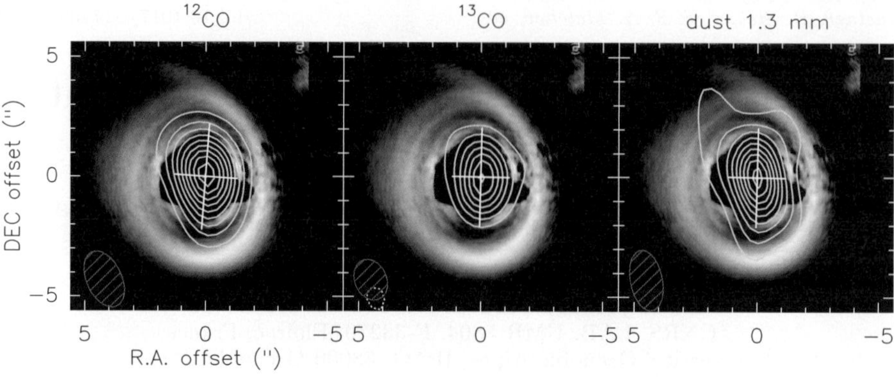

**Figure 1.** Integrated intensity of the CO and dust emission at 1.3 mm, superimposed to the *HST* scattered emission (Clampin *et al.* 2003). The cross indicates the position angle and aspect ratio as determined from gas modeling. Left: $^{12}$CO J=2-1 emission, contour spacing: $6\sigma$, i.e. $5.5\times10^{-1}$ Jy beam$^{-1}$ km s$^{-1}$. Beam size: $2.48 \times 1.45''$. Middle: $^{13}$CO J=2-1 emission, contour spacing: $3\sigma$, i.e. $7.8\times10^{-2}$ Jy beam$^{-1}$ km s$^{-1}$. Beam size: $1.76'' \times 1.32''$. Right: continuum emission, contour spacing: $3\sigma$, i.e. $2.0\times10^{-1}$ mJy beam$^{-1}$. Beam size: $2.57'' \times 1.31''$.

integrated intensity maps of the gas are displayed in Figure 1, as well as the continuum emission at 1.3 mm. We have modeled the data in the uv-plane using the code DiskFit (Piétu *et al.* 2007), based on a power-law description of the physical parameters, e.g. $T(r) = T_0(r/R_0)^{-q}$. Table 1 shows the parameters determined from this modeling for the $^{12}$CO and $^{13}$CO. The disk extends from $\sim$20 AU to 250 AU. From the $^{12}$CO/$^{13}$CO line ratio, the $^{12}$CO appears to be still optically thick while the $^{13}$CO is optically thin. The temperature is thus best determined from the $^{12}$CO modeling ($\sim$45 K at 100 AU, a typical value for an A star). The $^{13}$CO better probes the surface density, which is here $\sim$30 times less than around typical HAeBe disks, like MWC480.

HD141569A is thus a 'hybrid' disk with a large gas component, likely primordial, and an impressive evolved debris disk. The flux at 1.3 mm is 3.5±0.1 mJy, a low value in agreement with fast evolution of the dust. The links between gas and dust properties in this and other star/disk systems have to be studied in more detail, in particular to better understand the disk dissipation/evolution mechanisms which influence the shaping of young planetary systems.

# References


Augereau, J.-C., Lagrange, A. M., Mouillet, D. *et al.*, 1999, *A&A*, 350,51

Biller, B. A., Liu, M. C., Rice, K. *et al.*, 2015, *MNRAS*, 450, 4446

Boccaletti, A., Augereau, J.-C., Marchis, F. *et al.* 2003, *ApJ*, 585, 494

Dent, W. R. F., Greaves, J. S., Mannings, V. *et al.* 2005, *MNRAS*,277,25

Goto, M., Usuda, T., Dullemond, C. P. *et al.*, 2006, *ApJ*, 652, 758

Jonkheid, B., Kamp, I., Augereau, J.-C. *et al.*, 2006, *A&A*, 453,163

Merín, B., Montesinos, B., Eiroa, C. *et al.* 2004, *A&A*, 419, 301

Piétu, V., Dutrey, A., & Guilloteau, S., 2007, *A&A*, 467, 163

Sylvester, R. J., Skinner, C. J., Barlow, M. J. *et al.*, 1996, *MNRAS*, 279, 915

Thi, W.-F., Pinte, C., Pantin, E. *et al.*, 2014, *A&A*, 561, 50

van Leeuwen, F., 2007, *A&A*, 474, 653

Zuckerman, B., Forveille, T., & Kastner, J. H, 1995, *Nature*, 373, 494

*Young Stars & Planets Near the Sun*
*Proceedings IAU Symposium No. 314, 2015*
*J. H. Kastner, B. Stelzer, & S. A. Metchev, eds.*

© International Astronomical Union 2016
doi:10.1017/S1743921315005918

# Constraining X-ray-Induced Photoevaporation of Protoplanetary Disks Orbiting Low-Mass Stars

**Kristina M. Punzi[1], Joel H. Kastner[1], David Rodriguez[2], David A. Principe[3,4], and Laura Vican[5]**

[1] Laboratory for Multiwavelength Astrophysics, Rochester Institute of Technology
[2] Universidad de Chile
[3] Núcleo de Astronomía de la Facultad de Ingeniería, Universidad Diego Portales
[4] Millennium Nucleus Protoplanetary Disks
[5] University of California, Los Angeles

**Abstract.** Low-mass, pre-main sequence stars possess intense high-energy radiation fields as a result of their strong stellar magnetic activity. This stellar UV and X-ray radiation may have a profound impact on the lifetimes of protoplanetary disks. We aim to constrain the X-ray-induced photoevaporation rates of protoplanetary disks orbiting low-mass stars by analyzing serendipitous XMM-Newton and Chandra X-ray observations of candidate nearby (D < 100 pc), young (age < 100 Myr) M stars identified in the GALEX Nearby Young-Star Survey (GALNYSS).

**Keywords.** stars: low-mass, pre-main sequence; techniques: imaging spectroscopy; X-rays: stars

## 1. Introduction

Low-mass (M-type) stars represent some of the best targets for the discovery of potentially habitable exoplanets due to their low luminosities and the location of their habitable zones. Presently, only a small number of planets have been detected around M stars, with terrestrial planets being common and giant planets being rare, although these results may be affected by a selection bias (e.g., Mulders *et al.* 2015, Howard *et al.* 2012). This trend may be a consequence of the intense high-energy radiation fields of low-mass, pre-main sequence stars. A great deal of mass in protoplanetary disks is lost from the surface of the disk due to heating (photoevaporation) from high-energy radiation from the central star. The X-rays from young stars drive disk dissipation and chemistry, influencing the timescale and conditions for exoplanet formation. According to Owen *et al.* (2012), stellar X-ray luminosity alone sets the photoevaporation rate. However, Gorti *et al.* (2015) demonstrate that X-ray spectral hardness is also important. Hence, it is necessary to fully characterize the X-ray radiation fields incident on protoplanetary disks.

## 2. Serendipitously Detected X-ray Counterparts

To constrain the X-ray-induced photoevaporation rates of protoplanetary disks orbiting low-mass stars, we can examine stars from the GALEX Nearby Young-Star Survey (GALNYSS; Rodriguez *et al.* 2013) that have been serendipitously observed by either XMM-Newton or Chandra. GALNYSS combines ultraviolet (GALEX) and near-IR (WISE and 2MASS) photometry with kinematics to identify candidate nearby (D < 100 pc), young (age < 100 Myr), low-mass (M-type) stars. This survey has identified >2000 candidates, with most of the stars having spectral types in the range M3-M4.

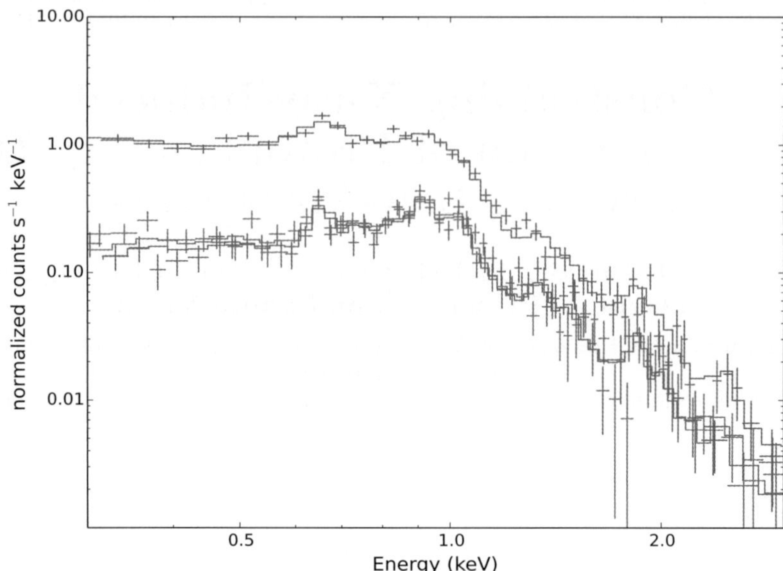

**Figure 1.** XMM-Newton EPIC extracted spectra (crosses) of J061313.30-274205.6 (spectral type M3, $\beta$ Pictoris Moving Group candidate – 99.24% likelihood) for pn (blue) and MOS (red and green) detectors. Overplotted are the best-fit models (histograms). Our model predicts plasma temperatures of $\sim 3.6$ and $\sim 12$ MK and an X-ray luminosity of $1.2 \times 10^{29}$ erg s$^{-1}$ at a distance of $\sim 25$ pc.

The XMM-Newton data for GALNYSS candidates are being reprocessed and analysed using the Scientific Analysis System following standard procedures†. Spectra are extracted from the EPIC detectors using circular regions centered on the 2MASS/WISE positions of the objects. Spectral modelling is performed with XSPEC. An example of the extracted spectra and resulting model fit is displayed in Figure 1. Our models consist of X-ray spectra that result from optically thin plasmas in collisional ionization equilibrium (vapec model in XSPEC‡). These emission spectra were combined with photoelectric absorption by using the XSPEC model wabs.

## 3. Conclusions and Future Work

Our preliminary results demonstrate that serendipitous XMM-Newton observations of GALNYSS stars are capable of producing useful constraints on the stellar X-ray temperature and luminosity and hence, on photoevaporation due to X-ray irradiation. These early results suggest that X-ray photoevaporation may not account for complete disk dispersal at ages $\sim$ 8-20 Myr. For the entire sample of serendipitously observed GALNYSS stars, we will determine X-ray temperatures and luminosities, the potential correlation between M star $L_X$ and residual disk mass, and the mass dependence of $L_X / L_{bol}$.

## References

Gorti, U., Hollenbach, D., & Dullemond, C. P., 2015, *ApJ*, 804, 29.
Howard, A. W., Marcy. G. W. Bryson, S. T., *et al.* 2012, *ApJS*, 201, 15.
Mulders, G. D., Pascucci, I., & Apai, D. 2015, *ApJ*, 798, 112.
Owen, J. E., Clarke, C., J., & Ercolano, B., 2012, *MNRAS*, 422, 1880.
Rodriguez, D. R., Zuckerman, B., Kastner, J. H., *et al.*, 2013, *ApJ*, 744, 101.

† See the SAS documentation at http://xmm.esac.esa.int/sas/current/documentation/threads.
‡ See the XSPEC documentation at https://heasarc.gsfc.nasa.gov/xanadu/xspec/manual/Models.htm for a description of the models.

*Young Stars & Planets Near the Sun*
*Proceedings IAU Symposium No. 314, 2015*
*J. H. Kastner, B. Stelzer, & S. A. Metchev, eds.*

© International Astronomical Union 2016
doi:10.1017/S1743921315006286

# The Age of Taurus: Environmental Effects on Disc Lifetimes

**J. M. Rees[1], T. Wilson[1], C. P. M. Bell[2], R. D. Jeffries[3] and T. Naylor[1]**

[1] School of Physics, University of Exeter, Exeter, EX4 4QL, UK
email: jon@astro.ex.ac.uk
[2] Dept. of Physics & Astronomy, University of Rochester, Rochester, NY 14627-0171, USA
[3] Astrophysics Group, Keele University, Staffordshire, ST5 5BG, UK

**Abstract.** Using semi-empirical isochrones, we find the age of the Taurus star-forming region to be 3-4 Myr. Comparing the disc fraction in Taurus to young massive clusters suggests discs survive longer in this low density environment. We also present a method of photometrically de-reddening young stars using $iZJH$ data.

**Keywords.** stars: evolution, stars: pre-main sequence, stars: fundamental parameters

## 1. Introduction

Taurus is a low-density star-forming region containing primarily low-mass stars and so represents an ideal laboratory for studying the environmental effects on circumstellar disc lifetimes (Kenyon *et al.* 2008). To investigate the impact of the low-density environment on the discs in Taurus, we used the Wide-Field Camera (WFC) on the 2.5m Isaac Newton Telescope (INT) on La Palma to obtain $griZ$ photometry of 40 fields in Taurus. Our fields are focused on the densest regions not covered by the Sloan Digital Sky Survey. The resultant INT-WFC survey mainly covers the L1495, L1521 and L1529 clouds.

We have augmented the WFC data with near-infrared $JHK$ data from 2MASS (Cutri *et al.* 2003). To determine the age of Taurus we compare the semi-empirical isochrones discussed in Bell *et al.* (2013, 2014) to the observed colour-magnitude diagrams (CMDs). For a brief description of these isochrones see Bell *et al.* (these proceedings).

## 2. De-reddening

The extinction in Taurus is spatially variable across the different clouds, and so we require a method of de-reddening the stars individually. We have found that in an $i - Z$, $J - H$ colour-colour diagram the reddening vectors are almost perpendicular to the theoretical stellar sequence (Fig. 1), whose position is almost independent of age, and so we can de-redden stars using photometry alone. We construct a grid of models over a range of ages (1 to 10 Myr) and binary mass ratios (single star to equal mass binary). We adopt the reddening law from Fitzpatrick (1999), apply it to the atmospheric models and fold the result through sets of filter responses to derive reddening coefficients in each photometric system. It is well known that for a fixed value of $E(B-V)$, extinction in a given filter will vary with $T_{\rm eff}$ (see e.g. Bell *et al.* 2013). To account for this we use extinction tables to redden the isochrones for a grid of $E(B-V)$ and $T_{\rm eff}$ values, and compare the reddened model grid to the data. We adopt a Bayesian approach and marginalise over binary mass, age and $T_{\rm eff}$. We take the extinction values from the model with the highest likelihood, and use this to de-redden the star.

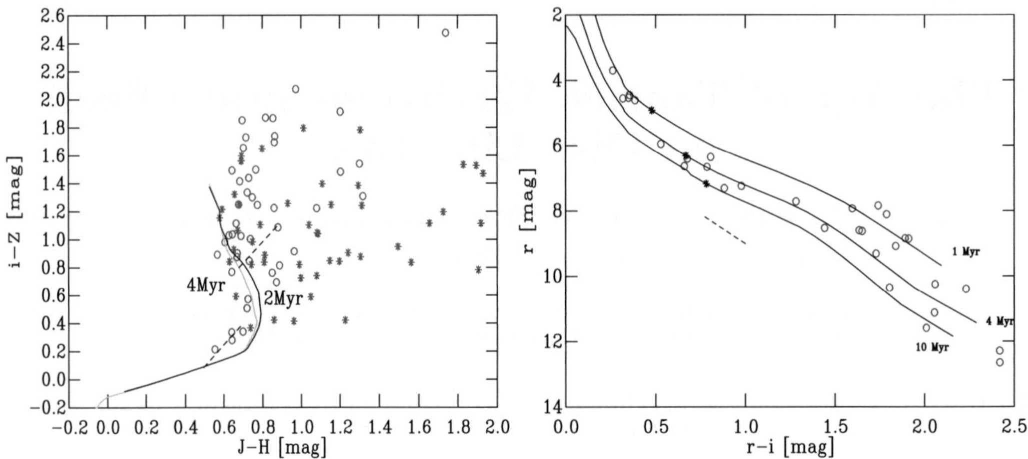

**Figure 1. Left:** $i - Z$, $J - H$ diagram for Taurus members. Asterisks are Class II sources, open circles are Class III sources. Overlaid as solid lines are a 2 and 4 Myr isochrone. The dashed lines are reddening vectors in this colour space. **Right:** $r$, $r - i$ diagram for Taurus members identified as Class III. Isochrones of 1, 4 and 10 Myr are overlaid. Asterisks indicate the position of a theoretical star with mass 0.75 $M_\odot$. The black dashed line shows a reddening vector for $A_V = 1$ mag.

## 3. Taurus

Plotting the de-reddened Taurus members in the $r$, $r - i$ CMD, we notice that a significant fraction of the Class II objects appear much fainter than the primary locus. This is likely an accretion effect, and if we were to fit for the age of these members we would derive an age that is erroneously old. To avoid this effect, we fit only the Class III sources. We note that those Class II sources that are not scattered below the sequence lie coincident with the Class III sources, and thus we believe the age derived from the Class III sources alone will be representative of the overall age. We plot our de-reddened Taurus members in an $r$, $r - i$ CMD to fit for the age (Fig. 1). We find that isochrones of 3–4 Myr (older than is commonly quoted in the literature) trace the observed stellar sequence well. To ensure consistency with the Bell *et al.* (2013) age scale we compare the position of a theoretical star with a mass of 0.75 $M_\odot$ to the middle of the observed sequence. We find consistency with the overall isochrone fitting, with an age of 3–4 Myr still providing a good fit. With a robust age for Taurus we then examined the disc fraction. Taurus has a disc fraction of 69% (Luhman *et al.* 2010). If we compare this to the other clusters in Bell *et al.* (2013), which are on the same age scale, we find that Taurus has the largest disc fraction in the sample, significantly higher than the group of young (2 Myr), massive clusters, suggesting that discs may have survived longer in the low density environment present in Taurus.

## References

Bell C. P. M., Naylor T., Mayne N. J., Jeffries R. D., & Littlefair S. P. 2013, *MNRAS*, 434, 806
Bell C. P. M. *et al.* 2014, *MNRAS*, 445, 3496
Cutri R. M. *et al.* 2003, *2MASS All Sky Catalog of point sources.*
Fitzpatrick E. L. 1999, *PASP*, 111, 63
Kenyon S. J., Gómez M., & Whitney B. A. 2008, *Handbook of Star Forming Regions, Vol. 1,*
    p.405
Luhman K. L., Allen P. R., Espaillat C., Hartmann L., & Calvet N. 2010, *ApJS*, 186, 111

*Young Stars & Planets Near the Sun*
*Proceedings IAU Symposium No. 314, 2015*
*J. H. Kastner, B. Stelzer, & S. A. Metchev, eds.*

© International Astronomical Union 2016
doi:10.1017/S174392131500616X

# A Molecular Disk Survey of Low-Mass Stars in the TW Hya Association

**David R. Rodriguez**[1], **Gerrit van der Plas**[1,8], **Joel H. Kastner**[2], **Adam C. Schneider**[3], **Jacqueline K. Faherty**[4,5], **Diego Mardones**[1], **Subhanjoy Mohanty**[6], and **David Principe**[7,8]

[1]Departamento de Astronomía, Universidad de Chile, Casilla 36-D, Santiago, Chile
[2]Center for Imaging Science, School of Physics & Astronomy, and Laboratory for Multiwavelength Astrophysics, Rochester Institute of Technology, Rochester, NY 14623, USA
[3]Department of Physics and Astronomy, University of Toledo, Toledo, OH 43606, USA
[4]Department of Terrestrial Magnetism, Carnegie Institution of Washington, Washington, DC 20015, USA
[5]Hubble Fellow
[6]Imperial College London, London SW7 2AZ, UK
[7]Núcleo de Astronomía, Facultad de Ingeniería, Universidad Diego Portales, Santiago, Chile
[8]Millennium Nucleus Protoplanetary Disks, Chile

**Abstract.** We have carried out an ALMA Cycle 2 survey of 15 confirmed or candidate low-mass ($< 0.2 M_\odot$) members of the TW Hya Association (TWA) with the goal of detecting line emission from CO molecular gas and continuum emission from cold dust. Our targets have spectral types of M4-L0 and hence represent the extreme low end of the TWA's mass function. The survey has yielded a detection of $^{12}$CO(2–1) emission around TWA 34. This newly discovered $\sim$10 Myr-old molecular gas disk lies just $\sim$50pc from Earth.

**Keywords.** open clusters and associations: individual (TWA) — protoplanetary disks — stars: evolution — stars: pre-main sequence

## 1. Introduction

Beginning with the identification of the TW Hya Association (TWA) several stellar associations with ages $\sim$8–200 Myr have been identified in close proximity to the Earth (Zuckerman & Song 2004, Torres *et al.* 2008). These young moving groups serve as excellent laboratories to explore the evolution of stellar and planetary properties. Of particular interest is the growing number of M dwarfs identified as members of these groups and how the lifetimes of protoplanetary disks orbiting such low-mass stars compare with those of higher mass members.

To this end, we have carried out an Atacama Large Millimeter Array (ALMA) survey of 15 recently identified very low-mass members or candidate members of the TWA. Our sample is drawn from Looper *et al.* (2007, 2010a,b), Looper (2011), Shkolnik *et al.* (2011), Rodriguez *et al.* (2011), Schneider *et al.* (2012), and Gagné *et al.* (2004). Several are known to host dusty, circumstellar disks as inferred from WISE and Herschel infrared excesses, but none had yet been observed with ALMA.

Our ALMA Cycle 2 program (2013.1.00457.S) consisted of observations of continuum dust emission at 230 GHz and observations of the $^{12}$CO(2–1) and $^{13}$CO(2–1) emission lines with a velocity resolution of 0.6 km/s. The requested angular resolution was 1.6″ as we did not aim to resolve any emission. We reached a sensitivity of 0.05 mJy/beam in the continuum and 5 mJy/beam per 0.6 km/s channel in $^{12}$CO and $^{13}$CO. Calibration and cleaning was performed by the ALMA staff with CASA version 4.2.2.

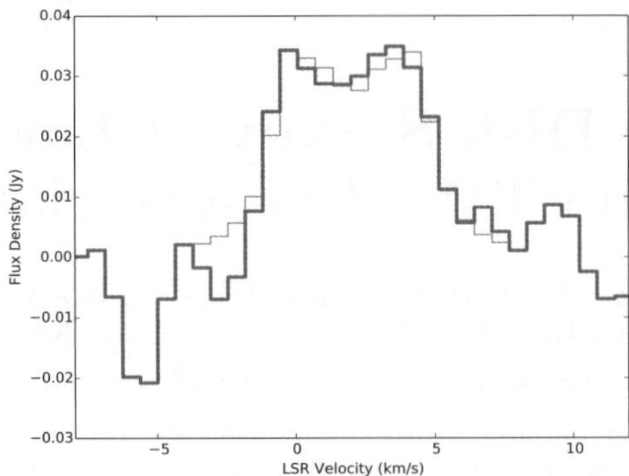

**Figure 1.** $^{12}$CO(2–1) emission line profile of the gas around TWA 34, the only target in our sample with detected CO gas. The red, thin line is a best-fit Keplerian profile.

## 2. Results

We report unambigious 1.3mm continuum detections of four objects (TWA 30B, 32, 33, 34). We only detected $^{12}$CO emission toward TWA 34. In Fig. 1, we show the integrated line profile of the CO emission, Hanning smoothed with a kernel size of 3 channels. To characterize this emission, we fit a parametrized Keplerian model as described in Kastner *et al.* (2008). We obtain an integrated intensity of $0.34 \pm 0.03$ Jy km/s. Combined with our 3-$\sigma$ upper limit on $^{13}$CO emission of $\sim$0.03 Jy km/s, we estimate a gas mass of $\sim$0.2 $M_E$, following the methods of Kastner *et al.* (2008) and assuming $T_{ex}$=40 K, a $^{12}$C:$^{13}$C ratio of 89, and a CO:H$_2$ ratio of $10^{-4}$.

With our best-fit Keplerian profile, we can estimate a systemic velocity for the CO emission. In the barycentric frame of reference, this corresponds to $13.3 \pm 0.1$ km/s. This is consistent with membership in TWA.

## 3. Conclusions

We have carried out an ALMA survey of 15 low-mass TWA members and candidates to search for molecular gas in the form of $^{12}$CO and $^{13}$CO as well as continuum dust emission. We detect continuum emission from four objects. Only one system, TWA 34, shows signatures of molecular gas in its disk in the form of $^{12}$CO (2–1) emission. TWA 34 is a new low-mass star hosting a molecular gas disk at just 50 pc from Earth. The systemic velocity we measure for the circumstellar CO around TWA 34 is consistent with membership in the TWA.

## References

Gagné, J., Faherty, J. K., Cruz, K., *et al.* 2014, *ApJ Letterse*, 785, LL14
Kastner, J. H., Zuckerman, B., & Forveille, T. 2008, *A&A*, 486, 239
Looper, D. L., Burgasser, A. J., Kirkpatrick, J. D., & Swift, B. J. 2007, *ApJ Letters*, 669, L97
Looper, D. L., Mohanty, S., Bochanski, J. J., *et al.* 2010, *ApJ*, 714, 45
Looper, D. L., Bochanski, J. J., Burgasser, A. J., *et al.* 2010b, *AJ*, 140, 1486
Looper, D. L. 2011, Ph.D. Thesis, University of Hawai'i
Rodriguez, D. R., Bessell, M. S., Zuckerman, B., & Kastner, J. H. 2011, *ApJ*, 727, 62
Roeser, S., Demleitner, M., & Schilbach, E. 2010, *AJ*, 139, 2440 754, 39
Schneider, A., Song, I., Melis, C., Zuckerman, B., & Bessell, M. 2012, *ApJ*, 757, 163
Shkolnik, E. L., Liu, M. C., Reid, I. N., Dupuy, T., & Weinberger, A. J. 2011, *ApJ*, 727, 6
Torres, C. A. O., *et al.* 2008, *Handbook of Star Forming Regions*, Volume II, 757
Zuckerman, B. & Song, I. 2004, *ARA&A*, 42, 685

*Young Stars & Planets Near the Sun*
Proceedings IAU Symposium No. 314, 2015
J. H. Kastner, B. Stelzer, & S. A. Metchev, eds.

© International Astronomical Union 2016
doi:10.1017/S1743921315006353

# Mid-infrared Variability and Accretion in NGC 2264 Protostars

**S. Terebey[1], A. M. Cody[2], L. M. Rebull[3], and J. R. Stauffer[3]**

[1]California State University Los Angeles; email: sterebe@calstatela.edu
[2]NASA Ames Research Center; [3]Spitzer Science Center; California Institute of Technology

**Abstract.** Variable mass accretion is thought to be an important aspect of protostar formation. Mid-infrared wavelength observations trace variations in accretion luminosity and thus can probe mass accretion on sub-AU scales. We present results from the Spitzer YSOVAR campaign towards Class I protostars in NGC 2264. The precise (0.02 mag) medium-cadence light curves at 3.6 and 4.5 microns show that young star variability is ubiquitous, with a variety of morphologies and time scales. A structure function analysis shows the light curves, on average, have a power-law behavior up to 30 days. The trend continues to longer timescales (years) for protostars (Class I), in contrast with the smaller brightness changes displayed by T Tauri stars (Class II). The power-law behavior suggests a stochastic process, such as turbulent mass accretion, drives the variability.

**Keywords.** stars:formation – stars:protostars – stars:variables: T Tauri

## 1. Introduction and Data

Young stars have long been recognized to be variable sources. The variability fits into a picture of a dynamic environment, where the inner disk feeds mass onto the star, and shocks dissipate energy thus leading to a variety of UV and optical phenomena. Protostars should share the same phenomena, although with lower stellar masses and higher accretion rates. However, protostars are enshrouded, so that optical and UV tracers are not accessible to study the inner disk, and thus motivate the need for mid-infrared studies. A key issue in star formation is understanding mass accretion. Variable mass accretion is important in providing a possible solution to the luminosity problem (e.g. Dunham *et al.* 2010).

In Orion, Morales-Calderon *et al.* (2011) found that 70% of stars with disks show mid-infrared variability. Extending to other clusters, and considering a data span of seven years, Rebull *et al.* (2014) find the mid-IR variability fraction scales with Class I content, which is used as a proxy for cluster age. For YSOs in NGC 2264, Cody *et al.* (2014) quantify a variety of light curve types, and demonstrate the kinds of variability that are fundamentally related to the presence of disks.

The current work focuses on characterizing the timescales of mid-infrared variability for protostars in NGC 2264. The analysis includes all available Spitzer 4.5 $\mu$m data, from the cryogenic era (Sung *et al.* 2009), from $\sim$ 40 days in 2010 (Rebull *et al.* 2014), and from $\sim$ 30 days in 2011 (Cody *et al.* 2014). This young (2–4 Myr) cluster is located at a distance of 760 pc, contains about 1000 known members, and a similar number of candidate members (Park *et al.* 2000, Sung *et al.* 2009). Our protostar sample includes 37 Class I plus flat SED sources. The criteria for inclusion were conservative, and required 70 micron flux detection (e.g. Forbrich *et al.* 2010). For a comparison sample, we used 82 Class II sources from Cody *et al.* (2014).

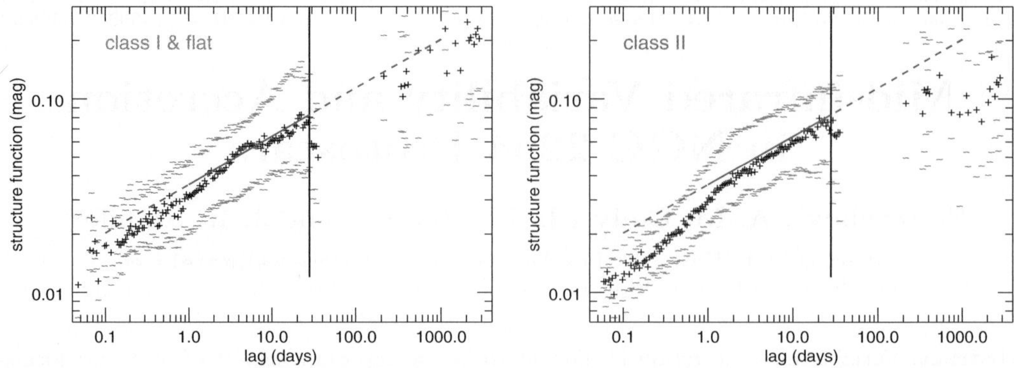


**Figure 1.** The 4.5 $\mu m$ ensemble Structure Function (crosses) shows the same power-law behavior (blue line) between 1 and 30 days for the Class I plus flat sample (left), and the Class II sample (right). At longer times ($\sim 1000$ days) the Class I plus flat sample approximately follows the same power-law trend (dashed blue line), whereas the Class II sample has 'turned over' at 0.1 mag, with the meaning that 0.1 mag is the typical variability of the Class II sources. Red dashes show the inner 50 percent quartile range for objects contributing to the ensemble SF.

## 2. Analysis and Conclusions

We use the structure function $SF(\Delta t)$ to characterize the light curves in a statistical sense. Similar to computing rms, it is an average of $\Delta mag$ values, but is restricted to data points that have the same time lag $\Delta t = t_i - t_j$. For example, a non-variable object exhibits a SF that is roughly constant versus time lag. The ensemble SF is made from the median of the SF curves for each object. Examples of the SF in the context of quasar variability are given in Kelly *et al.* (2009) and Kozlowski *et al.* (2010).

The ensemble SF (Fig. 1) shows that two magnitude measurements separated by 10 days will, on average, differ by about 0.06 magnitudes for both samples, Class I plus flat (left panel) and and Class II (right panel). Moreover, both samples exhibit the same power-law slope of 0.25 between 1 and 30 days. However, at 1000 days the samples diverge: a typical Class I will change by 0.2 mag, but a Class II by only 0.1 mag.

Time scales of $\sim$10 days at 4.5 $\mu m$ trace emission variability that occurs in the inner disk (0.1 AU). The power-law behavior exhibited between 1 and 30 days (Fig. 1) suggests a stochastic process such as turbulent mass-accretion is acting. The similar structure function found in the Class I and Class II samples implies similar physical processes are at work in the inner disk. At 1000 days, the increased variability of Class I objects suggests secular changes in the mass accretion rate. Future directions include the possibility of using mid-IR variability to determine *disk* mass accretion rates for protostars.

## References

Cody, A. M., Stauffer, J. R., Baglin, A. *et al.* 2014, *AJ*, 147, 82
Dunham, M. M., Evans, Neal J., II, Terebey, S., *et al.* 2010, *ApJ*, 710, 470
Forbrich, J., Tappe, A., Robitaille, T., *et al.* 2010, *ApJ*, 716, 1453
Kelly, B. C., Bechtold, J., & Siemiginowska, A. 2009, *ApJ*, 698, 895
Kozlowski, S., Kochanek, C. S., Stern, D., *et al.* 2010, *ApJ*, 716, 530
Morales-Calderon, M. , Stauffer, J. R., Hillenbrand, L. A. *et al.* 2011, *ApJ*, 733, 50
Park, B.-G., Sung, H, Bessell, M. S., & Kang, Y. H. 2000, *AJ*, 120, 894
Rebull, L. R., Cody, A. M., Covey, K. R., *et al.* 2014, *AJ*, 148, 92
Sung, H, Stauffer, J. R. & Bessell, M. S., 2009, *AJ*, 138, 1116

*Young Stars & Planets Near the Sun*
Proceedings IAU Symposium No. 314, 2015
J. H. Kastner, B. Stelzer, & S. A. Metchev, eds.

© International Astronomical Union 2016
doi:10.1017/S1743921315005980

# On the Possibility to Identify a Companion in a Protoplanetary Disk

## O. V. Zakhozhay

Main Astronomical Observatory, NAS of Ukraine, Kyiv, 03680, Ukraine
email: zakhozhay.olga@gmail.com

**Abstract.** In this paper, I investigate a possibility to detect a brown dwarf companion in a protoplanetary disk based on spectral energy distribution (SED) profile analysis. I present synthetic spectral energy distributions of protoplanetary disks with and without an embedded companion that clears a gap. The computations are performed for a star (0.8 $M_\odot$) and a substellar companion (30 $M_J$) at an age of 5 Myr embedded in a protoplanetary disk, located at a distance 100 pc from the Sun. Analysis of the SED profile shape indicates that the maximum difference between the fluxes of the systems with and without the companion is $\approx 0.43$ Jy at 34 $\mu$m.

**Keywords.** stars: formation, planetary systems: protoplanetary disks

## 1. Introduction

Planetary systems are formed in protoplanetary disks that surround young stellar and substellar objects. In the first hundred thousand years, such disks could contain hot and massive fragments that later may form giant planets and brown dwarf companions (e.g. Stamatellos & Whitworth 2009, Meru & Bate 2010). Surviving companions clear gaps along their orbital motion (e.g. Fouchet et a. 2010, Meru *et al.* 2014). Observational properties of protoplanetary and debris disks with embedded companions have been studied little so far (e.g. Varnière *et al.* 2006, Gonzalez *et al.* 2012).

In this work, I study the properties of synthetic SEDs, modeled for a stellar object surrounded by a protoplanetary disk with an embedded brown dwarf companion and compare it to an analogous system without a companion. I pay special attention to the fluxes differences and how they change with wavelength.

## 2. Results

The models have been computed for systems at 5 Myr with a 0.8 $M_\odot$ central object and a 30 $M_J$ substellar companion at 1 AU embedded in a protoplanetary disk (that is extended from 0.1 to 150 AU and is oriented face-on). The system is located at a distance of 100 pc from the Sun. To model the flux from the disk, I followed the approach of Andrews & Willams (2005) and by assuming that disk is composed of an inner region and an outer region. These two regions are separated by a gap cleared by the companion. I assume that there is no emitting material inside the gap and that the width of the gap cleared by the companion is the diameter of the Hill sphere. Fluxes from the star and companion were modeled using the black-body approximation and the physical parameters (temperatures and radii) from Baraffe *et al.* (1998).

Figure 1 shows SEDs of a system with a protoplanetary disk with a companion (black solid line) and without it (gray solid line). It also illustrates the contribution of the star (gray dashed line) and the companion (black dash-dotted line), disk inner region (black

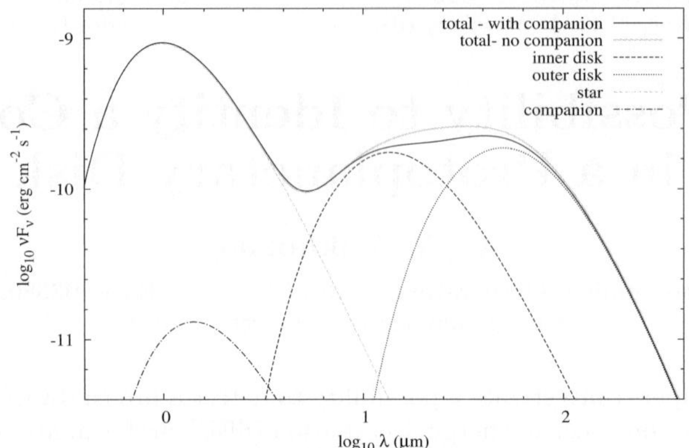


**Figure 1.** SED of modeled system with a protoplanetary disk and companion (black solid line) that consists of inner (black dashed line) and outer (black dotted line) parts. The flux from the companion is shown with the black dashed-dotted line and that from the star with gray dashed line. The flux from the system with the same parameters but without a companion is shown with the gray solid line.

**Table 1.** Fluxes from the systems with and without the companion.

| $\lambda$ ($\mu$m) | $\Delta F$ (Jy) | $F_{noC}$ (Jy) | $F_{withC}$ (Jy) |
|---|---|---|---|
| 10 | 0.02 | 0.57 | 0.55 |
| 20 | 0.29 | 1.63 | 1.35 |
| 34 ($\Delta F_{max}$) | 0.43 | 2.93 | 2.50 |
| 80 | 0.23 | 4.28 | 4.06 |
| 100 | 0.17 | 3.84 | 3.67 |

dashed line) and outer region (black dotted line). Visual examination of Figure 1 indicates that the gap cleared by the companion, located at 1 AU, in the protoplanetary disk introduces a flux depression at the wavelength interval 10–100$\mu$m. This flux depression is due to the absence of the emitting disk material between 0.77 AU and 1.23 AU from the star. The results of the calculations indicate that the maximum difference between SEDs from the system with and without the companion is $\approx 0.43$ Jy at 34 $\mu$m. The differences of the fluxes ($\Delta F$) from this system without ($F_{noC}$) and with the embedded companion ($F_{withC}$), as a function of wavelength ($\lambda$) are listed in the Table 1.

## 3. Acknowledgments

Z.O.V. acknowledges the travel support from the International Astronomical Union.

## References


Andrews, S. M. & Willams, J. P. 2005, *ApJ*, 631, 1134
Baraffe, I., Chabrier, G., Allard, F., *et al.* 1998, *A&A*, 337, 403
Fouchet, L., Gonzalez, J.-F., & Maddison, S. T. 2010, *A&A*, 518, A16
Gonzalez, J.-F., Pinte, C., Maddison, S. T., Ménard, F., & Fouchet, L. 2012, *A&A*, 547, A58
Meru, F. & Bate, M. R. 2010, *MNRAS*, 406, 2279
Meru, F., Quanz, S. P., Reggiani, M., Baruteau, C., & Pineda, J. E., *arXiv:1411.5366*
Stamatellos, D. & Whitworth, A. P. 2009, *MNRAS*, 392, 413
Varnière, P., Bjorkman, J. E., Frank, A. *et al.* 2006, *ApJ*, 637, L125

*Young Stars & Planets Near the Sun*
*Proceedings IAU Symposium No. 314, 2015*
*J. H. Kastner, B. Stelzer, & S. A. Metchev, eds.*
© International Astronomical Union 2016
doi:10.1017/S1743921315006560

# Direct Imaging of Exoplanets
# Living an Exciting Life

## G. Chauvin

UJF-Grenoble1/CNRS-INSU, Institut de Planétologie et d'Astrophysique de Grenoble UMR
5274, Grenoble, F-38041, France
email: `gael.chauvin@obs.ujf-grenoble.fr`

**Abstract.** With the development of high contrast imaging techniques and instruments, vast efforts have been devoted during the past decades to detect and characterize lighter, cooler and closer companions to nearby stars, and ultimately image new planetary systems. Complementary to other planet-hunting techniques, this approach has opened a new astrophysical window to study the physical properties and the formation mechanisms of brown dwarfs and planets. In this review, I will briefly describe the different observing techniques and strategies used, the main samples of targeted nearby stars, finally the main results obtained so far about exoplanet discoveries characterization of their physical properties, and study of their occurrence and possible formation and evolution mechanisms.

**Keywords.** planets and satellites: detection, formation, dynamical evolution and stability, atmospheres; techniques: image processing, spectroscopic, high angular resolution

---

## 1. Introduction

With two decades of exoplanet studies, observations regularly obtained key successes with the discovery of the Hot-Jupiters family, the detection and confirmation of about 2000 exoplanets today, the first glimpse of planetary atmopsheres and internal structures, the direct images of Super-Jupiters revolving around their host star as well as the determination of their rotating period, the discovery of Super-Earths in the Habitable Zone (where water is expected to be liquid) or the discovery Earth-mass planets. The five main hunting techniques currently used (radial velocity, transit, mirco-lensing, direct imaghing and astrometry) are complementary and can be combined to further constrain planet properties like density (using radial velocity and transit) or internal entropy (radial velocity and imaging). Radial velocity, transit, micro-lensing and astrometry enables the internal part study of exoplanetary systems, at less than 5–10 AU. Direct Imaging (DI) is here unique to explore the outer part at more than 10 AU to complete our view of the systems's architecture. In addition to the orbital properties, exoplanet's photons can be resolved and dispersed to probe the atmospheric properties of the imaged exoplanets. Being easier to detect at young ages, their atmospheres show low-gravity features, as well as the presence of clouds, and non-equilibrium chemistry processes. Moreover, DI enables to directly probe the presence of planets in their birth environment. Planet characteristics and disk spatial structures can then be linked to study the planet – disk interactions and system's stability.

In this review, I will briefly describe: i/ the observing challenges of DI, ii/ the samples of stars observed, iii/ the main surveys and discoveries published sofar, finally iv/ key results about the physics of exoplanets, the mass determination and accretion history, the orbit and architecture and the occurence and formation.

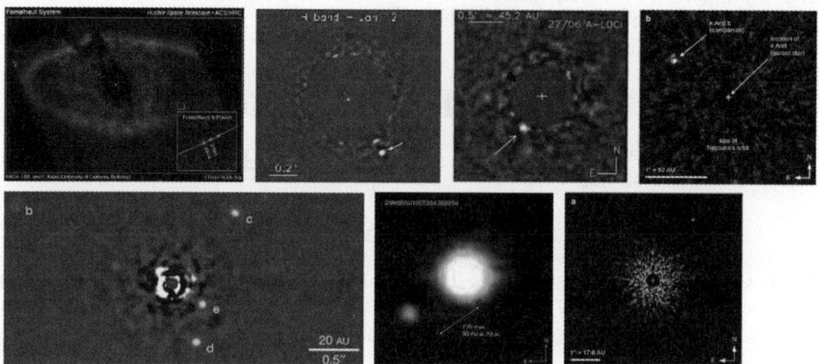

**Figure 1.** DI key discoveries: Fomalhaut b (Kalas *et al.* 2008), $\beta$ Pictoris b (Lagrange *et al.* 2009), HD 95086 b (Rameau et al. 2013a), $\kappa$ And b (Carson *et al.* 2013), HR 8799 bcde (Marois *et al.* 2010), 2M1207 b (Chauvin *et al.* 2005) and Gl504 b (Kuzuhara *et al.* 2013).

## 2. Observing challenges and techniques

To reach high contrast of $10^{-6} - 10^{-7}$ at separation as close as 200 mas for the discovery of young jupiters at $\sim$ 10 AU (for a young star at 50 pc), current planet imagers are designed with a 3-stages implemenation (four if we include the post-processing analysis):

(*a*) *High angular resolution.* From the ground, Extreme-Adaptive Optics (XAO) systems are used to to compensate for atmospheric, telescope and common path defects. The main impact is to access the diffraction limit of the telescope (40 mas in H-band), to concentrate the stellar flux in the coherent core of the PSF, then to boost the detection signal. Current XAO instruments like SPHERE (Beuzit *et al.* 2008) and GPI (Macintosh *et al.* 2008) rely on high-order deformable mirror, high temporal sampling frequency at more than 1.0 kHz, spatial filtering of the wavefront before sensing to achieve Strehl correction of 90% in H-band on bright stars.

(*b*) *Stellar-light attenuation.* Coronagraphy is the second stage to reduce by a factor $\sim$ 100 the intensity of the stellar diffracted-limited core, and to eliminate the diffraction features due to the pupil edges. Various coronagraphic concepts are currently implemented to optimize the access to small separations, the transmission throughput and the chromaticity-dependency. The most common ones are the classical and apodized Lyot coronographs and the phase-masks coronographs like the four-quadran-phase mask and the vortex coronographs (see Mawet *et al.* 2012).

(*c*) *Speckles subtraction.* After the first two stages, boiling atmospheric speckles and instrumental quasi-statics speckles still remain an important source of limitation. Differential techniques have been developped to correct either: i/ residual atmospheric speckles using spectral or polarimetric differential techniques (simultaneous calibration of the PSF) or ii/ quasi-static instrumental speckles using angular differential imaging technique (Marois *et al.* 2006).

(*d*) *Post-processing tools.* Finally, important progress in the past years has been made to develop innovative algorithms to optimally calibrate the PSF temporally and spectrally in order to suppress the stellar contribution and recover faint planetary signals as close as possible to the star (LOCI, Lafrenière et al. 2007; PCA, Soummer *et al.* 2012).

## 3. Targets, surveys & main discoveries

Young ($\leqslant$ 500 Myr), nearby ($\leqslant$ 100 pc) stars are very favourable targets for the direct detection of the lowest mass companions (see Zuckerman & Song 2004; Torres *et al.* 2008). The first planetary mass companions were detected at large distances ($\geqslant$ 100 AU) and/or

**Table 1.** Deep imaging surveys of young ($< 100$ Myr) and intermediate-old to old ($0.1 - 5$ Gyr), nearby ($< 100$ pc) stars. Imaging modes are indicated as: Cor-I (coronagraphic imaging), Sat-I (saturated imaging), SDI (spectral differential imaging), ADI (angular differential imaging), ASDI (angular and spectral differential imaging).

| Reference | Telescope | Instr. | Mode | Filter | FoV ($''\times''$) | # | SpT | Age (Myr) |
|---|---|---|---|---|---|---|---|---|
| Chauvin+03 | ESO3.6m | ADONIS | Cor-I | $H, K$ | $13 \times 13$ | 29 | GKM | $\lesssim 50$ |
| Neuhäuser+03 | NTT | Sharp | Sat-I | $K$ | $11 \times 11$ | 23 | AFGKM | $\lesssim 50$ |
| | NTT | Sofi | Sat-I | $H$ | $13 \times 13$ | 10 | AFGKM | $\lesssim 50$ |
| Lowrance+05 | HST | NICMOS | Cor-I | $H$ | $19 \times 19$ | 45 | AFGKM | $10 - 600$ |
| Masciadri+05 | VLT | NaCo | Sat-I | $H, K$ | $14 \times 14$ | 28 | KM | $\lesssim 200$ |
| Biller+07 | VLT | NaCo | SDI | $H$ | $5 \times 5$ | 45 | GKM | $\lesssim 300$ |
| | MMT | | SDI | $H$ | $5 \times 5$ | - | - | - |
| Kasper+07 | VLT | NaCo | Sat-I | $L'$ | $28 \times 28$ | 22 | GKM | $\lesssim 50$ |
| Lafrenière+07 | Gemini-N | NIRI | ADI | $H$ | $22 \times 22$ | 85 | | 10-5000 |
| Apai+08[a] | VLT | NaCo | SDI | $H$ | $3 \times 3$ | 8 | FG | 12-500 |
| Metchev+09 | Palomar | PHARO | Cor-I | $K$ | $25.2 \times 25.2$ | 266 | FK | 3-3000 |
| | Keck-II | NIRC2 | Cor-I | $K$ | $40.6 \times 40.6$ | - | - | - |
| Chauvin+10 | VLT | NaCo | Cor-I | $H, K$ | $28 \times 28$ | 88 | BAFGKM | $\lesssim 100$ |
| Heinze+10ab | MMT | Clio | ADI | $L', M$ | $15.5 \times 12.4$ | 54 | FGK | 100-5000 |
| Janson+11 | Gemini-N | NIRI | ADI | $H, K$ | $22 \times 22$ | 15 | BA | 20-700 |
| Vigan+12 | Gemini-N | NIRI | ADI | $H, K$ | $22 \times 22$ | 42 | AF | 10-400 |
| | VLT | NaCo | ADI | $H, K$ | $14 \times 14$ | - | - | - |
| Delorme+12 | VLT | NaCo | ADI | $L'$ | $28 \times 28$ | 16 | M | $\lesssim 200$ |
| Rameau+13c | VLT | NaCo | ADI | $L'$ | $28 \times 28$ | 59 | AF | $\lesssim 200$ |
| Yamamoto+13 | Subaru | HiCIAO | ADI | $H, K$ | $20 \times 20$ | 20 | FG | $125 \pm 8$ |
| Biller+13 | Gemini-S | NICI | Cor-ASDI | $H$ | $18 \times 18$ | 80 | BAFGKM | $\lesssim 200$ |
| Nielsen+13 | Gemini-S | NICI | Cor-ASDI | $H$ | $18 \times 18$ | 70 | BA | 50-500 |
| Wahhaj+13 | Gemini-S | NICI | Cor-ASDI | $H$ | $18 \times 18$ | 57 | AFGKM | $\sim 100$ |
| Janson+13 | Subaru | HiCIAO | ADI | $H$ | $20 \times 20$ | 50 | AFGKM | $\lesssim 1000$ |
| Brandt+14 | Subaru | HiCIAO | ADI | $H$ | $20 \times 20$ | 63 | AFGKM | $\lesssim 500$ |
| Chauvin+15 | VLT | NaCo | ADI | $H$ | $14 \times 14$ | 86 | FGK | $\lesssim 200$ |

with small mass ratio with their primaries, indicating a probable star-like or gravitational disk instability formation mechanism. The breakthrough discoveries of closer and/or lighter planetary mass companions like Fomalhaut b ($< 1$ $M_{\rm Jup}$ at 177 AU; Kalas et al. 2008), HR 8799 bcde (10, 10, 10 and 7 $M_{\rm Jup}$ at respectively 14, 24, 38 and 68 AU; Marois *et al.* 2010), $\beta$ Pictoris b (8 $M_{\rm Jup}$ at 8 AU; Lagrange *et al.* 2009) or more recently $\kappa$ And b ($14^{+25}_{-2}$ $M_{\rm Jup}$ at 55 AU; Carson *et al.* 2013), HD 95086 b ($4 - 5$ $M_{\rm Jup}$ at 56 AU; Rameau *et al.* 2013a,b) and GJ 504 b ($4^{+4.5}_{-1}$ $M_{\rm Jup}$ at 43.5 AU; Kuzuhara *et al.* 2013) indicate that we are just initiating the characterization of the outer part of planetary systems between typically $5 - 100$ AU (see Fig. 1). Vast efforts are now devoted to systematic searches of exoplanets in DI with an increasing number of large scale surveys (see Table 1; eight new surveys published between 2013 and 2015) and this will accelerate with the new generation of planet imagers.

## 4. Key astrophysical results

(*a*) *Physics of imaged planets.* Once an exoplanet is discovered and confirmed in DI (usually after a proper+parallactic motion test), the mass is derived from evolutionary models predictions. DI only enables to measure the photometry and luminosity of exoplanets, and not the mass. A key problem is that these models are not well calibrated at

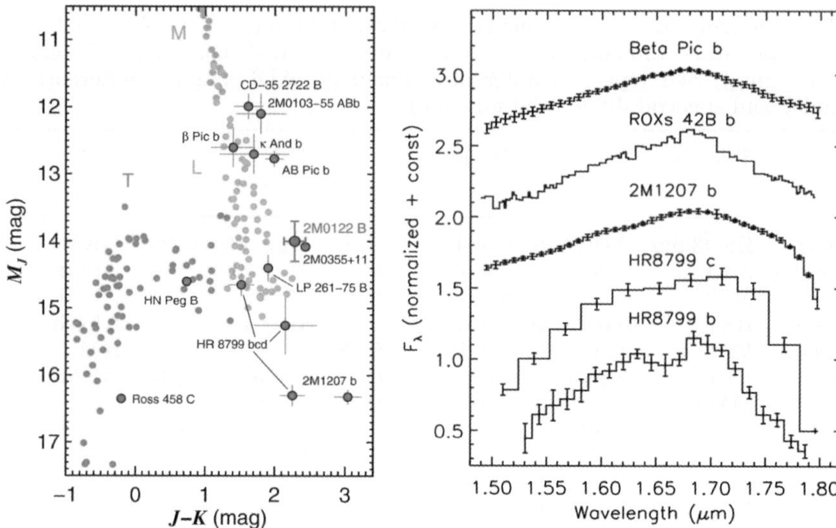

**Figure 2.** Left: $M_J$ vs. $J - K$ diagram for field M (orange), L (light green), and T (blue) dwarfs. Dark green circles show substellar companions with parallactic distances (either to the primaries or the secondaries themselves). For comparison are overplotted the measurements for young imaged planets and brown dwarfs. From Bowler *et al.* (2013). Right: H-band spectra of young, directly imaged planets. From Chilcote *et al.* (2015).

young ages. In addition to the system age, the predictions highly depend on the formation mechanisms and the gas-accretion phase that will form the exoplanetary atmosphere (see Chabrier *et al.* 2014 for review). In the ideal case of multi-wavelengths photometric study, the spectral energy distribution of the exoplanet can be constrained. In the case of $\beta$ Pic b, it shows atmospheric properties compatible with a young (low-gravity), dusty and early-L type and a planet's luminosity $log(L/L_\odot) = -3.90 \pm 0.07$ (see Bonnefoy et al. 2013; Fig. 6). Synthetic models give in addition a spectrum reproduced for $T_{eff} = 1650 \pm 150$ K and a $log(g) \leqslant 4.7$ dex and evolutionary models (hot-start) a mass between 8 and 13 $M_{Jup}$. Today, most imaged exoplanets are young, late-M or L-type exoplanets with dusty atmospheres. A striking feature is that some of them (2M1207 b, HR8799 b) appear redder and underluminous compared to field dwarfs of the same spectral type (see Fig. 2, Left). Methane absorption is also inhibited in some cases. The effect of low-gravity actually shifts the classical $T_{eff}$–SpT classification as highlighted by compartive studies between planets/young brown dwarfs and field brown dwarfs. Low-gravity conditions could also favour the formation of thick clouds causing photometric variability (Metchev *et al.* 2015).

(*b*) *Mass and accretion history.* As mentioned earlier, DI does not allow to measure the mass of exoplanets, but the luminosity. The so-called hot-, warm- and cold-start models of planetary formation and evolution, that describe different states of the gas-accretion shock during the planetary atmosphere formation, predict luminosities that are spread over several order of magnitudes for young, massive giants (see Marley *et al.* 2007, Mordasini *et al.* 2012). The combination of DI with radial velocity is here extremely precious to derive the dynamical mass of imaged planets. Such a combination between HARPS and NACO/GPI observations is illustrated for the case of $\beta$ Pic b (Bonnefoy *et al.* 2014) and showed that the planet evolutive state may result from an inefficient accretion shock and/or a planetesimal density at formation higher than in the classical core-accretion model. Future systematic simultaneous determination of mass and

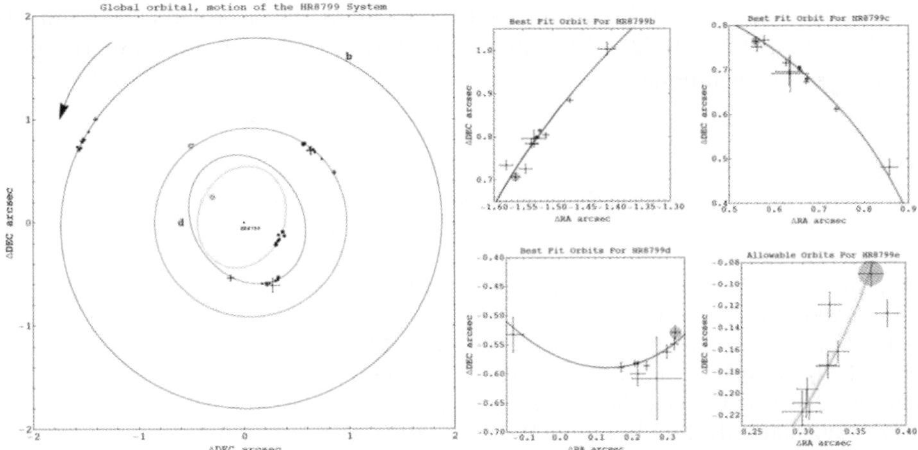

**Figure 3.** On-sky projection of the best fit orbits for HR8799 bcde (Pueyo *et al.* 2015).

luminosity will help calibrating the models. Much can be expected from the synergy between GAIA and SPHERE/GPI.

(*c*) *Orbit and architecture.* The third interesting question addressed by DI studies concerns the architecture of exoplanetary systems. In the case of multiple planetary system like HR 8799, a long-term astrometric monitoring enabled to characterize the planet orbits, to check for coplanarity, and ultimately to verify the dynamical stability of the whole system. Pueyo *et al.* (2015) recently showed that the HR 8799 d planet seems to be misaligned by 15–20 degrees and more eccentric ($\sim 0.3$) than the other planets in the system, suggesting remaining imprints of dynamical interactions (see Fig. 3). They also confirm that mean motion resonances are likely to be present to stabilize the system. For HR 8799 and HD 95086, the planet location(s) can be compared to the dust spatial distribution and architecture, i.e for both cases with the presence of an inner warm belt and an outer cold belt surrounding the planets. The case of $\beta$ Pic is even more striking in terms of planet–disk interactions, as the simultaneous characterization of the planet orbit and peculiar disk geometry confirm that the planet is responsible for the inner warped disk morphology, and that the orbital parameters are consistent with the transit-like event observed in November 1981, and the cometary activity observed for decades in that system (Chauvin *et al.* 2012; Lagrange *et al.* 2012).

(*d*) *Occurrence and formation.* Theories of planetary formation have drastically transformed in the last two decades and now favor the formation of planets within a protoplanetary disk by accretion of solids, building up a 10 to 15 $M_{\oplus}$ core followed by rapid agglomeration of gas. Whereas physical conditions and timescales favor core accretion in the inner disk ($\leqslant 10$ AU), gravitational instability or gravo-turbulent fragmentation could be alternative solutions to form massive planetary mass companions at wider separations ($\geqslant 10$ AU) in the earliest phase of the disk's lifetime (Boley 2009). The planets could either migrate inward, toward or outward, from the star by disk-planet interactions or during planet-planet interactions, which will alter the original semi-major axis distribution. One key element addressed by DI surveys concerns the occurence of giant planets beyond 10 AU to actually test these theories (Nielsen *et al.* 2013; Rameau *et al.* 2013b). Observed occurrences tend to reject gravitational instability or gravo-turbulent fragmentation as a dominant mechanism to form wide-orbit giant planets (Janson *et al.* 2013). For what concerns core accretion, the bulk of this population is probably still not explored by DI. However, SPHERE and GPI systematic surveys sensitive to physical separations as close as 5–10 AU (compared to 20–40 AU with previous instruments like NaCo, NICI, HiCIAO, NIRC2) will probably reach that bulk and allow a more significant overlap with other techniques.

## References


Apai D., Janson M., Moro-Martín A. *et al.* 2008, *ApJ*, 672, 1196

Beuzit J.-L., Feldt M., Dohlen K. *et al.* 2008, *SPIE*, 7014, 18

Biller B. A., Close L. M., Masciadri E. *et al.* 2007, *ApJS*, 173, 143

Biller E., Liu M., Wahhaj Z. *et al.* 2013, *ApJ*, 777, 160

Boley A. C. 2009, *ApJ*, 695, L53

Bowler B. P., Liu M., Shkolnik *et al.* 2013, *ApJ*, 774, 55

Bonnefoy M., Boccaletti A., Lagrange A. M. *et al.* 2013a, *A&A*, 555, 107

Bonnefoy M., Marleau G., Galicher R. *et al.* 2014, *A&A*, 567, L9

Brandt T., Kuzuhara M., McElwain M. W. *et al.* 2014, *ApJ*, 794, 159

Chabrier, G., Johansen A., JAnson M. & Rafikov R. 2014, PPVI proc., 914 pp., p.619-642

Chauvin G., Thomson M., Dumas C. *et al.* 2003, *A&A*, 404, 157

Chauvin G., Lagrange A.-M., Dumas C. *et al.* 2005, *A&A*, 438, L25

Chauvin G., Lagrange A.-M., Bonavita M. *et al.* 2010, *A&A*, 509, 52

Chauvin G., Lagrange A.-M., Beust H. *et al.*, 2012, *A&A*, 542, 41

Chauvin G., Vigan A., Bonnefoy M. *et al.* 2015, *A&A*, 573, 27

Chilcote J., Barman T., Fitzgerald M. P. *et al.* 2015, *ApJ*, 798, L3

Carson J., Thalmann C., Janson M. *et al.* 2013, *ApJ*, 763, L32

Delorme P., Lagrange A.-M., Chauvin G. *et al.* 2012, *A&A*, 539, 72

Heinze A. N., Hinz P. M., Sivanandam S. *et al.* 2010a, *ApJ*, 714, 1551

Heinze A. N., Hinz P. M., Kenworthy M. *et al.* 2010b, *ApJ*, 714, 1570

Janson M., Bonavita M., Klahr H. *et al.* 2011, *ApJ*, 736, 89

Kalas P., Graham J. R., Chiang E. *et al.* 2008, *Science* 322, 1345

Kasper M., Apai D., Janson M. & Brandner W. 2007, *A&A*, 472, 321

Kuzuhara M., Tamura M., Kudo T. *et al.* 2013, *ApJ*, 774, 11

Lafrenière D., Doyon R., Marois C. *et al.* 2007, *ApJ*, 670, 1367

Lagrange A.-M., Gratadour D., Chauvin G. *et al.*, 2009, *A&A*, 506, L972

Lagrange A.-M., Boccaletti A., Milli J., *et al.*, 2012, *A&A*, 542, L40

Lowrance P. J., Becklin E. E., Schneider G. *et al.* 2005, *AJ*, 130, 1845

Macintosh B., Graham J. R., Palmer D. *et al.* 2008, *SPIE*, 7015, 18

Marley M., Fortney J., Hubickyj O. *et al.* 2007, *ApJ*, 655, 541

Marois C., Lafrenière D., Doyon R., Macintosh B. & Nadeau D. 2006, *ApJ*, 641, 556

Marois C., Zuckerman B., Konopacky Q. M. *et al.* 2010, *Nature* 468, 1080

Masciadri E., Mundt R., Henning Th. & Alvarez C. 2005, *ApJ*, 625, 1004

Mawet D., Pueyo L., Lawson P. *et al.* 2012, *SPIE*, 8442, 04

Metchev S. & Hillenbrand L. 2009, *ApJS*, 181, 62

Metchev S. A., Heinze, A., Apai, D. *et al.* 2015, *ApJ*, 799, 154

Mordasini C., Alibert Y., Georgy C. 2012, *A&A*, 547, 112

Neuhäuser R., Guenther E. W., Alves J. *et al.* 2003, *AN*, 324, 535

Nielsen E., Liu M., Wahhaj Z. *et al.* 2013, *ApJ*, 776, 4

Pueyo L., Soummer R., Hoffman J. *et al.* 2015, *ApJ*, 803, 31

Rameau J., Chauvin G., Lagrange A.-M. *et al.* 2013a, *ApJ*, 772, L15

Rameau J., Chauvin G., Lagrange A.-M. *et al.* 2013b, *A&A*, 553, 60

Soummer R., Pueyo L. & Larkin J. 2012, *ApJ*, 755, L28

Torres C.A.O., Quast G.R., Melo C.H.F. 2008 Handbook of Star Forming Regions Vol. II 757

Vigan A., Patience J., Marois C. *et al.* 2012, *A&A*, 544, 9

Wahhaj Z., Liu M., Nielsen E. L. *et al.* 2013, *ApJ*, 773, 179

Yamamoto K., Matsuo T., Shibai H. *et al.* 2013, *PASJ*, 65, 90

Zuckerman B. & Song I. 2004, *ARA&A*, 42, 685

## Discussion

QUESTION-1: Can we make any difference between imaged planets and imaged brown dwarfs?

ANSWER-1: Currently, no. We cannot access the formation history.

QUESTION-2: In addition to core accretion and gravitational instability, we should also consider stellar formation mechanism such as core fragmentation to form planetary mass objects (isolated or as companion).

ANSWER-2: Yes, I fully agree. It was refered as gravo-turbulent fragmentation in the slides.

*Young Stars & Planets Near the Sun*
*Proceedings IAU Symposium No. 314, 2015*
*J. H. Kastner, B. Stelzer, & S. A. Metchev, eds.*

© International Astronomical Union 2016
doi:10.1017/S1743921315006572

# Mapping the Distributions of Exoplanet Populations with NICI and GPI

**Eric L. Nielsen[1,2], Michael C. Liu[3], Zahed Wahhaj[4], Beth A. Biller[5], Thomas L. Hayward[6], Laird M. Close[7], the Gemini NICI Planet-Finding Campaign Team, Bruce Macintosh[2], Dmitry Savransky[8], Jason J. Wang[9], James R. Graham[9], Robert J. De Rosa[9], Abhijith Rajan[10], and the GPIES Consortium**

[1]SETI Institute, 189 Bernardo Ave. Suite 100, Mountain View, CA 94043. email: `enielsen@seti.org` [2]Kavli Institute for Particle Astrophysics and Cosmology, 452 Lomita Mall, Stanford, CA 94305. [3]Institute for Astronomy, University of Hawaii at Manoa. [4]European Southern Observatory. [5]Institute for Astronomy, University of Edinburgh. [6]Gemini Observatory. [7]Steward Observatory, University of Arizona. [8]Cornell University. [9]University of California, Berkeley. [10]Arizona State University.

**Abstract.** While more and more long-period giant planets are discovered by direct imaging, the distribution of planets at these separations ($\gtrsim 5$ AU) has remained largely uncertain, especially compared to planets in the inner regions of solar systems probed by RV and transit techniques. The low frequency, the detection challenges, and heterogeneous samples make determining the mass and orbit distributions of directly imaged planets at the end of a survey difficult. By utilizing Monte Carlo methods that incorporate the age, distance, and spectral type of each target, we can use all stars in the survey, not just those with detected planets, to learn about the underlying population. We have produced upper limits and direct measurements of the frequency of these planets with the most recent generation of direct imaging surveys. The Gemini NICI Planet-Finding Campaign observed 220 young, nearby stars at a median H-band contrast of 14.5 magnitudes at 1", representing the largest, deepest search for exoplanets by the completion of the survey. The Gemini Planet Imager Exoplanet Survey is in the process of surveying 600 stars, pushing these contrasts to a few tenths of an arcsecond from the star. With the advent of large surveys (many hundreds of stars) using advanced planet-imagers we gain the ability to move beyond measuring the frequency of wide-separation giant planets and to simultaneously determine the distribution as a function of planet mass, semi-major axis, and stellar mass, and so directly test models of planet formation and evolution.

**Keywords.** instrumentation: adaptive optics, techniques: high angular resolution, planetary systems

---

## 1. Introduction

While over 1000 extrasolar planets have been found to date, only about a dozen have been directly imaged. Given the small angular separations ($\lesssim 1$"), high contrasts between planets and stars ($10^4$–$10^7$ for young giant planets), and low occurrence rate of intermediate-separation giant planets ($\lesssim 5\%$, Nielsen & Close 2010), large-scale surveys with advanced adaptive optics systems at large telescopes are required to find and characterize these planets. We have recently entered the era of multi-hundred star high contrast planet searches with surveys like NICI (Liu *et al.* 2010), IDPS (Vigan *et al.* 2012), SEEDS (Tamura *et al.* 2014), LEECH (Skemer *et al.* 2014), GPI (Macintosh *et al.* 2014a), and SPHERE (Beuzit *et al.* 2010).

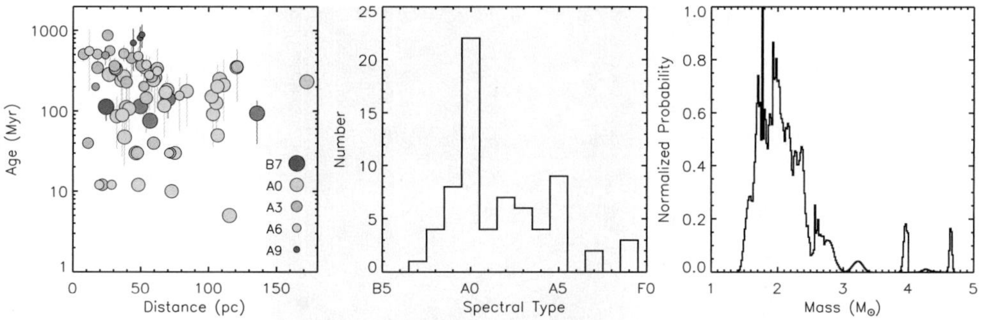

**Figure 1.** B and A stars observed by the Gemini NICI Planet-Finding Campaign. (Left) stars without error bars are members of known moving groups, while the rest have ages determined by our Bayesian method (Nielsen *et al.* 2013). (Center and right) spectral type and mass distributions of the B and A stars in the sample.

## 2. The Gemini NICI Planet-Finding Campaign

Using the Near Infrared Coronagraphic Imager at Gemini South, the Gemini NICI Planet-Finding Campaign surveyed 220 stars between 2008 and 2012 (Liu *et al.* 2010, Wahhaj *et al.* 2013a, Nielsen *et al.* 2013, Biller *et al.* 2013, Wahhaj *et al.* 2013b, Hayward *et al.* 2014). With median contrasts of 14.5 magnitude at 1", NICI represented the largest, deepest imaging search for planets conducted at the time. The target list was made up of a balance of spectral types from late to mid-M, with an emphasis on the youngest, nearest stars available in the southern sky. In addition to discovering 4 new brown dwarf companions (Biller *et al.* 2010, Wahhaj *et al.* 2011, Nielsen *et al.* 2012, Nielsen *et al.* 2013), the Campaign also charactered debris disks (Wahhaj *et al.* 2014, Biller *et al.* 2015) and the exoplanet $\beta$ Pic b (Males *et al.* 2014, Nielsen *et al.* 2014).

### 2.1. *The Ages of B and A stars*

Since the luminosity of giant planets is a strong function of age, determining ages is of key importance in carrying out direct imaging surveys, required to choose a target list, to properly characterize detected objects, and to evaluate survey completeness. B and A stars in particular are challenging, as the most common methods for evaluating age (e.g. calcium, lithium, X-ray, and rotation) aren't applicable for this range of mass. As part of the Gemini NICI Planet-Finding Campaign we developed a new Bayesian technique to produce accurate ages and uncertainties for these early-type stars given their position and errors in the color-magnitude diagrams, finding in many cases significantly older ages than have been previously reported in the literature (Nielsen *et al.* 2013). Figure 1 gives the derived ages and masses of our high-mass sample, as well as the distance and spectral type distribution.

## 3. Completeness to Planets and Brown Dwarfs

In order to evaluate survey completeness we utilize Monte Carlo simulations, as detailed in (Nielsen *et al.* 2005, 2008, and 2010). These simulations create an ensemble of simulated planets around each target star with full orbital parameters, and determine what fraction can be detected given the measured contrast curve for that star. By repeating over a grid of mass and semi-major axis we produce a completeness map for a single star, giving what fraction of planets, as a function of mass and orbital distance, can be detected with our observations. In the case of multiple epochs of contrast curves, the simulated planets are advanced forward in their orbits and compared to each contrast curve for that star.

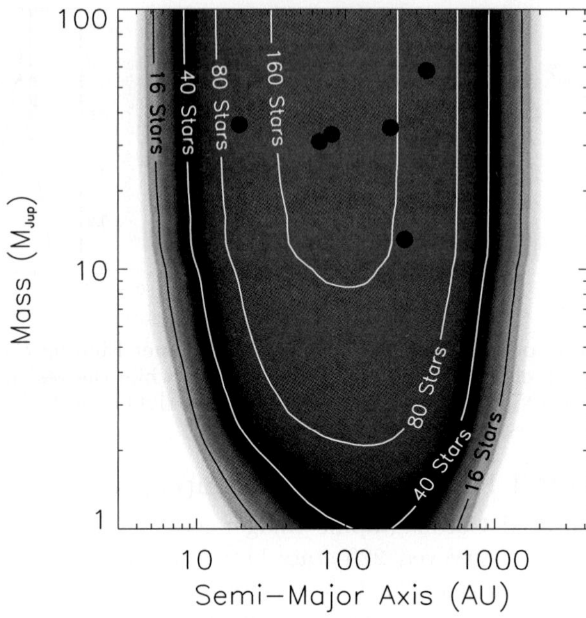


**Figure 2.** Completeness to planets and brown dwarfs from the NICI Campaign. Contours give the ranges of semi-major axis and companion mass for which the survey is complete to planets around the given number of stars. Black dots show detected brown dwarfs.

By summing these completeness maps for all stars we produce the completeness for the entire survey, as shown in Figure 2. The contours represent the number of stars for which we are 100% complete to planets of the given parameters (alternatively, if all stars had a planet of that mass and semi-major axis, the contour values give the number of planets we would expect to detect). Since the distances of target stars varied by over an order of magnitude, and no star is 100% complete to any type of planet (there are always some fraction of orbits that place the companion either behind or in front of the star, or off the detector), the survey completeness never reaches the total number of stars in the survey, 220. Nevertheless we show we were able to detect planets down to a few Jupiter masses at semi-major axes ∼20–800 AU for a significant number of target stars.

By taking our survey completeness shown in Figure 2, expanding it into a four-dimensional completeness map to include stellar mass and projected separation, we can use a Bayesian method to compare the NICI-detected brown dwarfs to a model of brown dwarf populations. Using a series of power-law fits to brown dwarf mass, stellar host mass, and semi-major axis we find distributions consistent (at the $1\sigma$ level) with similar fits to radial velocity giant planets presented by Cumming *et al.* (2008) and Johnson *et al.* (2010), despite the wide variation in mass and orbital separation between the two sets of populations. We do find that small-separation giant planets (<2 AU) are 2-3 times more frequent than wide-separation (10–1000 AU) brown dwarfs, however.


## 4. The Gemini Planet Imager Exoplanet Survey

The Gemini Planet Imager (GPI) represents the latest technology for reaching high contrasts (∼15 magnitudes) close to the star (<1"), and so is able to image giant planets at the equivalent orbital distances of the giant planet region of our own solar system. The GPI Exoplanet Survey (GPIES) is currently searching for planets around 600 young, nearby stars (Macintosh *et al.* 2014a, 2014b, Perrin *et al.* 2014, 2015, Chilcote

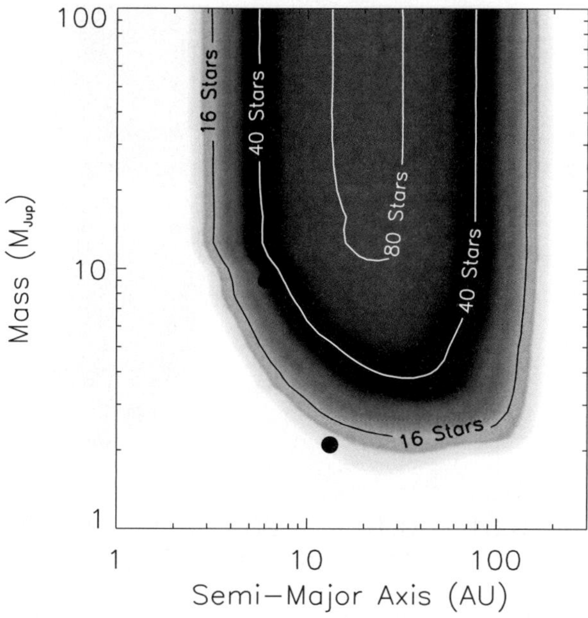


**Figure 3.** Completeness to planets and brown dwarfs from the first 93 stars of the GPIES Campaign. Note that the semi-major axis range of this plot has changed compared to Figure 2 to focus on the area of GPIES sensitivity, closer to the star compared to NICI.

*et al.* 2015). As of this writing, 93 stars have been observed by the GPIES campaign, and the first planet discovery is reported in Macintosh *et al.* (2015). We present the completeness to planets in Figure 3 for the GPIES Campaign. While the NICI instrument was able to reach deeper contrasts given its larger field of view (reaching contrasts of 17–18 magnitudes for some stars beyond 4"), GPI is probing smaller orbital distances compared to NICI. The new parameter space unseen by NICI but being studied by GPI runs from ∼5 to 20 AU. Additionally, NICI's completeness reflects the fact that the NICI Campaign observed 220 stars compared to 93 for GPI, as the GPIES campaign continues the completeness will become deeper and move inward and outward in semi-major axis.

## 5. Early Measurements of Exoplanet Populations

In order to assess the constraints on planet populations we apply the same Bayesian technique we used for brown dwarf populations discussed in Section 3, this time for extrasolar giant planets. We combine the completeness from the NICI and GPIES results, as well as known and new planet detections. Our results are again broadly consistent with the RV results, finding similar power law distributions for 5-1000 AU giant planets as seen for <2 AU giant planets. Our frequency is consistent with that for RV planets as well, but may be slightly lower than that for the closer-in planets. Further observations and detections will allow us to make a more definitive comparison between these two sets of giant planet populations, and determine if they are in fact the same population, or if a break is evident that may trace evolution or formation mechanisms.

## 6. Conclusions

GPI's unprecedented sensitivity to small-separation giant planets is moving closer to the known radial velocity giant planets (Figure 4). At the current rate of observations

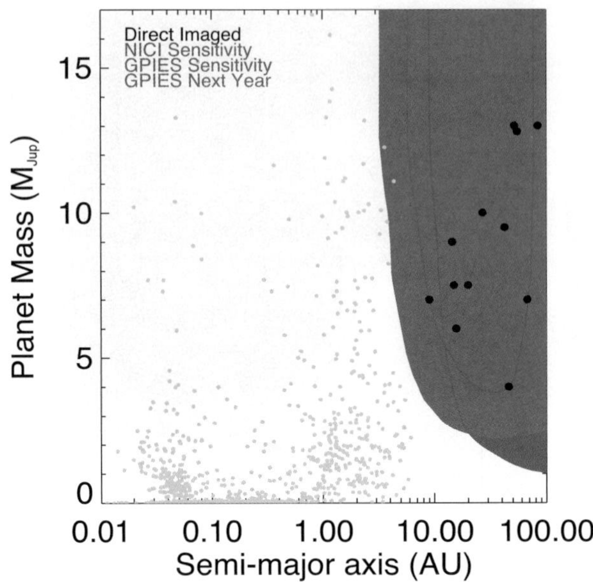



**Figure 4.** Known giant extrasolar planets from exoplanets.eu, and completeness from NICI and GPIES. Also shown is the projected completeness from the GPIES Campaign by June 2016, given the current rate of observations.

the number of stars observed by GPIES will rise from 93 to 300 in about a year from this writing, in June 2016. We can project out to that date by increasing the current completeness by a factor of three, which is the purple shaded region in Figure 4. This region, corresponding to half of GPIES, overlaps known radial velocity giant planets, and suggests we will soon be closing the gap between RV and imaged planets. Within a year we can begin to measure giant exoplanets from the surface of the star out to thousands of AU.

# References


Beuzit, J. L., Boccaletti, A., Feldt, M., Dohlen, K., Mouillet, D., *et al.* 2010, *ASPC*, 430, 231.
Biller, B. A., Liu, M. C., Wahhaj, Z., Nielsen, E. L., Close, L. M., *et al.* 2010, *ApJ*, 720, 82.
Biller, B. A., Liu, M. C., Wahhaj, Z., Nielsen, E. L., Hayward, T. L., *et al.* 2013, *ApJ*, 777, 160.
Biller, B. A., Liu, M. C., Rice, K., Wahhaj, Z., Nielsen, E. L., *et al.* 2015, *MNRAS*, 450, 4446.
Cumming, A., Butler, R. P., March, G. W., Vogt, S. S., *et al.* 2008, *PASP*, 120, 531.
Chilcote, J., Barman, T., Fitzgerald, M. P., Graham, J. R., *et al.* 2015, *ApJ*, 798, 3.
Liu, M. C., Wahhaj, Z., Biller, B. A., Nielsen, E. L., Chun, M., *et al.* 2010, *SPIE*, 7736, 1.
Hayward, T. L., Biller, B. A., Liu, M. C., Nielsen, E. L., *et al.* 2014, *PASP*, 126, 1112.
Johnson, J. A., Aller, K. M., Howard, A. W., & Crepp, J. R. 2010, *PASP*, 122, 905.
Macintosh, B., Graham, J. R., Ingraham, P., Konopacky, Q., *et al.* 2014, *PNAS*, 1111 2661.
Macintosh, B., Anthony, A., Atwood, J., Bauman, B., Cardwell, A., *et al.* 2014b, *SPIE*, 9148 0.
Macintosh, B., Graham, J. R., Barman, T., De Rosa, R. J. *et al.* 2015, *Science*, in press.
Males, J. R., Close, L. M., Morzinski, K. M., Wahhaj, Z., Liu, M. C., *et al.* 2014, *ApJ*, 786, 32.
Nielsen, E. L., Close, L. M., Guirado, J. C., Biller, B. A., Lenzen, R., *et al.* 2005, *AN*, 326, 1033.
Nielsen, E. L., Close, L. M., Biller, B. A., Masciadri, E., & Lenzen, R. 2008, *ApJ*, 674, 466.
Nielsen, E. L. & Close, L. M. 2010, *ApJ*, 717, 878.
Nielsen, E. L., Liu, M. C., Wahhaj, Z., Biller, B. A., Hayward, T. L., *et al.* 2012, *ApJ*, 750, 53.
Nielsen, E. L., Liu, M. C., Wahhaj, Z., Biller, B. A., Hayward, T. L., *et al.* 2013, *ApJ*, 776, 4.
Nielsen, E. L., Liu, M. C., Wahhaj, Z., Biller, B. A., Hayward, T. L., *et al.* 2014, *ApJ*, 794, 158.

Perrin, M. D., Maire, J., Ingraham, P., Savransky, D., *et al.* 2014, *SPIE*, 9147, 3.
Perrin, M. D., Duchene, G., Millar-Blanchaer, M., Fitzgerald, M. P., *et al.* 2015, *ApJ*, 799, 182.
Skemer, A. J., Hinz, P., Esposito, S., Skrutskie, M. F., Defrere, D., *et al.* 2014, *SPIE*, 9148, 0.
Tamura, M., *et al.* 2014, *IAUS*, 299, 12.
Vigan, A., Patience, J., Marois, C., Bonavita, M., De Rosa, R. J., *et al.* 2012, *A&A*, 544, 9.
Wahhaj, Z., Liu, M. C., Biller, B. A., Clarke, F., Nielsen, E. L., *et al.* 2011, *ApJ*, 729, 139.
Wahhaj, Z., Liu, M. C., Nielsen, E. L., Biller, B. A., *et al.* 2013a, *ApJ*, 773, 179.
Wahhaj, Z., Liu, M. C., Biller, B. A., Nielsen, E. L., Hayward, T. L., *et al.* 2013b, *ApJ*, 779, 80.
Wahhaj, Z., Liu, M. C., Biller, B. A., Nielsen, E. L., Hayward, T. L., *et al.* 2014, *ApJ*, 567, 34.

## Discussion

QUESTION: Does GPI really require astrometric follow-up given it's an IFS?

ANSWER: It's true that we have spectra for many of our candidate companions and thus have high confidence from a single observation that we're viewing background stars. However for completeness we follow up these objects with second epoch astrometric observations to confirm their nature as background objects as opposed to common proper motion companions. Unlike previous surveys with large fields of view (the NICI detector is 20"x20"), the GPI detector is 3" on a side, so the number of background objects detected in our survey is small enough that getting second epochs is non prohibitive. Additionally, for the faintest candidates (corresponding to the lowest mass planets, if real) the candidates are seen only when all the wavelength channels are combined to increase S/N. For these candidates only astrometric follow-up can definitively establish their nature.

QUESTION: Isn't it a concern that RV and direct imaging probe different ages of target stars, and so planets?

ANSWER: That's correct, RV surveys are largely targeting old, quiet stars, though some recent surveys have begun to search for planets around younger targets. As GPI and SPHERE find planets in the overlap region between direct imaging and RV we can begin to examine if orbital evolution between $\sim$10–100 Myr and several Gyr is a significant factor in planet populations, if the distributions of planets in the same region of parameter space found by the different techniques is significantly different for different ages.

*Young Stars & Planets Near the Sun*
*Proceedings IAU Symposium No. 314, 2015*
*J. H. Kastner, B. Stelzer, & S. A. Metchev, eds.*

© International Astronomical Union 2016
doi:10.1017/S1743921315006225

# What Do Young Brown Dwarfs Tell Us About Exoplanets?

**Katelyn N. Allers[1], Michael C. Liu[2] and Trent J. Dupuy[3]**

[1] Department of Physics and Astronomy, Bucknell University, Lewisburg, PA 17837, USA
email: `k.allers@bucknell.edu`
[2] Institute for Astronomy, University of Hawaii, 2680 Woodlawn Drive, Honolulu, HI 96822, USA
[3] Department of Astronomy, The University of Texas at Austin, 2515 Speedway C1400, Austin, TX 78712, USA

**Abstract.** In recent years, all-sky surveys have uncovered a new and interesting population of young ($\approx$10–200 Myr), nearby substellar objects. Many of these objects have inferred masses and temperatures that overlap those of directly imaged exoplanets. These young brown dwarfs provide valuable analogs to young, dusty exoplanets in a context where detailed spectroscopic observations across a broad range of wavelengths and at high S/N are possible. How do the temperatures inferred by atmospheric models and evolutionary models compare? Can we determine the formation mechanism of a young planetary-mass object? How well do we understand the role that disequilibrium chemistry and dust clouds play in the atmospheres of these objects? We review the successes and challenges in determining the fundamental properties (mass, $\log(g)$, effective temperature) of young substellar objects, both brown dwarfs and gas-giant exoplanets.

**Keywords.** (stars:) brown dwarfs, infrared:stars, infrared:planetary systems

## 1. Introduction

Studies of directly-imaged exoplanets (e.g., Marois *et al.* 2008; Lagrange *et al.* 2009), planetary-mass companions (e.g., Gauza *et al.* 2015; Artigau *et al.* 2015) and free-floating planetary-mass objects (e.g.; Liu *et al.* 2013b; Gizis *et al.* 2012) share a common goal: determination of fundamental properties (mass, $\log(g)$, effective temperature).

The spectra of young, directly-imaged exoplanets are well matched by young field brown dwarfs (e.g., Chilcote *et al.* 2015). Recently, young brown dwarfs with similar colors and magnitudes as directly-imaged exoplanets have been discovered (Liu *et al.* 2013b, Gauza *et al.* 2015). Given that detailed followup spectroscopy is possible for a large population of young field brown dwarfs, to what extent can these objects be used as analogs to extrasolar planets? Can we develop a unified picture of young, planetary-mass objects that includes both free-floating objects, wide companions, and exoplanets?

## 2. Near-IR Spectroscopic Age Determinations

Given the mass-luminosity-age degeneracy for substellar objects (brown dwarfs and planetary-mass objects), determination of age is critical for estimating their fundamental properties. In various presentations at this symposium, methods for determining the ages of young stars have been discussed. For low-mass stars, the detection of lithium and/or the position of an object above the zero age main sequence on an H-R diagram are the standard methods for determining age. Neither of these methods, however, can determine the age of brown dwarfs or directly-imaged exoplanets. First, most brown dwarfs ($<65\ M_{\rm Jup}$) and all planetary-mass objects do not achieve high enough core

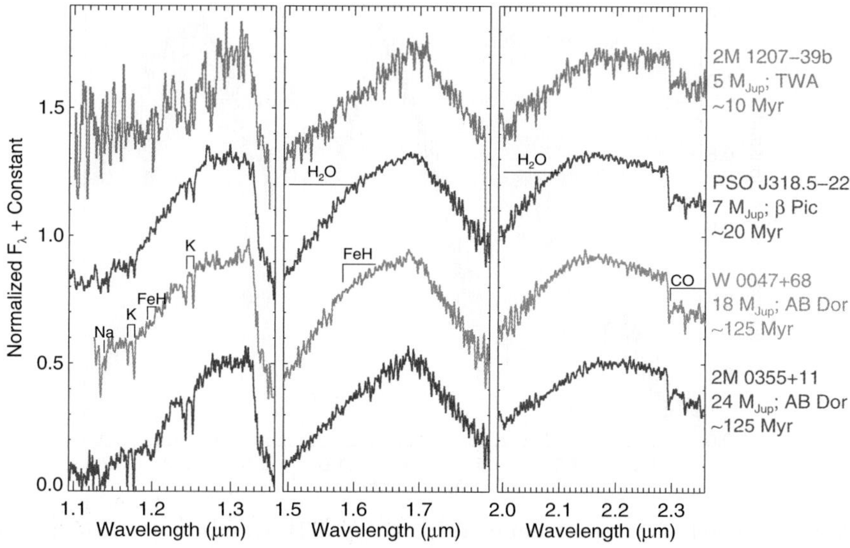


**Figure 1.** Comparison of the spectra of young brown dwarfs and planetary-mass objects that are kinematically linked to known young moving groups. Despite having similar masses and ages, the spectra of PSO J318.5−22 (Liu *et al.* 2013b) and 2M 1207-39b (Patience *et al.* 2010) are not the most similar. The free-floating brown dwarfs, 2M 0355+11 and W 0047+68 are both members of the ∼125 Myr old AB Dor moving group (Liu *et al.* 2013a, Gizis *et al.* 2015), yet have quite disparate spectra (Allers & Liu 2013), despite their similar masses.

temperatures to destroy lithium. Second, recent studies indicate that young, ultracool brown dwarfs are fainter than expected, having similar luminosities as their older field dwarf counterparts of the same spectral type (e.g. Liu *et al.* 2013a).

In contrast to most directly imaged planets, young brown dwarfs are accessible to detailed spectroscopic observations. Cruz *et al.* (2009) developed a classification scheme for the optical spectra of young brown dwarfs. Allers & Liu (2013) presents a corresponding classification system using near-IR spectra. Both of these studies classify the gravity-sensitive spectral features into rough age bins of $\lesssim$30 Myr (labeled as $\gamma$ in the optical and VL-G in the near-IR) and 30–200 Myr ($\beta$ in the optical and INT-G in the near-IR). Both studies emphasize that objects of the same spectral classification can have varying $J - K$ colors. Allers & Liu (2013) also note that objects of the same spectral type that are tied to same YMG can display significant differences in their gravity-sensitive (youth) spectral features in the near-IR. An illustration of this is shown in Figure 1. In terms of mass and age, 2M 1207−39b and PSO J318.5−22† are remarkably similar. When looking at the spectral shapes and gravity sensitive features, PSO J318.5−22 is much more similar to W 0047+68, a markedly older object. Likewise, the spectrum 2M 1207-39b most closely resembles 2M 0355+11 (Faherty *et al.* 2013), an older, more massive object, which is over 2 magnitudes brighter than 2M 1207−39b in $J$ band.

Even objects that lie at the same position on a CMD can display disparate spectra. The recently discovered planetary-mass object VHS 1256−1257B (Gauza *et al.* 2015) has colors and magnitudes that agree (to within the uncertainties) with the directly-imaged exoplanet, HR 8799b. Figure 2 shows the spectra of these two objects. Despite having incredible similarity in color and magnitude, these two objects have quite different spectral shapes. *Overall, observations of young brown dwarfs and directly-imaged exoplanets*

† The recently-measured radial velocity of PSO J318.5−22 confirms that it is a member of the ∼20 Myr-old $\beta$ Pictoris moving group (Allers *et al.*, in prep).

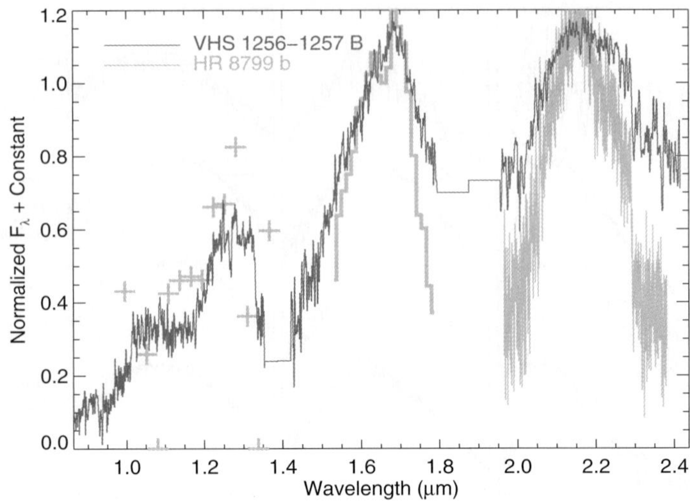


**Figure 2.** Comparison of the spectrum of VHS 1256-1257B (Gauza *et al.* 2015) and HR 8799b (Barman *et al.* 2011, 2015; Oppenheimer *et al.* 2013). The $J - H$ color and absolute $J$-band magnitude of VHS 1256−1257B agree (to within the uncertainties) with photometry of the exoplanet, HR 8799b. The spectra, however, have very different spectral shapes. In particular, the $K$-band spectrum of HR 8799b shows deeper $H_2O$ and CO absorption than seen in the spectrum of VHS 1256−1257B. In addition, HR 8799b has (weak) methane absorption present in its $K$-band spectrum, whereas VHS 1256−1257B does not.

*point to diversity in near-IR spectral morphologies, even among objects of the same color, luminosity, and/or age.*

## 3. Atmospheric and Evolutionary Models

Based on the spectral diversity seen for objects of similar masses and ages, it is clear that one cannot determine accurate ages and masses for planetary-mass objects from spectra alone. For the handful of planetary-mass objects that are members of young moving groups, one can estimate their masses using the model isochrone for the age of the group. One caution is that evolutionary models have not been tested at such young ages by observations. Indeed, the few available tests at older ages ($\approx$0.5-1.0 Gyr) find systematic errors in the model-predicted luminosities and masses at the level of a factor of $\approx$2 and $\approx$25%, respectively, likely due to the effect of cloud evolution (Dupuy *et al.* 2009, 2014).

Effective temperatures ($T_{\mathrm{eff}}$) for ultracool dwarfs can be computed using evolutionary models, given the objects' measured bolometric luminosities and estimated ages (e.g., Golimowski *et al.* 2004). Early discoveries of ultracool companions to young stars suggested a systematic difference between the temperatures of young objects compared to old objects of the same spectral type (e.g., Metchev & Hillenbrand 2006; Luhman *et al.* 2007; Dupuy *et al.* 2009). The HR 8799 planets have brought this effect to the forefront, given the sharp discrepancy between their cool temperatures ($\approx$1000 K) inferred from evolutionary models and their very red, methane-poor spectral energy distributions characteristic of hotter field objects (e.g., Marois *et al.* 2008; Bowler *et al.* 2010; Barman *et al.* 2011; Marley *et al.* 2012). This discrepancy is now seen in free-floating objects in the field as well, as shown by the young ($\approx$10-20 Myr) late-L dwarf PSO J318−22 (Liu *et al.* 2013). Figure 3 compiles all the temperature results to date derived from evolutionary models, including a greatly expanded sample of temperatures from the Hawaii

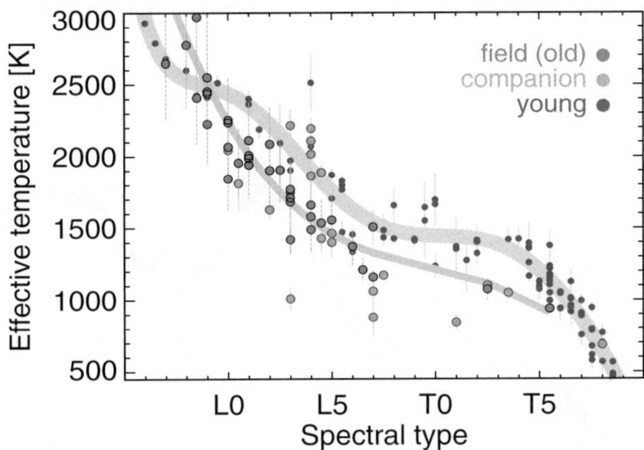


**Figure 3.** Summary of $T_{\text{eff}}$ as a function of spectral type for ultracool dwarfs. Field (old) objects (*grey*) and young companions (*blue*) are from the calculations and compilation by Bowler *et al.*(2013), with additions from Naud *et al.*(2014) and Gauza *et al.*(2015). Young field objects (*orange*) are primarily from Liu *et al.*(in prep), with additional objects from Liu *et al.*(2013), Gizis *et al.*(2015), and Gagne *et al.*(2015). Polynomial fits to the old and young samples are shown as thick colored lines. The young sequence for L and T dwarfs runs $\approx$200–300 K cooler than the old sequence.

Infrared Parallax Program (Dupuy & Liu 2012) for young objects (Liu *et al.*, in prep). We see now the young field L and T dwarfs have a systematic offset relative to the field objects of $\approx$200–300 K cooler, while the young late-M dwarfs tend to run hotter than the field objects. A similar offset appears to be present between young companions and old field objects though less distinct (see also Bowler *et al.* 2013), perhaps reflecting the more heterogeneous spectral typing of companion discoveries or (more speculatively) the possibility that young companion and young field objects behave differently.

How do the $T_{\text{eff}}$'s determined by fitting model atmospheres compare to the $T_{\text{eff}}$'s determined from evolutionary models? Figure 4 shows the best fit atmospheric models to the near-IR spectrum of PSO J318.5−22. As discussed in Liu *et al.* (2013b) and Gizis *et al.* (2015), current atmospheric models for young brown dwarfs and planetary-mass objects with L-type spectra overpredict their $T_{\text{eff}}$'s . Indeed, if young, planetary-mass objects had the $T_{\text{eff}}$'s inferred by atmospheric models, their radii would necessarily be implausibly small ($< 1$ R$_{\text{Jup}}$) to match their observed luminosities. The $T_{\text{eff}}$'s derived from evolutionary and atmospheric models become more divergent for younger and lower-mass objects (Gizis *et al.* 2015). Recent advances in atmospheric models, including treatment of clouds and disequilibrium chemistry are likely to provide better fits (see contribution by M. Marley). The combination of issues associated with current model atmospheres as well as the diversity seen in the observed spectra of young planetary-mass objects suggests that $T_{\text{eff}}$'s determined from fitting model atmospheres to near-IR spectra are unreliable and corresponding surface gravity determinations even moreso.

## 4. Conclusions

- Spectral diversity is the norm in the young substellar regime, even for objects of similar, mass, age, and/or position on a color-magnitude diagram.
- Young substellar objects (brown dwarfs and exoplanets) have cooler effective temperatures than field brown dwarfs of the same spectral type, both for low-gravity objects in the field and those found as companions to young stars.

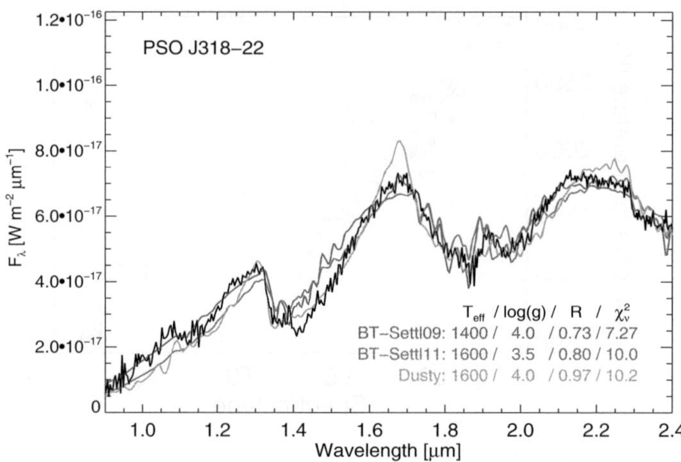


**Figure 4.** Comparison of the spectrum of PSO J318.5−22 (black; Liu *et al.* 2013b) to best fit model atmospheres (Allard *et al.* 2001, 2012), as determined from $\chi^2$ minimization. Since PSO J318.5−22 has a known distance, the scaling factor needed to match the flux level of the model to the observed spectrum ($= R^2/d^2$ where $R$ is the radius and $d$ is the distance) provides an estimate of PSO J318.5−22's radius. The best-fit models yield implausibly small radii. As a member of the ≈10–20 Myr old $\beta$ Pictoris moving group, evolutionary models predict a radius of 1.56 $R_{\rm Jup}$ and effective temperature of 1190 K for PSO J318.5−22, significantly cooler than the temperatures of the best-fit atmospheric models.

• The effective temperatures of young brown dwarfs and directly-imaged exoplanets are overpredicted by atmospheric models for L-type objects.

## References


Allard, F., Hauschildt, P. H., Alexander, D. R., Tamanai, A., & Schweitzer, A. 2001, *ApJ*, 556, 357

Allard, F., Homeier, D., Freytag, B., & Sharp, C. M. 2012. *Atmospheres From Very Low-Mass Stars to Extrasolar Planets*. EAS Publications Series 57, 3-43

Allers, K. N. & Liu, M. C. 2013, *ApJ*, 772, 79

Artigau, É., Gagné, J., Faherty, J., *et al.* 2015, *ApJ*, 806, 254

Barman, T. S., Konopacky, Q. M., Macintosh, B., & Marois, C. 2015, *ApJ*, 804, 61

Barman, T. S., Macintosh, B., Konopacky, Q. M., & Marois, C. 2011, *ApJ*, 733, 65

Bowler, B. P., Liu, M. C., Dupuy, T. J., & Cushing, M. C. 2010, *ApJ*, 723, 850

Bowler, B. P., Liu, M. C., Shkolnik, E. L., & Dupuy, T. J. 2013, *ApJ*, 774, 55

Chilcote, J., Barman, T., Fitzgerald, M. P., *et al.* 2015, *ApJL*, 798, L3

Cruz, K. L., Kirkpatrick, J. D., & Burgasser, A. J. 2009, *AJ*, 137, 3345

Dupuy, T. J., Liu, M. C., & Ireland, M. J. 2009, *ApJ*, 699, 168

Dupuy, T. J., Liu, M. C., & Ireland, M. J. 2014, *ApJ*, 790, 133

Dupuy, T. J. & Liu, M. C. 2012, *ApJS*, 201, 19

Faherty, J. K., Rice, E. L., Cruz, K. L., Mamajek, E. E., & Núñez, A. 2013, *AJ*, 145, 2

Gagné, J., Burgasser, A. J., Faherty, J. K., *et al.* 2015, arXiv:1506.04195

Gauza, B., Béjar, V. J. S., Pérez-Garrido, A., *et al.* 2015, *ApJ*, 804, 96

Gizis, J. E., Allers, K. N., Liu, M. C., *et al.* 2015, *ApJ*, 799, 203

Gizis, J. E., Faherty, J. K., Liu, M. C., *et al.* 2012, *AJ*, 144, 94

Golimowski, D. A., Leggett, S. K., Marley, M. S., *et al.* 2004, *AJ*, 127, 3516

Lagrange, A.-M., Gratadour, D., Chauvin, G., *et al.* 2009, *A&A*, 493, L21

Liu, M. C., Dupuy, T. J., & Allers, K. N. 2013a, *Astronomische Nachrichten*, 334, 85

Liu, M. C., Magnier, E. A., Deacon, N. R., *et al.* 2013b, *ApJL*, 777, L20

Luhman, K. L., Patten, B. M., Marengo, M., *et al.* 2007, *ApJ*, 654, 570
Marley, M. S., Saumon, D., Cushing, M., *et al.* 2012, *ApJ*, 754, 135
Marois, C., Macintosh, B., Barman, T., *et al.* 2008, *Science*, 322, 1348
Metchev, S. A. & Hillenbrand, L. A. 2006, *ApJ*, 651, 1166
Naud, M.-E., Artigau, É., Malo, L., *et al.* 2014, *ApJ*, 787, 5
Oppenheimer, B. R., Baranec, C., Beichman, C., *et al.* 2013, *ApJ*, 768, 24
Patience, J., King, R. R., de Rosa, R. J., & Marois, C. 2010, *A&A*, 517, A76


## Discussion

J. GAGNÉ: You show some very red objects with "peaky" H-band spectra that are not young. Do you have an explanation for why these objects appear the way they do?

AUTHOR: There is some speculation in the literature that this could be a metallicity or inclination effect.

S. METCHEV: Please get to the bottom of these dust vs. youth indicators. Also, how useful is VO as a youth indicator?

AUTHOR: In Allers & Liu (2013), we present a classification system that can distinguish between young and dusty objects. As for VO, it is an excellent indicator of youth for objects with spectral types of L0–L4.

B. BOWLER: Could some of the youth effects be caused by patchy clouds? i.e. could it be that gravity effects are actually caused by seeing emergent spectra from different cloud layers in the atmosphere?

AUTHOR: Patchy clouds could certainly do interesting things to the spectra (see contribution by M. Marley), but the youth that we see isn't likely driven by cloud patchiness. We see strong indicators of youth in objects with late M and early L spectral types which shouldn't have the patchy clouds that are associated with the L/T transition.

*Young Stars & Planets Near the Sun*
*Proceedings IAU Symposium No. 314, 2015*
*J. H. Kastner, B. Stelzer, & S. A. Metchev, eds.*
© International Astronomical Union 2016
doi:10.1017/S1743921315006493

# A Young Planetary Mass Companion to the Nearby M Dwarf VHS J125601.92-125723.9

B. Gauza[1], V. J. S. Béjar[1], A. Pérez-Garrido[2], M. R. Zapatero Osorio[3], N. Lodieu[1], R. Rebolo[1], E. Pallé[1] and G. Nowak[1]

[1]Instituto de Astrofísica de Canarias (IAC), Calle Vía Láctea s/n, E-38200 La Laguna, Tenerife, Spain
email: bgauza@iac.es vbejar@iac.es
[2]Dpto. Física Aplicada, Universidad Politécnica de Cartagena, Campus Muralla del Mar, Cartagena, Murcia E-30202, Spain
[3]Centro de Astrobiología (CSIC-INTA), Ctra. Ajalvir km 4, 28850, Torrejón de Ardoz, Madrid, Spain

**Abstract.** We have recently identified a young, very red ($J - K_s = 2.47$ mag) late L-type companion at 8.06" $\pm\,0.03$" ($\sim$102 AU) from a previously unrecognized M dwarf. We determined the parallactic distance of the system to be 12.7$\pm$1.0 pc. Non-detection of lithium and the kinematics of the primary allowed us to constrain the age of the system in the range of 150–300 Myr. By comparison with theoretical evolutionary models we derived a mass of $73^{+20}_{-15}$ $M_{\rm Jup}$ for the primary, at around the substellar mass regime and $11.2^{+9.7}_{-1.8}$ $M_{\rm Jup}$ for the secondary, near the deuterium burning mass limit.

**Keywords.** stars: brown dwarfs – stars: imaging – infrared: planetary systems – stars: individual (VHS J125601.92-125723.9)

## 1. Introduction

Young L dwarfs have recently been identified, either free-floating (Peña Ramírez et al. 2012; Zapatero Osorio *et al.* 2014) or as companions to stars (e. g., Wahhaj *et al.* 2011; Bowler *et al.* 2013; Chauvin *et al.* 2015). They were found to share similar photometric and spectral properties that are distinct from those of typical late-type objects of the field population. Some of these peculiarities have been attributed to low surface gravities and cloudy atmospheres, as expected at early evolutionary stages, less than several hundred million years (Cruz *et al.* 2009; Allers & Liu 2013). They have very red colors ($J - K_s > 2$ mag), $J$-band absolute magnitudes fainter than their old, field counterparts, and show distinctive spectral features such as, for example, a sharply peaked, triangular-shaped continuum in the $H$-band and weaker Na I and K I lines.

Recent studies have revealed a strong resemblance between the young L dwarfs and directly imaged planetary mass companions (e. g., 2MASS 1207-39 b, Chauvin *et al.* 2005, HR 8799 bcde, Marois *et al.* 2010, GJ 504 b Kuzuhara *et al.* 2013). They have similar near-infrared (near-IR) colors and absolute magnitudes, overlapping effective temperature regimes of $\sim$1000–1500 K and masses of a few to a few tens of Jupiter masses Spectroscopic and photometric studies of young substellar objects can provide information on the physical properties of gas giant exoplanets found by transit and radial velocity surveys, in particular, on the characteristics and composition of their complex atmospheres. Here we describe the physical properties of the recently identified nearby, young binary system VHS 1256-1257: spectral types of the two components, distance, age, luminosities, masses, and effective temperatures.

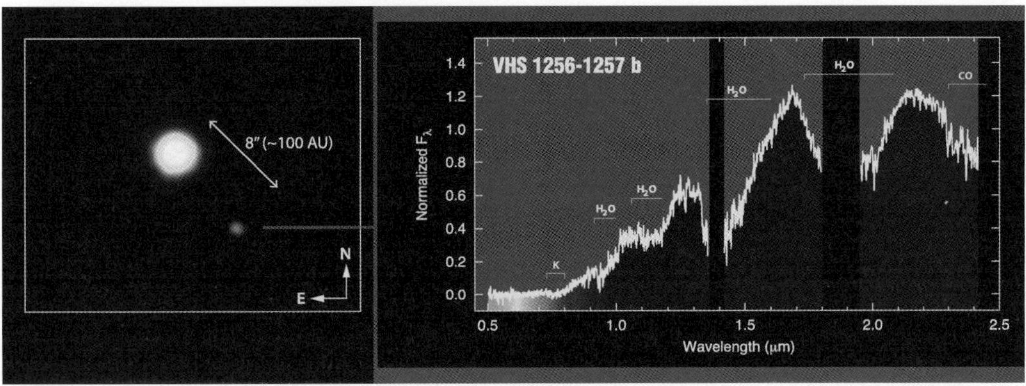


**Figure 1.** *Left*: image of the planetary mass companion at 8.06" ± 0.03" (~102 AU) from the low-mass star VHS 1256. This false color image was obtained from the $Y$, $J$ and $K_s$ bands images taken with the VISTA telescope. *Right*: Low-resolution optical and near-infrared spectra of the companion.

**Table 1.** Selected measurements and parameters of VHS 1256-1257 components.

| | SpType | $\mu_\alpha \cos\delta$ [mas/yr] | $\mu_\delta$ [mas/yr] | $J$ [mag] | $J - K_s$ [mag] | $\log (L_{\rm bol}/L_\odot)$ | Mass $[M_{\rm Jup}]$ | $T_{\rm eff}$ [K] |
|---|---|---|---|---|---|---|---|---|
| Primary | M7.5±0.5 | -281.5±5.3 | -205.5±15.2 | 11.02±0.02 | 0.97±0.04 | -3.14±0.10 | $73^{+20}_{-15}$ | 2620±140 |
| Secondary | L7.0±1.5 | -275.4±5.3 | -198.4±15.2 | 17.14±0.02 | 2.47±0.03 | -5.05±0.22 | $11.2^{+9.7}_{-1.8}$ | $880^{+140}_{-110}$ |

## 2. Identification of the system

Using the VISTA Hemisphere Survey (VHS, McMahon *et al.* 2013) data and the 2MASS (Skrutskie *et al.* 2006), we carried out a search for high proper motion objects, by cross-matching both catalogs. The search focused on objects that had moved at least 2 arcsec and a maximum of 30 arcsec from 2MASS to VHS. The time baseline between the two surveys is typically about 12 yr, which gives proper motions of approximately 0.15–3.0 arcsec yr$^{-1}$. Over the common area between 2MASS and VHS we found more than 6000 objects with $J = 11$–17 mag and proper motion higher than 150 mas yr$^{-1}$. We have searched for common proper motion pairs and multiples among these objects with proper motions consistent within 40 mas yr$^{-1}$ in both $\mu_\alpha$ and $\mu_\delta$. VHS 1256-1257 system was one of the identified candidates.

A VISTA image of the pair is shown in Fig. 1. The companion is located at a projected angular separation of 8.06 ± 0.03 arcsec. This separation corresponds to a projected orbital separation of 102 ± 9 AU at the estimated distance of the system. The two components share a common proper motion, which significantly differs from the proper motion of background stars. The $\mu_\alpha \cos\delta$ and $\mu_\delta$ measured from the VHS and 2MASS positions of the sources and other selected parameters of the two components are listed in Table 1.

## 3. Physical properties of the components

To determine the spectral types, we used low-resolution optical and near-IR spectra obtained using GTC/OSIRIS and NTT/SofI instruments, respectively. We classified the objects through visual comparison of the spectra with a set of field dwarf spectral templates, in the optical and near-IR separately, with known young L dwarfs in the near-IR, and using the spectral indices established by Allers & Liu (2013). We adopted an M7.5 ± 0.5 and L7.0 ± 1.5 spectral types for the primary and secondary, respectively.

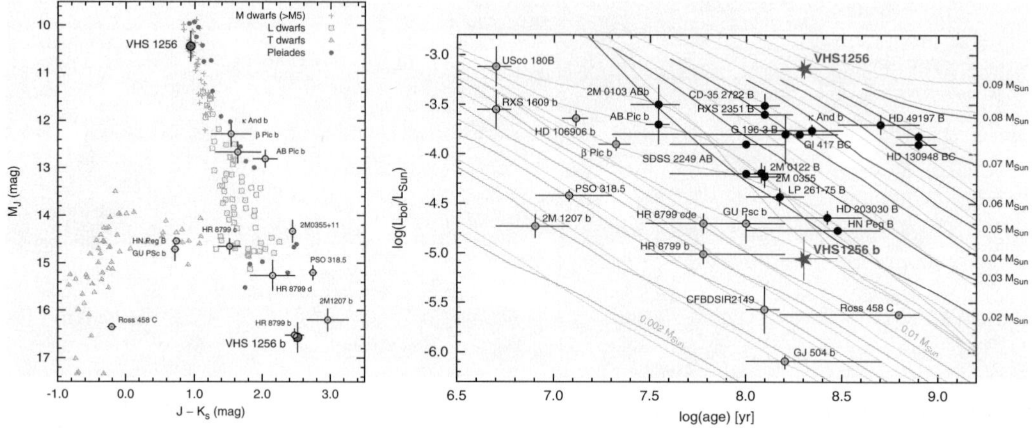

**Figure 2.** *Left*: $M_J$ vs. $J - K_s$ color-magnitude diagrams comparing the two components of VHS 1256-1257 with field M, L, and T dwarfs with parallax measurements, known young substellar objects and the least-massive Pleiades members. *Right:* Luminosity and age of the two components compared with evolutionary tracks from the cloudy atmosphere models of Saumon & Marley (2008), the BT-Settl models (Allard & Barman 2004) and known young substellar objects. The Saumon & Marley models are plotted in color, gray curves correspond to BT-Settl.

The near-IR spectrum of the secondary (right panel of Fig. 1) shows a peaked triangular shape of the $H$-band continuum, and other spectral features recognized as hallmarks of low surface gravity and youth, like weak Na I and K I lines. The optical spectrum of the primary also shows weaker alkali lines than of the field counterparts, but no Li I at 670.82 nm at a limit of pEW $< 30$ mÅ.

Using the VISTA data and a set of follow-up images collected from March to December, 2014 using NTT/SofI, IAC80/CAMELOT and WHT/LIRIS instruments, we have obtained a first measurement of the trigonometric parallax. The absolute parallax is $\pi = 78.79 \pm 6.4$ mas, which translates into a distance of the system of $12.7 \pm 1.0$ pc. To measure the heliocentric radial velocity of VHS 1256-1257 we employed high spectral resolution ($R \sim 40\,000$) UVES data and the cross-correlation method against the M6V star GJ 406, which has a known, constant radial velocity of $v_r = 19.5 \pm 0.1$ km s$^{-1}$ (Nidever *et al.* 2002). GJ 406 was also observed with the VLT/UVES instrument with similar spectral resolution, data is available in the ESO Archive. The measured heliocentric radial velocity is $v_r = -1.4 \pm 5.0$ km s$^{-1}$, Having the measurements of proper motion, parallax, and radial velocity we could determine the three components of the Galactic space velocity, $U$, $V$, and $W$, obtaining -9.4±2.0, -16.4±3.0 and -8.5±3.9 km s$^{-1}$, for each component, respectively.

The non-detection of Li I imposes a lower limit on the mass of the primary of 0.055–0.060 $M_\odot$ and also imposes a lower limit on the age of the system, since objects of similar spectral type in the Pleiades cluster (age $\sim 120$ Myr) have fully preserved this element. Theoretical evolutionary models predict that objects with $T_{\rm eff} \sim 2600$ K (which is the corresponding effective temperature of an M7.5 dwarf) have preserved their Li content at ages $< 150$ Myr. The galactic $UVW$ velocities indicate that the system probably belongs to the Local Association, whose members have estimated ages of 10–300 Myr. In conclusion, based on the absence of Li I in the primary and the likely membership in the Local Association, we adopt a range of 150–300 Myr for the age of the VHS 1256-1257 system.

In Fig. 2 we compare $M_J$ versus $J - K_s$ and luminosities at the adopted range of ages of the two components of the pair with field M, L and T dwarf sequence (on the color-mag diagram), with the Saumon & Marley (2008) and BT-Settl models evolutionary tracks (on the right panel) and with known young substellar objects. On the $M_J$, $J - K_s$ CMD, the secondary, VHS 1256 b is located at almost the same position as the HR 8799 b planet. Using bolometric corrections we derive luminosities and compared with theoretical models to derive masses, temperatures and log $g$. For the primary, we obtained a mass of $73^{+20}_{-15}$ $M_{\rm Jup}$, close to the boundary between stars and brown dwarfs. The effective temperature and log $g$ found from the models are $2620 \pm 140$ K and $5.05 \pm 0.10$ dex, respectively. For the companion, we obtained a mass of $11.2^{+9.7}_{-1.8}$ $M_{\rm Jup}$, indicating that the object is near the mass limit at which the onset of deuterium fusion occurs. Given the uncertainty in the mass determination, it is currently unclear whether the object is above or below it. The effective temperature and log $g$ of the companion obtained from the evolutionary models are $880^{+140}_{-110}$ K and $4.24^{+0.35}_{-0.10}$ dex.

## 4. Final remarks

VHS 1256-1257 b is among the nearest currently known planetary mass companions detected by direct imaging. Moreover, it is one of the very few young, extremely red L dwarfs with age constrained within a narrow range, given by the likely belonging to the Local Association and the absence of Li I in the primary. The $T_{\rm eff}$ of $\sim 900$ K determined from evolutionary models based on the luminosity does not seem to be consistent with the expected $T_{\rm eff}$ range of field dwarfs of similar spectral type and with the absence of methane, which is expected to appear in atmosphere cooler than 1400 K. Following Barman *et al.* (2011) the formation of clouds with substantial vertical thickness and non-equilibrium chemistry in a low-gravity object like VHS 1256-1257 b could provide an explanation of the apparent high atmospheric temperature ($>1500$ K) as compared to cooling track effective temperature predictions (900–1000 K). Since it is a relatively nearby and bright object near the deuterium-burning limit, it becomes one of the most promising targets to study the application of the deuterium test. From the masses and separation of the components we estimate the orbital period to be about 3900 yr. Assuming a circular orbit with a face-on orientation, the displacement caused by the orbital motion would be from 4 to 13 mas/yr, which will become feasible to measure in the next few years using precise astrometric observations.

## References


Allard, F. & Barman, T. S. 2004, *IAU Symposium*, Vol. 213, 119
Allers, K. N. & Liu, M. C. 2013, *ApJ*, 772, 79
Barman, T. S., Macintosh, B., Konopacky, Q. M., & Marois, C. 2011, *ApJL*, 735, L39
Bowler, B. P., Liu, M. C., Shkolnik, E. L., & Dupuy, T. J. 2013, *ApJ*, 774, 55
Chauvin, G., Lagrange, A. M., Dumas, C., *et al.*, 2005, *A&A*, 438, L25
Chauvin, G., Vigan, A., Bonnefoy, M., *et al.*, 2015, *A&A*, 573, 127
Cruz, K. L., Kirkpatrick, J. D., & Burgasser, A. J. 2009, *AJ*, 137, 3345
Deacon, N. R. & Hambly, N. C 2007, *A&A*, 468, 163
Eggen, O. J. 1992, *AJ*, 104, 2141
Eisenbeiss, T., Ammler-von Eiff, M., Roell, T., *et al.*, 2013, *A&A*, 556, 53
Gauza, B., Béjar, V. J., Pérez-Garrido, A., *et al.*, 2015, *ApJ*, 804, 96
Kuzuhara, M., Tamura, M., Kudo, T., *et al.*, 2013, *ApJ*, 774, 11
Marois, C., Zuckerman, B., Konopacky, Q. M., *et al.*, 2010, *Nature*, 468, 1080
McMahon, R. G., Banerji, M., Gonzalez, E., *et al.*, 2013, *The Messenger*, 154, 35
Peña Ramírez, K., Béjar, V. J. S.., Zapatero Osorio, M. R., *et al.*, 2012, *ApJ*, 754, 30
Saumon, D. & Marley, M. S. 2008, *ApJ*, 689, 1327S

Skrutskie, M. F., Cutri, R. M., Stiening, R., *et al.*, 2006, *AJ*, 131, 1163
Nidever, D. L., Marcy, G. W., Butler, R. P., Fischer, D. A., & Vogt, S. S. 2002, *ApJS*, 141, 503
Wahhaj, Z., Liu, M. C., Biller, B. A., *et al.*, 2011, *ApJ*, 729, 139
Zapatero Osorio, M. R., Gálvez Ortiz, M. C., Bihain, G., *et al.*, 2014, *A&A*, 568, A77

## Discussion

AUTHOR: A question by Eric Mamajek was regarding the age upper limit. He asked if the restriction of 300 Myr which we adopt comes only from the probable membership of the system to the Local Association.

AUTHOR: Having measured the proper motion, parallax and radial velocity of the primary, we could determine its $UVW$ galactic velocities. The kinematics are consistent with a range where several young moving groups overlap, but some of them like the $\beta$ Pictoris, Taurus-Auriga or TW Hya can be rejected since they are younger than 100 Myr, and this age is incompatible with the non-detection of lithium in the atmosphere of the primary. However, the kinematic and spectroscopic properties of VHS 1256-1257 appear consistent with membership in Local Association Group, a coherent kinematic stream of young stars, all below 300 Myr (Eggen 1992). The Local Association is considered by some authors as having much wider spread of ages, up to 1 Gyr, we thus agree that this upper age constraint is not as strong as the lower limit of 150 Myr coming from the non-detection of Li I. Nevertheless, we see various spectral features of both of the components suggesting an age considerably younger than 1 Gyr. The weak alkali lines, peculiar red $J - K_s$ color and peaked, triangular shape of the H-band continuum are recognized as hallmarks of very young ages, below 500 Myr. We would like to highlight, that in case of VHS 1256 its not the case since we know the system is older than 150 Myr.

Additionally, from tests using the LACEwING moving group identification code, Adric Riedel has found (private communication) that the system has a significant probability of being a member of the Hercules-Lyra moving group (250 Myr, Eisenbeiss *et al.* 2013), imposing an age constraint consistent with our estimates.

AUTHOR: As mentioned in the original paper (Gauza *et al.* 2015) the primary and the secondary were cataloged in 2MASS with designations 2MASS J125602.15-125721.7 and 2MASS J125601.83-125727.6, respectively. The primary was also listed as SIPS 1256-1257 in the sample of low-mass stars with $\mu > 0.1\,\mathrm{arcsec\,yr}^{-1}$ from Deacon & Hambly (2007). Eric has also noticed, what we didn't realize before, that the primary was also detected by Luyten as a HPM star and listed in his catalog as LP736-15.

*Young Stars & Planets Near the Sun*
Proceedings IAU Symposium No. 314, 2015
J. H. Kastner, B. Stelzer, & S. A. Metchev, eds.

© International Astronomical Union 2016
doi:10.1017/S1743921315006675

# Planetary Evaporation and the Dynamics of Planet Wind/Stellar Wind Bow Shocks

**A. Frank[1], B. Lui,[1] J. Carroll-Nellenback,[1], A. C. Quillen[1], E. G. Blackman[1], J. Kasting[2] and I. Dobbs-Dixon[3]**

[1] University of Rochester
[2] Penn State University
[3] New York University Abu Dhabi

**Abstract.** We present initial results of a new campaign of simulations focusing on the interaction of planetary winds with stellar environments using Adaptive Mesh Refinement methods. We have confirmed the results of Stone & Proga (2009) that an azimuthal flow structure is created in the planetary wind due to day/night temperatures differences. We show that a backflow towards the planet will occur with a strength that depends on the escape parameter. When a stellar outflow is included, we see unstable bow waves forming through the outflow's interaction with the planetary wind.

**Keywords.** Exo-planets. Evaporative Flows, Bow Shocks

## 1. Introduction

Planetary blow-off occurs when irradiation from the central star, especially in the extreme ultraviolet (EUV), heats of the upper layers of the atmosphere to produce an extended envelope of gas which transitions into an wind. Such flows are believed to occur in Hot Jupiters (i.e. HD 209458b, Ballister *et al.*2007). A characteristic measure of the strength of the wind is the ratio of gravitational potential to thermal energy at the top of the atmosphere. This is usually called the hydrodynamic escape parameter,

$$\lambda = \frac{GM_p\mu}{(R_p k T_p)} \tag{1.1}$$

where $M_p$ and $R_p$ are the mass and radius of the planet, and $T_p$ and $\mu$ are the temperature and mean mass per particle in the atmosphere. For $\lambda >> 10$, the atmosphere is too tightly bound for a hydrodynamic wind to form. Note that weaker outflows may be produced via non-thermal processes, e.g., Hunten (1982). For $\lambda \sim 10$, a Parker-type, thermally driven hydrodynamic wind is expected (note that $\lambda \sim 15$ for the sun with its $T \sim 10^6$ K corona). Note also that the composition of the atmosphere must also be considered in detailed models of wind launching.

When these atmospheric winds occur, they are expected to have a wide range of applications and observational consequences in exoplanet studies. In particular many open questions exist about how these winds will change when additional physics are added. Physical processes that could affect the wind structure and outflow rates include planetary magnetic fields (Owen and Adams 2014), time dependent EUV flux (Lecavelier des Etangs *et al.* 2012), atmospheric circulation (Teyssandier *et al.* 2015) and the interaction between stellar and planet winds (Murray-Clay *et al.* 2009, Stone & Proga 2009). Many of these processes have only recently begun to be incorporated into existing simulations (Schneiter *et al.* 2007, Cohen *et al.* 2011, Bisikalo *et al.* 2013). The recent work of

A. Frank et al.

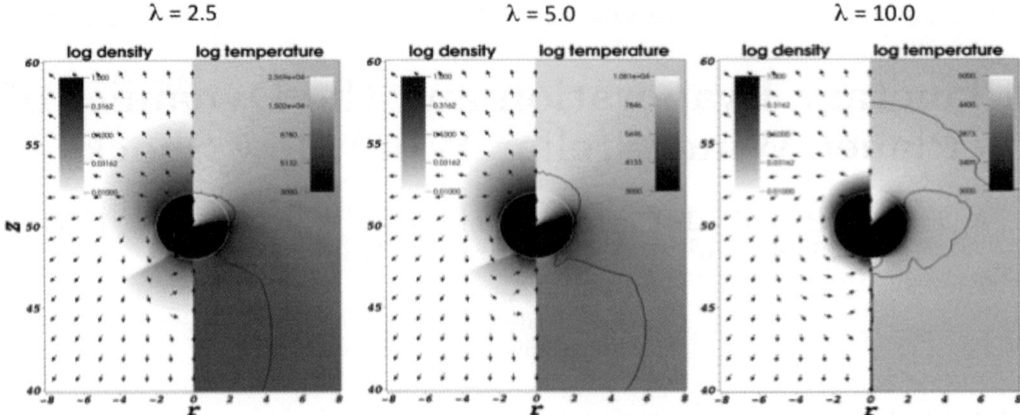

**Figure 1.** Steady state planetary wind solutions computed at high resolution in 2.5D using AstroBEAR for different escape parameter $\lambda$ values. Flow and density are shown on the left and thermal structure and $M = 1$ contours are shown on the right. Note the flow from the day-side (top) to the night-side passing through a shock (apparent at $\theta \sim 135^o$ from vertical) and leading to backflow on to the planets night-side.

Matsakos *et al.* 2015, for example, includes a variety of processes in a 3-D MHD study that tracked the full orbital dynamics of the wind.

In this contribution we present initial results of a new campaign of simulations focusing on the interaction of planetary winds with stellar environments using Adaptive Mesh Refinement methods. The goal of this initial study is to explore the multi-dimensional wind structure from photo-evaporating planets and the time-dependence of planetary wind/stellar wind interactions.

## 2. Code and Set-up

In this study we present first results of our adaptive mesh refinement (AMR) simulations of planetary wind launching and planet-wind/stellar wind interactions using the AstroBEAR code (Cunningham *et al.*2009). AstroBEAR is a fully parallelized AMR MHD multi-physics code which currently includes modules for the treatment of self-gravity, ionization dynamics, chemistry, heat conduction, viscosity, resistivity and radiation transport via flux-limited diffusion (or plane-parallel ray tracing). For these initial simulations we explore only hydrodynamic interactions in axisymmetry, (2.5D), with a polytropic equation of state (the polytropic index $\gamma$ is an input parameter and we assume isothermal conditions $\gamma = 1.01$).

Our initial set up is similar to that of Stone & Proga (2009) . We initialize the outer boundary of the planet (assumed to be the wind launch point) with azimuthally variable temperature $T(\theta) = T_o sin(\theta)$ where $\theta = 90^o$ is the sub solar point and $T_o = 10^4$ K. The high temperatures on the day side of the planet launch a strong Parker-type thermal wind. On the night side the lower temperatures imply a lower value of $\lambda$ leading to an aspherical flow from the planet. In some of our models we also drive a plane parallel stellar wind into the grid. Given the small orbital radius for planets on "hot" orbits, it is possible that the planet will still be in the stellar wind acceleration region and we consider both subsonic ($M = 0.8$) and supersonic ($M = 5.0$) stellar winds.

## 3. Planetary Wind Structure: Nightside Backflow

On order to asses the multi-dimensional structure of the planetary evaporative flow we have run a series of simulations at high resolution with different initial conditions for the planetary parameters. The simulation domain has a resolution of 64 zones per planetary radii at the highest level of refinement. We choose our planetary parameters to sample three values of the escape parameter ($\lambda = 2.5, 5, 10$). Figure 1 shows the wind solutions (density, temperature, velocity field, Mach surface) for these runs.

In all three cases we see a strong Parker type thermal wind being launched from the day side of the planet (top hemisphere). Consideration of the $M = 1$ mach surface (black contour) demonstrates that in all 3 cases, the wind reaches supersonic velocities close the planet. The distance to the sonic surface however depends on the escape parameter increasing with the value of $\lambda$. The flow from the planet is not however purely radial. As first shown by Proga & Stone (2009), azimuthal pressure gradients in the sub-sonic and trans-sonic regions of the planetary flow drive material from the hot day side towards the colder night side of the planet. As this flow converges towards the night side, a conical shock forms at approximately latitude $\theta = 135^\circ$ (relative to the substellar point). This shock weakens with increasing $\lambda$ and is not seen in our $\lambda = 10$ simulation because the flow is entirely subsonic at these radii. We note that the converging flow on the nightside does more than just redirect outflowing material. In our simulations we see some gas fall back towards the planetary boundary. We find the relative strength of this backflow also decreases for the largest values of our $\lambda$ series.

## 4. Stellar Wind Interactions

The high resolution of our simulations translates into higher Reynolds numbers and hence a greater ability to capture instabilities associated with stellar wind/planetary wind interactions. As the stellar wind sweeps past the planet, it encounters the planetary wind and, depending on the ratio of ram pressures (or thermal pressures for subsonic flows), a bow wave may form. Such a bow wave/bow shock may be prone to a variety of instabilities including non-linear thin shell modes (NTSI) if the bow shock cools effectively. Since instabilities will generate vorticity one can expect density inhomogeneities to form which will be swept downstream on a timescale of $t_s \sim R_p/V_{bs}$ where $V_{bs}$ represents the flow behind the bow shock.

Because the stellar wind may not have reached its final supersonic velocity we have carried out simulations representing both subsonic and supersonic stellar flows impinging on a planetary wind. In figure 2 we show flow conditions for a $M_* = 0.8$ (left) and a $M_* = 5.0$ (right) wind flowing downward from the upper boundary to interact with a $\lambda = 5$ planetary wind (see figure 1). In both the subsonic and supersonic cases we see the bow wave is not smooth but breaks up due to instabilities. The morphology of the two cases is however quite different. In the subsonic stellar wind case the bow consists of one shock facing back into the planetary wind. We see that the instabilities are triggered at the contact discontinuity between the $M < 1$ stellar wind and the $M > 1$ planetary wind in a manner similar to the thin shell instability. In the supersonic stellar wind case, the bow consists of two shocks facing upstream and downstream and here we see larger vortices forming in the unstable downstream flow.

## 5. Conclusions

In this contribution we have presented first results from a new campaign of AMR simulations designed to explore the interactions of an evaporative planetary wind with its stellar environment. We have confirmed the results of Stone & Proga (2009) showing that azimuthal structure of the wind forms due to day/night temperature differences. Going further we have also found that such structures depend on the value of the escape parameter $\lambda$. When a stellar outflow is included we see an unstable bow wave forming

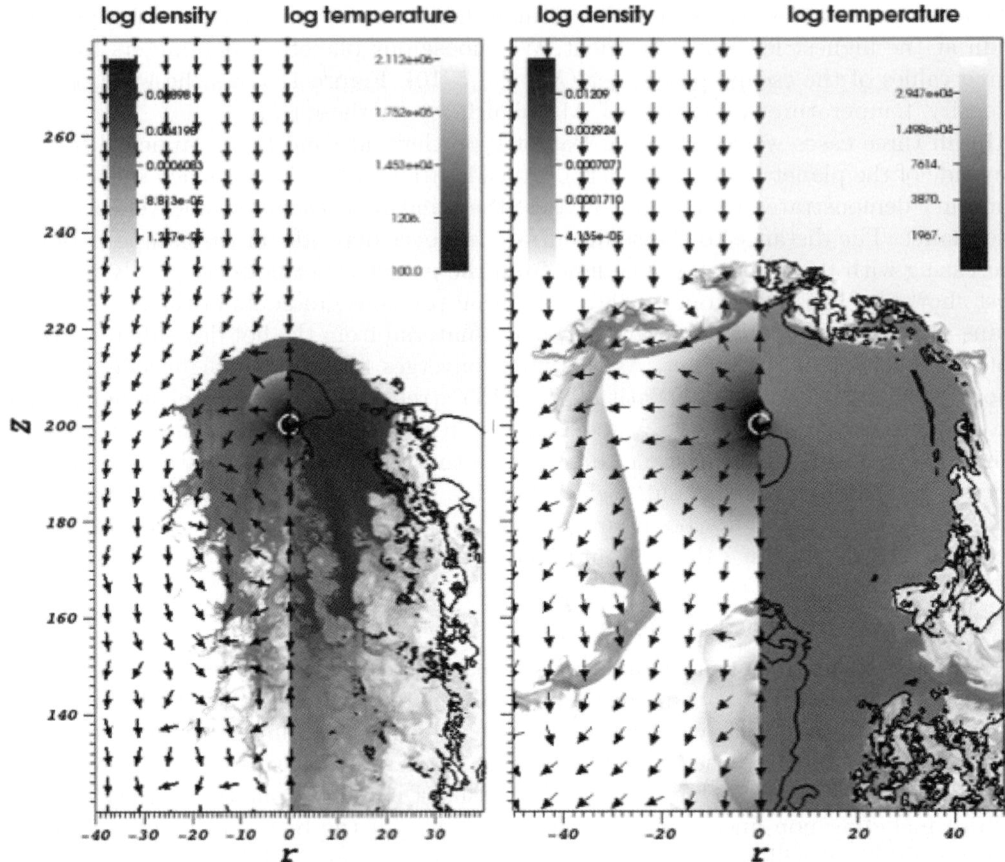

**Figure 2.** Stellar wind/planetary wind simulations. (Left) $M_* = 0.8$ and (right) $M_* = 5.0$ wind flowing downward from the upper boundary to interact with a $\lambda = 5$ planetary wind.

between it and the planetary wind. The bow wave is unstable in both sub-sonic and supersonic stellar outflow cases. The morphology of the resulting flows, however, differs in the two cases. Thus our work implies a time-dependent nature of the interaction region (with variations expected on the crossing time $t_s \sim R_p/V_{bs}$). Future studies will focus on carrying the simulations out in the co-rotating frame.

# References


Ballister, G., King, D., & Herbert, 2007, *Nature*, 445, 511

Bisikalo, D., *et al.* 2013. *ApJ* 764, 19

Cohen, O., Kashyap, V. L., Drake, J. J., Sokolov, I. V., & Gombosi, T. I. 2011. *ApJ* 738, 166

Cunningham A., Frank, A., Varniere, P., Mitran, S., & Jones, T. W., 2009, *ApJ*, 182, 519

Hunten, D. M., Pepin, R. O., & Walker, J. C. G., 1987, *Icarus*, 69, 532549

Lecavelier des Etangs, A., Ehrenreich, D., Vidal-Madjar, A., *et al.* 2010, *A&A*, 514, 72

Matsakos, T., Uribe, A., & Königl, A. 2015, ArXiv e-prints arXiv:1503.03551

Owen, J. E. & Adams, F. 2014, *MNRAS* 444, 3761

Schneiter, E. M., *et al.* 2007, *The Astrophysical Journal* 671, L57

Teyssandier, J., Owen, J. E., Adams, F. C., & Quillen, A. C. 2015, submitted to *MNRAS*, http://arxiv.org/abs/1504.01680

Stone, J. M. & Proga, D, 2009, *Astrophys. J.*, 694, 205

*Young Stars & Planets Near the Sun*
*Proceedings IAU Symposium No. 314, 2015*
*J. H. Kastner, B. Stelzer, & S. A. Metchev, eds.*

© International Astronomical Union 2016
doi:10.1017/S1743921315006638

# Can the Dustiest Main Sequence Stars Tell Us About the Rocky Planet Formation Process?

## Carl Melis

Center for Astrophysics and Space Sciences, University of California, San Diego, California
92093-0424, USA
email: `cmelis@ucsd.edu`

**Abstract.** Main sequence stars hosting extreme quantities of inner planetary system debris are likely experiencing transient dust production events. The nature of these events, if they can be unambiguously attributed to a single process, can potentially inform us on the formation and/or early evolution of rocky Earth-like planets. In this contribution I examine some of the dustiest main sequence stars known and three processes that may be capable of reproducing their observed properties. Through this activity I also make an estimate for the likelihood of an A-type star to have an asteroid belt-like planetesimal population.

**Keywords.** planets and satellites: formation, circumstellar matter, planetary systems

## 1. Introduction

Understanding how rocky terrestrial planets form and evolve is paramount to the question of how life forms on planets and whether or not other sentient beings can exist in the universe. Currently the best means of identifying stars undergoing terrestrial planet formation or collisional evolution events is discovery and characterization of dusty debris disks that orbit in the inner planetary system of their host star (e.g., Kenyon & Bromley 2006; Melis *et al.* 2010; Jackson & Wyatt 2012; Meng *et al.* 2014, 2015; Genda *et al.* 2015). Dust at a distance of $\sim$1 AU from a Sun-like star will be heated to Earth-like temperatures, $\sim$300 K, and will emit in the mid-infrared, thus most surveys for inner planetary system dust look for excess emission at wavelengths of 10-20 $\mu$m.

Investigations into how mid-infrared excess stars provide insight into the rocky planet formation process are being actively pursued by several groups. Statistical works conducted to date typically do not discriminate on how dusty a given debris disk system is when including it in their analysis (e.g., Meyer *et al.* 2008; Jackson & Wyatt 2012; Genda *et al.* 2015). It is worth noting that it need not be the case that all mid-infrared excesses are linked to rocky planet formation events. Indeed, another hypothesis is that weaker excesses can be attributed to dust produced through collisional evolution of asteroid belt-analogs. Such a suggestion seems quite appropriate for the Meyer *et al.* (2008) sample which covers an age range of 0.3-3.0 Gyr where one would not expect rocky planet formation collisional events to be occurring. As such, one might interpret the Meyer *et al.* (2008) statistics as telling us that 19-32% or up to 62% of Sun-like stars host asteroid belt analogs. Under this banner, I estimate the occurrence rate for asteroid belts around A-type stars.

Examination of the incidence rate of such systems at young ages (10-30 Myr old), when collisional cascade dust production is at its height (e.g., Wyatt 2008), should enable a reasonable estimate of the fraction of stars that host asteroid-belt analogs. To ensure a robust estimate, it is important to use a statistical sample of stars observed in a blind survey where it is clear that the mid-infrared excess emission originates from inner planetary system material. Morales *et al.* (2009) provide just such a data set by combining *Spitzer* MIPS and IRS measurements for relatively luminous stars. Their data set, which probes stars of spectral type A and late-B that were previously determined to have excess emission at $24\,\mu$m, shows that 10 out of 14 (or 71%) of 24 and $70\,\mu$m detected stars have blackbody-fit dust temperatures $\gtrsim 200$ K (only those stars with single-temperature blackbody fits that are detected at 24 and $70\,\mu$m are selected to ensure that the dust temperature — and hence physical location — is well constrained). All of these 10 sources have a fractional infrared luminosity that is $<2\times 10^{-4}$ and thus can be interpreted as hosting active asteroid belt analogs (similar to the conclusion reached by Morales *et al.* 2009). Rieke *et al.* (2005) and Su *et al.* (2006) present unbiased $24\,\mu$m excess statistics, and each find that roughly 40-50% of A-type stars with ages between 10-30 Myr host $24\,\mu$m excess emission. When combined with the finding from Morales *et al.* (2009) that $\approx 71\%$ of A-type stars with $24\,\mu$m excess emission have dust in their inner planetary system, it is found that at least 33% of A-type stars with age in the range of 10-30 Myr should host an active planetesimal belt in their inner planetary system. This occurrence rate is in good agreement with the statistics of polluted white dwarf stars discussed by Zuckerman *et al.* (2003) and Zuckerman *et al.* (2010), indicating that such asteroid belt-analogs go on to eventually pollute their dead stars' atmospheres.

Rather than working with all mid-infrared excess main sequence stars, one might instead opt to examine only those hosting extreme quantities of inner planetary system material as they are likely experiencing transient dust production events. To date, only a handful of main sequence stars are known to have sufficiently high levels of mid-infrared excess emission (and hence inner planetary system dust) that points to an unambiguous origin in transient processes that may be relevant to terrestrial planet formation or evolution (e.g., Song *et al.* 2005, Rhee *et al.* 2007, Rhee *et al.* 2008, Zuckerman *et al.* 2012, Melis *et al.* 2012, Olofsson *et al.* 2012, Schneider *et al.* 2013, Melis *et al.* 2013, Kennedy & Wyatt 2014). Melis *et al.* (2014) describe how to sort inner planetary system debris disks into two general categories: those that result from collisional grinding down of a population of rocky planetesimals (an active asteroid belt-analog) and those that require a transient dust production event (stochastic or transient collisions involving planetary-scale objects). They begin with the assumption that all dust disks are the product of collisions of numerous small rocky bodies in an active planetesimal belt. It is then estimated what total mass of parent bodies — $M_{PB}$ — is necessary to reproduce the observed parameters for the small dust grains and compare $M_{PB}$ to the model simulations of Kenyon & Bromley (2006) for the formation of terrestrial planets to see if this amount of mass should have coalesced to form planetary embryos or planets. Should the indicated mass of parent bodies be sufficient to expect the formation of planetary-scale bodies ($M_{PB}\gtrsim 1\,\mathrm{M}_{Earth}$), then it is concluded that the active planetesimal belt hypothesis is false (as any planetesimal population should have developed into planetary embryos or planets) and that the observed dust results from transient dust-production involving large rocky objects.

## 2. Extreme debris disk systems

All main sequence stars currently known and confirmed to be hosting substantial quantities of inner planetary system dust — and thus likely to be undergoing transient dust production events ($\mathrm{M}_{PB}\gtrsim 1\,\mathrm{M}_{Earth}$) — are presented in Figure 1. Examination of the properties of these systems reveals that those with early-type host stars are mostly observed when the star is $\sim 10$ Myr old. By contrast, this epoch extends later for Sun-like

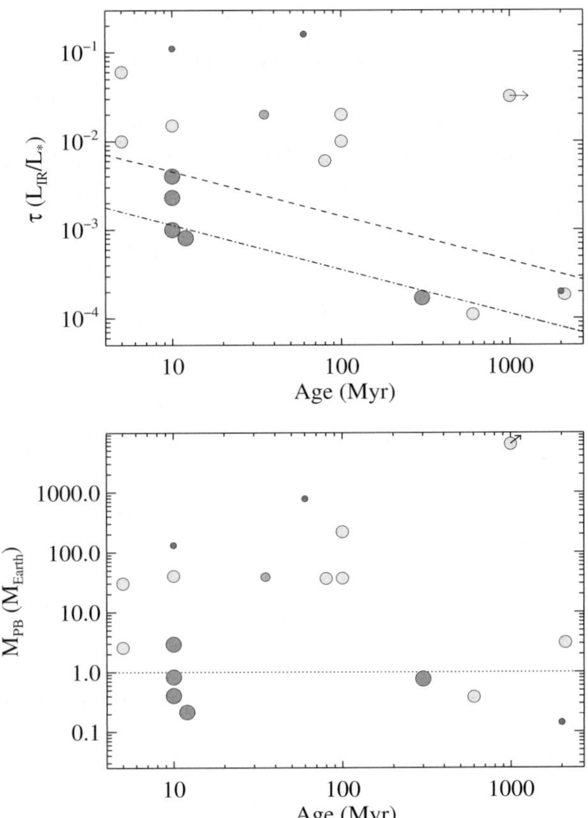

**Figure 1.** *Both Panels:* Stars hosting large quantities of inner planetary system material likely generated in transient dust production events ($M_{PB} \gtrsim 1\,M_{Earth}$; see Melis *et al.* 2014). Objects down to $0.1\,M_{Earth}$ are still considered as candidates but could be explained by collisional grinding down of an active planetesimal belt; some representative systems are shown. Sources are drawn from Melis *et al.* (2010), Melis *et al.* (2013), and references therein. Small, red circles are K-type stars. The medium-small, gold circle is a G-type star. Medium-large, yellow circles are F-type stars. Large, blue circles are A-type stars. The object with a limit arrow and age of 1000 Myr is BD+20 307 as its age is a lower limit. *Upper Panel:* The dashed line represents $\tau$ necessary to realize a parent body mass of $1\,M_{Earth}$ for a star of solar mass, radius, and temperature that is orbited by 300 K dust. The dash-dotted line represents the $\tau$ necessary to realize a parent body mass of $1\,M_{Earth}$ for a star of $1.7\,R_\odot$, $2.0\,M_\odot$, and 8180 K effective temperature that is orbited by 300 K dust. *Lower Panel:* The dotted line represents the cutoff above which the observed dust is expected to be the result of transient processes (see Melis *et al.* 2014).

stars with observed specimens typically having ages $\leqslant 100$ Myr. The incidence rate of the dustiest Sun-like stars, determined by Melis *et al.* (2010) to be roughly 1 in 300 for the age range of 30 to 100 Myr, is comparable to that for early-type stars of approximately 1 in 200 for the age range of 10 to 20 Myr (Melis *et al.* in preparation). If all stars undergo an exceptionally dusty phase of inner planetary system evolution, then the observable signature of it should persist for about $1.5 \times 10^5$ years for Sun-like stars (Melis *et al.* 2010) and $\sim 7 \times 10^4$ years for early-type stars (Melis *et al.* in preparation).

To be consistent with the data presented in Figure 1, any model should reproduce the high level of dust observed around each star, the dominant timescale in which this dust is observed, and the observed incidence rate. Three possibilities are presented below.

## 2.1. *Giant Impacts*

Rhee *et al.* (2008) and Melis *et al.* (2010) interpret the incidence rate of the dustiest main sequence stars with the aid of colliding planetary embryo models developed in Agnor *et al.* (1999), Agnor & Asphaug (2004), and Asphaug *et al.* (2006). In brief, following a giant impact of two rocky planetary embryos or planets, fragmented debris covering a range of sizes is released. The collision time for the smallest dust particles produced in the ensuing collisional cascade will be about 1 year divided by $80 \times L_{IR}/L_{bol}$ (Melis *et al.* 2014 and references therein). If the largest initial fragments ejected during a giant impact have radii $\sim$100 m (see work by Agnor, Asphaug, and colleagues – note that this is in contrast to the suggestion by Jackson & Wyatt (2012) and Genda *et al.* (2015) who suggest ejecta up to sizes of 500 km in diameter), then their collisional lifetime is $\sim 5 \times 10^4$ years for Sun-like stars and $\sim 2 \times 10^4$ years for early-type stars (see Melis *et al.* 2010). However, the formation of rocky terrestrial planets involves more than a single giant impact event (e.g., Stewart & Leinhardt 2012). Through the analysis of Stewart & Leinhardt (2012) and the simulations referenced therein it is found that a typical terrestrial planet will experience $\sim$6 giant impact-type events before it is fully formed. If half of these result in the production of large monolith fragments (as suggested by Stewart & Leinhardt 2012), then the model-estimated observable transient dust lifetime could be a few times more than the above estimates and hence in reasonably good agreement with the values given at the beginning of this section. If accurate, this suggests that on average there is one terrestrial planet for every star of Solar-mass and above (this does not necessarily mean all stars have terrestrial planets, as some may have multiple and others none). This is in good agreement with the recent determination of the prevalence of rocky bodies orbiting other stars from the *Kepler* data set (e.g., Burke *et al.* 2015).

## 2.2. *Leftover Planetesimals*

Radiometric age-dating of Solar system rocky bodies provides a detailed account of the impact history in the inner parts of our planetary system. Study of impact-reset $^{40}$Ar–$^{39}$Ar ages in lunar samples and meteorites shows the well-known increase in impactor flux between 3.4 and 4.1 Gyr ago (usually attributed to a Late-Heavy Bombardment phase), but meteorite samples alone also show a spike of reset activity around $\approx$4.5 Gyr ago (e.g., see Figure 1 of Marchi *et al.* 2013 and references therein). As discussed in Marchi *et al.* (2013), these radiometric reset events are often attributed to cooling of the parent body, but substantial evidence suggests that impacts are also important in generating this spike. Impacts in this time frame (ages of 10-100 Myr for the Sun), if indeed sufficiently energetic to reset radiometric geochronometers, likely would have produced significant quantities of dust. Such impacts could in theory be capable of explaining the data presented in Figure 1.

Marchi *et al.* (2013) postulate that impactors capable of producing the $\approx$4.5 Gyr old $^{40}$Ar–$^{39}$Ar spike could come from a class of "leftover planetesimals" excited to high orbital eccentricity and inclinations by forming rocky protoplanets (see also Bottke *et al.* 2007). Bottke *et al.* (2007) show that a scattered planetesimal population would rapidly deplete, with $\approx$90% of the sample being removed from the planetary system by the time its host star is $\sim$100 Myr old. The implication is that dust production from this depletion would be dominated by planetesimal-asteroid collisions, although some material should impact the rocky planetary bodies that expelled it. Large rocky planetary bodies are necessary in generating the dust, although they will typically lack an immediate connection to it. As an interesting aside, it is prudent to wonder if such objects could also be responsible for late accretion or veneer (e.g., Day *et al.* 2012; Schlichting *et al.* 2012).

### 2.3. *Dynamical Instability*

Solar system formation models (e.g., the Nice model; Gomes *et al.* 2005; Morbidelli *et al.* 2005; Tsiganis *et al.* 2005) predict a phase of outer planetary system reconfiguration wherein the inner planetary system experienced an enhanced influx of planetesimals. This "Late Heavy Bombardment" — which occurred $\approx 600$ Myr after the Sun was born — has observational evidence in the cratering record of various inner Solar system rocky objects. It is enticing to consider a similar phase of dynamical instability as capable of driving the exceptionally dusty states of the stars considered in Figure 1 (indeed, Fujiwara *et al.* 2009, 2012 advance dynamical instabilities to explain the dusty stars they study).

Some studies have shown that it is possible for dynamical instabilities to produce the observed incidence rate of the dustiest main sequence stars (e.g., Bonsor *et al.* 2013, 2014 and references therein). These simulations also suggest the incidence rate of dusty systems resulting from instabilities and the level of dust being produced by them tend to decrease with increasing stellar age, again in reasonable agreement with the data in Figure 1. But, it is not clear if the level of dust observed can be reproduced as instability models have not yet tracked collisional and dynamical evolution in tandem. New codes should allow coupled simulations and direct tests of these models through predictions for direct observables (e.g., Stark & Kuchner 2009; Kral *et al.* 2013).

The implication of this model should it be appropriate is that these systems would host giant planets and planetesimal belts (typically in the outer planetary system) that are experiencing dramatic reconfigurations; rocky Earth-like planets are not immediately necessary.

## 3. Conclusion

There is strong evidence that transient dust production events occur in the inner planetary systems of other stars. Examination of the dustiest main sequence stars as a subset of the debris disk population provides a safe route to selecting those systems where it is clear transient events are occurring and thus to indirectly probing terrestrial planet formation and evolution through the statistics of infrared excess stars. However, more observations, theory development, and simulations are necessary before we can robustly attach these systems to specific phases of planetary system formation or evolution.

## References

Agnor, C. & Asphaug, E. 2004, *ApJL*, **613**, L157
Agnor, C. B., Canup, R. M., & Levison, H. F. 1999, *Icarus*, **142**, 219
Asphaug, E., Agnor, C. B., & Williams, Q. 2006, *Nature*, **439**, 155
Bonsor, A., Raymond, S. N., & Augereau, J.-C. 2013, *MNRAS*, **433**, 2938
Bonsor, A., Raymond, S. N., Augereau, J.-C., & Ormel, C. W. 2014, *MNRAS*, **441**, 2380
Bottke, W. F., Levison, H. F., Nesvorný, D., & Dones, L. 2007, *Icarus*, **190**, 203
Burke, C. J., *et al.* 2015, *ApJ*, **809**, 8
Day, J. M. D., Walker, R. J., Qin, L., & Rumble, III, D. 2012, *Nature Geoscience*, **5**, 614
Fujiwara, H., *et al.* 2009, *ApJL*, **695**, L88
— 2012, *ApJL*, **749**, L29
Genda, H., Kobayashi, H., & Kokubo, E. 2015, *ArXiv e-prints*
Gomes, R., Levison, H. F., Tsiganis, K., & Morbidelli, A. 2005, *Nature*, **435**, 466

Jackson, A. P. & Wyatt, M. C. 2012, *MNRAS*, **425**, 657

Kennedy, G. M. & Wyatt, M. C. 2014, *MNRAS*, **444**, 3164

Kenyon, S. J. & Bromley, B. C. 2006, *AJ*, **131**, 1837

Kral, Q., Thébault, P., & Charnoz, S. 2013, *A&A*, **558**, A121

Marchi, S., *et al.* 2013, *Nature Geoscience*, **6**, 303

Melis, C., Zuckerman, B., Rhee, J. H., & Song, I. 2010, *ApJL*, **717**, L57

Melis, C., Zuckerman, B., Rhee, J. H., Song, I., Murphy, S. J., & Bessell, M. S. 2012, *Nature*, **487**, 74

— 2013, *ApJ*, **778**, 12

— 2014, in *IAU Symposium*, edited by N. Haghighipour, vol. 293 of *IAU Symposium*, 273–277

Meng, H. Y. A., *et al.* 2014, *Science*, **345**, 1032

— 2015, *ApJ*, **805**, 77

Meyer, M. R., *et al.* 2008, *ApJL*, **673**, L181

Morales, F. Y., *et al.* 2009, *ApJ*, **699**, 1067

Morbidelli, A., Levison, H. F., Tsiganis, K., & Gomes, R. 2005, *Nature*, **435**, 462

Olofsson, J., Juhász, A., Henning, T., Mutschke, H., Tamanai, A., Moór, A., & Ábrahám, P. 2012, *A&A*, **542**, A90

Rhee, J. H., Song, I., & Zuckerman, B. 2007, *ApJ*, **671**, 616

— 2008, *ApJ*, **675**, 777

Rieke, G. H., *et al.* 2005, *ApJ*, **620**, 1010

Schlichting, H. E., Warren, P. H., & Yin, Q.-Z. 2012, *ApJ*, **752**, 8

Schneider, A., Song, I., Melis, C., Zuckerman, B., Bessell, M., Hufford, T., & Hinkley, S. 2013, *ApJ*, **777**, 78

Song, I., Zuckerman, B., Weinberger, A. J., & Becklin, E. E. 2005, *Nature*, **436**, 363

Stark, C. C. & Kuchner, M. J. 2009, *ApJ*, **707**, 543

Stewart, S. T. & Leinhardt, Z. M. 2012, *ApJ*, **751**, 32

Su, K. Y. L., *et al.* 2006, *ApJ*, **653**, 675

Tsiganis, K., Gomes, R., Morbidelli, A., & Levison, H. F. 2005, *Nature*, **435**, 459

Wyatt, M. C. 2008, *ARA&A*, **46**, 339

Zuckerman, B., Koester, D., Reid, I. N., & Hünsch, M. 2003, *ApJ*, **596**, 477

Zuckerman, B., Melis, C., Klein, B., Koester, D., & Jura, M. 2010, *ApJ*, **722**, 725

Zuckerman, B., Melis, C., Rhee, J. H., Schneider, A., & Song, I. 2012, *ApJ*, **752**, 58

*Young Stars & Planets Near the Sun*
*Proceedings IAU Symposium No. 314, 2015*
*J. H. Kastner, B. Stelzer, & S. A. Metchev, eds.*

© International Astronomical Union 2016
doi:10.1017/S1743921315006390

# A Possible Dynamical History for the Fomalhaut System

## Virginie Faramaz

Instituto de Astrofísica - Pontificia Universidad Católica de Chile
email: `vfaramaz@astro.puc.cl`

**Abstract.** Fomalhaut b was long thought to shape the eccentric debris belt in the Fomalhaut system, but its orbit was found to be too eccentric for it to be the dominant belt-shaping perturber. This indicates that Fomalhaut b is Earth-sized at most and that the belt-shaping perturber, hereafter named Fomalhaut c, remains to be discovered. In addition, since its orbit more or less crosses that of Fomalhaut b, it also indicates that the current configuration of the system is transient and was reached recently. In this talk, we show that this current configuration can be explained if Fomalhaut c is Saturn- to Neptune-sized, and Fomalhaut b originates from a mean-motion resonance with Fomalhaut c.

**Keywords.** Stars: Fomalhaut, Planetary systems, Methods: numerical, Celestial mechanics

---

## 1. Introduction

Fomalhaut ($\alpha$ Psa) is a 440 Myr old A3V star, located at 7.7 pc (van Leeuwen 2007; Mamajek 2012). Fomalhaut is surrounded by an eccentric dust ring (e = 0.11 ± 0.01) (Kalas *et al.* 2005). This eccentric shape hinted at the presence of a massive body orbiting inside the belt on an eccentric orbit, dynamically shaping the belt (Quillen 2006; Deller & Maddison 2005). This hypothesis was apparently confirmed by the direct detection of a companion near the inner edge of the belt, Fomalhaut b (hereafter Fom b; Kalas *et al.* 2008). However, orbital fitting for this perturber has revealed a highly eccentric orbit that crosses the belt. In addition, it is close to apsidal alignement with the belt. Since such an eccentric orbit would create a significant apsidal misalignment, Fom b cannot be responsible for the disk shaping (Kalas *et al.* 2013; Beust *et al.* 2014; Pearce *et al.* 2015). The most straightforward solution to this apparent paradox is to suppose the presence of a yet undetected body in the system, hereafter named Fom c, which would be much more massive than Fom b. Consequently, Fom c would be dynamically predominant and responsible for the belt shaping. This is supported by recent dynamical or photometric studies which suggest that Fom b is no more than Earth- or Super-Earth sized (Beust *et al.* 2014; Janson *et al.* 2012; Galicher *et al.* 2013). However, in this configuration, which is illustrated in the bottom panel of Figure 1, the orbit of the putative belt-shaping planet Fom c would be crossed by that of Fom b. Such a two-planet system is highly unstable, and would indicate that Fom b must have been perturbed recently, potentially by Fom c (Beust *et al.* 2014).

## 2. Investigating the dynamics of a two-planet system

Investigation of the dynamics of this two-planet system in Faramaz *et al.* (2015), where Fom c is a massive belt-shaping body and Fom b a much less massive body originating from the inner parts of the system, has revealed a possible three-step dynamical scenario

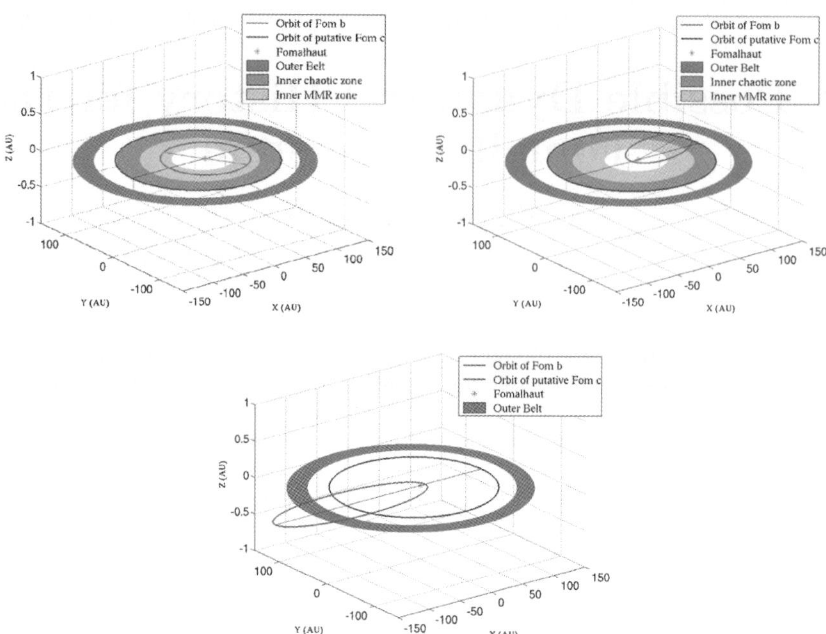

**Figure 1. Top left:** Probable initial configuration of the Fomalhaut system. Fom b is in MMR with the belt-shaping eccentric Fom c. **Top right:** Probable intermediate configuration of the Fomalhaut system. MMRs with an eccentric perturber generate very eccentric orbits, which leads Fom b to cross the chaotic zone of Fom c and be scattered on its current orbit. **Bottom:** Probable current configuration of the Fomalhaut system.

which can explain both why Fomalhaut b is on such an eccentric orbit and why it was set on it recently:

(*a*) *Mean-motion resonances between Fom b and the suspected Fom c :* Fom b is likely to have formerly resided in an inner mean-motion resonance (MMR) with the additional planet, as illustrated in the top left panel of Figure 1. MMRs with an eccentric perturber such as the belt-shaping Fom c induce a gradual eccentricity increase, which can cause Fom b to cross the chaotic zone of Fom c, where it can then be scattered by Fom c on its current orbit (top right panel of Fig. 1). The dynamical timescale involved in this process, that is, the typical time necessary for Fom b to reach a sufficient orbital eccentricity from its MMR position and be scattered on its current orbit, strongly depends on the mass of the putative Fom c. In particular, the scattering event can be delayed on timescales comparable to the age of the system with a Neptune- or Saturn-sized Fom c, which would explain why Fom b was recently set on its orbit.

(*b*) *Close encounter with the suspected Fom c:* inspection of the close encounters between Fom b and Fom c reveals that these can set Fom b on an orbit with semi-major axis compatible with that of Fom b, but that they also preferentially produce orbits which are not eccentric enough to be compatible with that of the observed one ($a = 81 - 415\,\mathrm{AU}$ and $e = 0.69 - 0.98$, at the 95% confidence level Beust *et al.* 2014).

(*c*) *Secular evolution with the suspected Fom c:* an additional eccentricity increase can be provided when Fom b is under the secular influence of the eccentric Fom c, which is indeed mainly expected at semi-major axes with $a = 81 - 415\,\mathrm{AU}$. However, this eccentricity increase is accompanied by an apsidal alignment with the belt-shaping Fom c, and thus with the belt, which may explain the tendency for the observed orbit to be apsidally aligned with the belt.

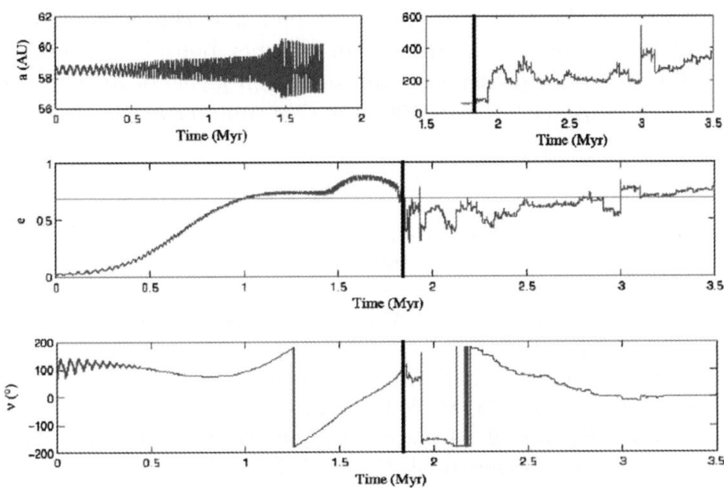

**Figure 2. Example of the three-step process that may have placed Fom b on its current orbit.** We display the evolution in time of the semi-major axis $a$, eccentricity $e$, and longitude of periastron $\nu$ of a massless test particle initially in 5:2 MMR with a $3M_{\rm Jup}$ Fom c, with semi-major axis 108.6 AU and orbital eccentricity 0.1. Note that this process can be generated via several other MMRs. The test particle endures a three-step dynamical evolution, starting with resonant evolution, where its semi-major axis suffers small oscillations around the exact resonant location, and its eccentricity largely increases, while co-evolving with the longitude of periastron. The vertical black line at $\sim 2\,{\rm Myr}$ indicates the second step of the process, that is, a close encounter with Fom c when the highly eccentric orbit of the test particle crosses the chaotic zone of Fom c. Note that this delay of several Myr with a Jupiter-sized Fom c increases up to several 100 Myr with a Saturn- or Neptune-sized Fom c. The semi-major axis of the test particle is comparable with this of Fom b after the close encounter, but its eccentricity remains smaller than 0.69 *(horizontal red line)*, and thus is incomparable with that of Fom b. The third step consists mainly in a secular evolution of the test particle with the eccentric Fom c, although its orbit endures small chaotic variations. This secular evolution allows the eccentricity to increase and become greater than 0.69, which occurs when there is an apsidal alignment between the perturber and the test particle, that is, when the longitude of periastron of the test particle is close to zero.

The whole process is summarized and illustrated in Figure 2.

In addition, when an eccentric planet such as Fom c coexists with km-sized planetesimals, orbits such as that of Fom b can be expected to be generated with great efficiency via the three-step mechanism detailed above (Faramaz *et al.* 2015). Therefore, one should probably expect the Fomalhaut system to contain a broad population of solid bodies on highly eccentric orbits that closely approach the star at periastron. These bodies would feed the inner parts of the system with dust, creating hot or warm inner belts. This is extremely interesting in the context of the Fomalhaut system, since both a warm and a hot inner belt were detected (Lebreton *et al.* 2013).

## 3. Conclusions

The study of the Fomalhaut system has revealed a robust process by which orbits such as that of Fom b naturally result from interactions between an eccentric massive perturber such as the suggested Fom c—assumed to shape the outer belt of this system – and a much less massive body, such as a smaller planet or a planetesimal. In addition, this process involves a delay in the production of Fom b-like orbits, which can be greater than 100 Myr if the eccentric massive perturber is Neptune- to Saturn-sized. This can provide an explanation both for the shape of the outer belt and the current dynamical status of Fom b. In addition, it may be at the origin of inner belts in the Fomalhaut system, and provide a solution to the presence of unusual high levels of dust in the vicinity of a significant number of stars with age $> 100\,\mathrm{Myr}$ (Faramaz *et al.*, in prep).

## References


Beust, H., Augereau, J.-C., Bonsor, A., *et al.* 2014, *A&A*, 561, A43
Deller, A. T. & Maddison, S. T. 2005, *ApJ*, 625, 398
Faramaz, V., Beust, H., Augereau, J.-C., Kalas, P., & Graham, J. R. 2015, *A&A*, 573, A87
Galicher, R., Marois, C., Zuckerman, B., & Macintosh, B. 2013, *ApJ*, 769, 42
Janson, M., Carson, J. C., Lafrenière, D., *et al.* 2012, *ApJ*, 747, 116
Kalas, P., Graham, J. R., Chiang, E., *et al.* 2008, *Science*, 322, 1345
Kalas, P., Graham, J. R., & Clampin, M. 2005, *Nature*, 435, 1067
Kalas, P., Graham, J. R., Fitzgerald, M. P., & Clampin, M. 2013, *ApJ*, 775, 56
Lebreton, J., van Lieshout, R., Augereau, J.-C., *et al.* 2013, *A&A*, 555, A146
Mamajek, E. E. 2012, *ApJ*, 754, L20
Mamajek, E. E., Bartlett, J. L., Seifahrt, A., *et al.* 2013, *AJ*, 146, 154
Neuhäuser, R., Hohle, M. M., Ginski, C., *et al.* 2015, *MNRAS*, 448, 376
Pearce, T. D., Wyatt, M. C., & Kennedy, G. M. 2015, *MNRAS*, 448, 3679
Quillen, A. C. 2006, *MNRAS*, 372, L14
Shannon, A., Clarke, C., & Wyatt, M. 2014, *MNRAS*, 442, 142
van Leeuwen, F., ed. 2007, *Astrophysics and Space Science Library*, Vol. 350, Hipparcos, the New Reduction of the Raw Data


## Discussion

E. MAMAJEK: Fomalhaut A is part of a triple star system. Did you consider one of the stellar companions could be responsible for the shape of the debris disk instead of an unseen Fom c?

V. FARAMAZ: Indeed, Shannon *et al.* (2014) argued that Fom C could excite the eccentricities of the disk at the observed value. However increasing eccentricities is not sufficient to create a well-defined eccentric ring, which is an extended structure. Providing first that Fom C is coplanar to the debris disk of Fom A, which is unknown, the process takes several precession timescales. With a separation of $\sim 200\,\mathrm{kAU}$ between Fom A and Fom C (Mamajek *et al.* 2013), this precession timescale is of the order of Gyr, that is, much longer than the age of the Fomalhaut system. This is why the possibility that Fom C is responsible for the shape of the debris disk of Fom A is discarded by Faramaz *et al.* (2015).

UNIDENTIFIED CONFERENCE PARTICIPANT: What do you think about the scenario by Neuhäuser *et al.* (2015) in which Fom b is actually a background neutron star?

V. FARAMAZ: It is not a possibility to exclude completely, however, this scenario fails to explain the fact that the orbit of Fom b was found to be nearly coplanar with the debris ring, as found first by Kalas *et al.* (2013), and confirmed since by Beust *et al.* (2014) and Pearce *et al.* (2015). The neutron star scenario was also motivated by the fact that the lack of giant planets in the inner parts of the system ruled out a scenario in which Fom b was scattered on its current orbit. However, as shown by Faramaz *et al.* (2015), there is no need for such giant planets for Fom b to be scattered, and the inferred belt-shaping planet itself, which can be as small as a Neptune, can explain the current configuration of the system.

*Young Stars & Planets Near the Sun*
Proceedings IAU Symposium No. 314, 2015
J. H. Kastner, B. Stelzer, & S. A. Metchev, eds.

© International Astronomical Union 2016
doi:10.1017/S1743921315005967

# Collisions of Planetesimals and Formation of Planets

Rudolf Dvorak[1], Thomas I. Maindl[1], Áron Süli[2], Christoph M. Schäfer[3], Roland Speith[4], and Christoph Burger[1]

[1] Department of Astrophysics, University of Vienna, Austria
email: rudolf.dvorak@univie.ac.at
[2] Eötvös University, Department of Astronomy, Budapest, Hungary
[3] Institut für Astronomie und Astrophysik, Eberhard Karls Universität Tübingen, Germany
[4] Physikalisches Institut, Eberhard Karls Universität Tübingen, Germany

**Abstract.** We present preliminary results of models of terrestrial planet formation using on the one hand classical numerical integration of hundreds of small bodies on CPUs and on the other hand—for comparison—the results of our GPU code with thousands of small bodies which then merge to larger ones. To be able to determine the outcome of collision events we use our smooth particle hydrodynamics (SPH) code which tracks how water is lost during such events.

**Keywords.** methods: n-body simulations; methods: numerical; minor planets, asteroids; planets and satellites: formation; solar system: formation

## 1. Introduction and scenarios

The outcome of numerical simulations of planet formation is extremely sensitive to initial conditions like disk surface density, total mass, the initial distribution of planetesimals and planetary embryos after the gas phase in the protoplanetary disk, and the dynamical model. Hence, the resulting planetary system models and even statistical statements are highly biased by the choice of these parameters. While many results on formation of planets and their water content have been achieved, several open questions remain (e.g., Raymond et al. 2014).

Here, we present preliminary results for a dynamical model including two gas giants, Jupiter and Saturn. By SPH simulation of collision events we can deduce the outcome of collisions in terms of merging/fragmentation and water loss depending on mass and size of the bodies, collision velocity, and impact angle (Fig. 2b and Maindl et al. 2013, 2014), thereby addressing the yet unclear effect of including water transport probabilities in planet formation simulations (cf. Izidoro et al. 2014).

We take the outcome of the last phases of the Grand Tack scenario (Walsh et al. 2011) as initial conditions for our simulations. Jupiter and Saturn are in their actual positions in the Solar System and we distribute small bodies in the mass range $3 \times 10^{-9}\,M_\odot < m < 3 \times 10^{-7}\,M_\odot$ and semi-major axes $0.4\,\mathrm{AU} < a < 2.7\,\mathrm{AU}$ with small inclinations and small eccentricities. Bodies initially outside 2 AU are assumed to have a water content of 10 percent.

## 2. Preliminary simulation results

Figure 1 shows a typical result out of 30 different runs after an integration time of 5 Myr using the Lie integration method with adaptive step size (Eggl & Dvorak 2010). We are able to include thousands of gravitating bodies instead of hundreds by using our new massively parallelized GPU code (a descendant of the one described in Süli 2013) which is up to two orders of magnitude faster than our CPU n-body integrator (cf.

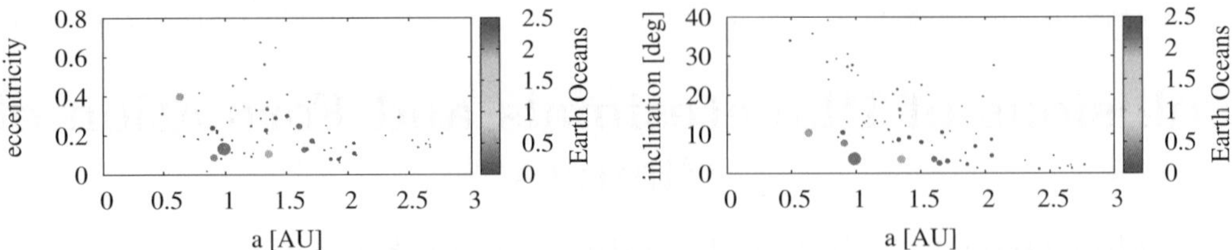

**Figure 1.** Eccentricities and inclinations for a typical scenario. The water content is color-coded, circle sizes denote the bodies' masses after 5 Myr. The largest bodies are $\sim 30$ lunar masses.

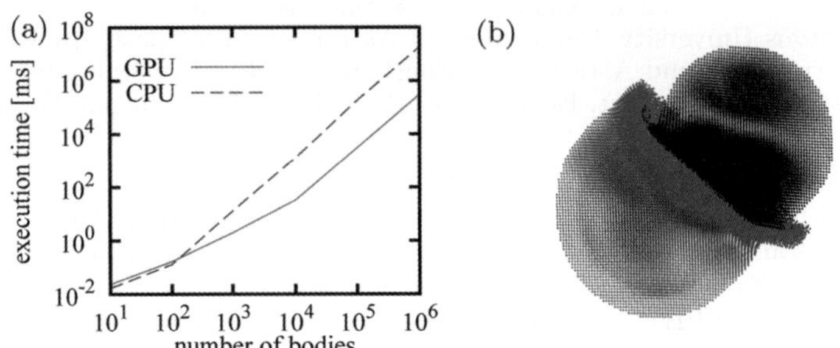

**Figure 2.** a) The GPU-over-CPU computational gain first grows with the number of bodies involved and settles at a factor of approximately 100. b) Low-resolution SPH simulation of two Ceres-mass bodies colliding at 1 km/s and a $30°$ impact angle (50k SPH particles). Blue dots: 30 wt-% water ice mantle on the target; black dots: solid basalt projectile.

Fig. 2a). This large number of bodies can handle dynamical friction with a precision not previously achieved (e.g., O'Brien et al. 2006).

## 3. Conclusions

While there exist several results from different groups with different underlying assumptions (e.g., Izidoro et al. 2014; Hansen 2009), we believe that independent computations including phenomena like dynamical friction and realistic collisional water transfer are desirable for answering the key questions of terrestrial planet formation.

## Acknowledgements

We acknowledge support from the FWF Austrian Science Fund project S 11603-N16.

## References

Eggl, S., Dvorak, R. 2010, in *Lecture Notes in Physics*, Berlin Springer Verlag, Vol. 790, Lecture Notes in Physics, Berlin Springer Verlag, ed. J. Souchay & R. Dvorak, 431–480

Hansen, B. M. S. 2009, *ApJ*, 703, 1131

Izidoro, A., Haghighipour, N., Winter, O. C., & Tsuchida, M. 2014, *ApJ*, 782, 31

Maindl, T. I., Dvorak, R., Schäfer, C., & Speith, R. 2014, in *IAU Symposium*, Vol. 310, IAU Symposium, 138–141

Maindl, T. I., Schäfer, C., Speith, R., *et al.* 2013, *Astronomische Nachrichten*, 334, 996

O'Brien, D. P., Morbidelli, A., & Levison, H. F. 2006, *Icarus*, 184, 39

Raymond, S. N., Kokubo, E., Morbidelli, A., Morishima, R., & Walsh, K. J. 2014, *Protostars and Planets VI*, 595

Süli, Á. 2013, Astronomische Nachrichten, 334, 1000

Walsh, K. J., Morbidelli, A., Raymond, S. N., O'Brien, D. P., & Mandell, A. M. 2011, *Nature*, 475, 206

*Young Stars & Planets Near the Sun*
*Proceedings IAU Symposium No. 314, 2015*
*J. H. Kastner, B. Stelzer, & S. A. Metchev, eds.*

© International Astronomical Union 2016
doi:10.1017/S174392131500650X

# Young Stars and Planets Near the Sun: Explosive Phenomena from Falling Evaporating Bodies

### Subhon Ibadov[1,2] and Firuz S. Ibodov[1]

[1]Lomonosov Moscow State University, Sternberg Astronomical Institute, 119991 Moscow, Russia
email: mshtf@sai.msu.ru
[2]Institute of Astrophysics, Tajik Academy of Sciences, 734042 Dushanbe, Tajikistan
email: ibadovsu@yandex.ru

**Abstract.** Impacts of falling evaporating bodies (FEBs) with stars and planets at velocities $V \gtrsim 10 - 20$ km/s will be accompanied, due to aerodynamic effects such as crushing and transversal expansion of the crushed mass, by the FEB's "explosion" and the generation of a strong "blast" wave, resulting in FEB-generated explosive/flare phenomena. Multiwavelength monitoring of nearby young stars (and exoplanets) with dense protoplanetary disks rich in FEB's is hence of interest for identifying such FEB-related mechanisms possibly underlying their variability.

**Keywords.** Young stars, falling evaporating bodies, FEB-generated explosive/flare phenomena

---

## 1. Introduction

Spectral observations of nearby, young stars like $\beta$ Pictoris indicate the presence of fluxes of comet-like falling evaporating bodies (FEBs; e.g., Lagrange *et al.* 1987). More than 2000 sungrazing comets (perihelion distances $\lesssim 0.01$ AU) were discovered during the last three decades (Weissman 1983†). Fast thermalization of the kinetic energy of these bodies leads to high-temperature explosive phenomena (cf. Ibadov 1990; 2011; Ibadov *et al.* 2009). One such giant meteor observed was the June 30, 1908 Tunguska fireball, which exploded at heights range 5-10 km with an energy release of about 20 megatons of TNT.

An analytic approach to the explosion of bolides in planetary atmospheres was developed by Prof. Grigorian during 1976-1979 by modifying the basic equations of the physical theory of meteors. This approach was further developed in connection with the 100th anniversary of the Tunguska phenomenon. The complete analytic theory of bolide explosions was presented and used to explain the February 15, 2013 Chelyabinsk event (Grigorian *et al.* 2013 and references therein).

Fortunately, Tunguska-like planetary explosive phenomena in our Solar System have small probabilities, i.e., about one event per 200-300 years. Thus, young stars (and exoplanets) located close to the Sun, for which FEB explosive phenomena should be far more frequent, are unique natural laboratories to study the early evolution of the Sun/stars and planets, including their habitability. Here, we present some theoretical results concerning evolution of FEBs in the atmospheres of young stars and planets, derived on the basis of the above-mentioned theory.

† See also http://sungrazer.nrl.navy.mil

## 2. Explosions of FEBs in Stellar and Planetary Atmospheres

Calculations show that intense aerodynamic crushing of FEBs along with transversal expansion occurs within the stellar/solar chromosphere. Disintegration of FEBs within the chromosphere has a meteor-like character and can cause anomalously high abundances of atoms of refractory elements, such as Fe and Si. Moreover, additional emission due to such metal atoms can be detected above the background of solar coronal radiation, if the FEBs' radii are $\gtrsim 100$ m (cf. Ibadov $et$ $al.$ 1999).

It should be noted that comet nuclei/FEBs can also be completely destroyed within the solar corona, but that destruction may be due to another processes, such as collisions with other circumsolar bodies. Comet outbursts are potential observational manifestations of such phenomena (cf. Schrijver $et$ $al.$ 2012; Ibadov 2012 and references therein).

The equation for the kinetic energy loss of aerodynamically fully crushed and flattening FEBs allows the determination of all parameters related to the place of maximum energy release (due to FEB explosion) in an atmosphere (Grigorian $et$ $al.$ 2013). The characteristic time for thermalization of the FEB's kinetic energy within the "exploding" layer is $\sim 0.3$ sec, at a layer thickness of roughly $0.7H = 140$ km (where $H$ is height scale of photosphere). FEBs similar to comet Halley 1986 III would release explosive energy close to the energies of large solar flares, $10^{32}$ erg, while FEBs similar to comet Hale-Bopp 1995 OI (radius $\gtrsim 30$ km) should lead to stellar super-flares (Grigorian $et$ $al.$ 2000; Ibadov $et$ $al.$ 2009; Eichler & Mordecai 2012). Using similar arguments, we have explained the heights of plumes observed with HST during the collision of comet SL-9 with Jupiter in July 1994 and have predicted the ejection of plumes from young stars due to FEBs (Ibodov & Ibadov 2014 and references therein).

## 3. Conclusions

Nearby young stars and planets are unique objects for understanding the evolution of our Solar System during the epoch of intense impacts with FEBs. Impacts of FEBs should produce hot ($\sim 10^7$ K) plasma, strong shock waves, plasma plumes, and anomalously high abundances of metal atoms. Multi-wavelength observations/monitoring of nearby young stars and exoplanets are of interest for studying/identifying such potential mechanisms underlying their flare activity (e.g., Teets $et$ $al.$ 2011).

### Acknowledgements

The authors are grateful to Prof. J. H. Kastner and Dr. S. Lepine for their invitation to IAUS 314, SAI MSU for hospitality, the editor for constructive comments, and E. Shimanovskaya and N. Gostev for technical support.

### References

Eichler, D. & Mordecai, D. 2012, $ApJL$, 761, L27
Grigorian, S. S., Ibadov, S., & Ibodov, F. S. 2000, $D$-$Ph$, 45, 463
Grigorian, S. S., Ibodov, F. S., & Ibadov, S. I. 2013, $Sol.$ $Sys.$ $Res.$, 47, 268
Ibadov, S. 1990, $Icarus$, 86, 283
Ibadov, S. 2011, in $Proc.IAUS$ $283$, A.Manchado, L.Stanghellini, & D.Schonberner, eds., p.392
Ibadov, S. 2012, $Adv.Space$ $Res.$, 49, 467
Ibadov, S, Ibodov, F. S. & Grigorian, S. S. 1999, $Rom.$ $As.$ $J$ $Suppl$, 9, 125
Ibadov, S. $et$ $al.$ 2009, in $Proc.$ $IAUS$ $257$, N. Gopalswamy & D. F. Webb, eds., p. 341
Ibodov, F. S. & Ibadov, S. 2014, in $Proc.$ $IAUS$ $300$, B. Schmieder $et$ $al.$, eds., p. 509.
Lagrange, A. M., Ferlet, R., & Vidal-Madjar, A. 1987, $A\&A$, 173, 289
Schrijver, C. J., Brown, J. C., Battams, K., $et$ $al.$ 2012, $Science$, 335, 324
Teets, W. K., Weintraub, D. A., Grosso, N., $et$ $al.$ 2011, $ApJ$, 741, 83
Weissman, P. R. 1983, $Icarus$, 55, 448

*Young Stars & Planets Near the Sun*
*Proceedings IAU Symposium No. 314, 2015*
*J. H. Kastner, B. Stelzer, & S. A. Metchev, eds.*

© International Astronomical Union 2016
doi:10.1017/S1743921315006110

# Near-infrared Spectroscopy of Brown Dwarf and Planetary-Mass Members in Upper Scorpius

## Nicolas Lodieu [1,2]

[1] Instituto de Astrofísica de Canarias (IAC), Calle Vía Láctea s/n, E-38200 La Laguna, Tenerife, Spain
email: nlodieu@iac.es
[2] Departamento de Astrofísica, Universidad de La Laguna (ULL), E-38206 La Laguna, Tenerife, Spain

**Abstract.** In these proceedings, I present new VLT/X-shooter near-infrared spectroscopy of brown dwarf and planetary-mass candidates with masses below 30 Jupiter masses identified in a deep VISTA $ZYJ$ survey of 13.5 square degrees in the Upper Scorpius (USco) association. These spectra represent new benchmarks at 5–10 Myr to compare with known and future discoveries of members in nearby moving groups and other young regions.

**Keywords.** Brown dwarfs — planetary-mass objects — surveys — photometry — spectroscopy

## 1. The Upper Scorpius association

USco is part of the nearest OB association to the Sun, Scorpius Centaurus, located at 145 pc from the Sun (de Bruijne *et al.* 1997). The combination of its close distance, young age (5–10 Myr; Preibisch & Zinnecker 1999; Pecaut *et al.* 2012; Song *et al.* 2012), and mean proper motion distinct from field stars makes it an ideal ground to search for cool brown dwarfs. Various groups have examined the association in X-rays and at optical and infrared wavelengths to identify bona-fide members. Spectroscopy of hundreds of stars and substellar members is now available, allowing us to characterise in-depth their binary and disk properties. The mass function of the association is very similar to the field mass function, from high-mass stars all the way to the substellar regime (Preibisch & Zinnecker 2002; Lodieu *et al.* 2011).

## 2. Photometric selection of brown dwarf member candidates

We conducted a deep $ZYJ$ survey of 13.5 square degrees of USco with VISTA (Visible and Infrared Survey Telescope for Astronomy) in April/May 2012 in service mode under good conditions (Lodieu *et al.* 2013b). Our VISTA survey is 100% complete down to 22.0, 21.2, and 20.5 mag in $Z$, $Y$, and $J$, respectively. We complemented this dataset with additional $H$ and $K$ photometry as well as proper motions from the UKIDSS Galactic Clusters Survey (Lawrence *et al.* 2007). We identified 67 bona-fide $ZYJ$ candidates in multiple colour-magnitude diagrams (Fig. 1) as well as five $YJ$-only candidates with $J$-band magnitudes in the 14–20 mag range, corresponding to masses between 30 and 5 Jupiter masses according to state-of-the-art atmospheric models. We concluded that the USco mass function in the planetary-mass regime might be flat or decreasing by counting the number of potential members in the 20–10 and 10–5 Jupiter mass bins.

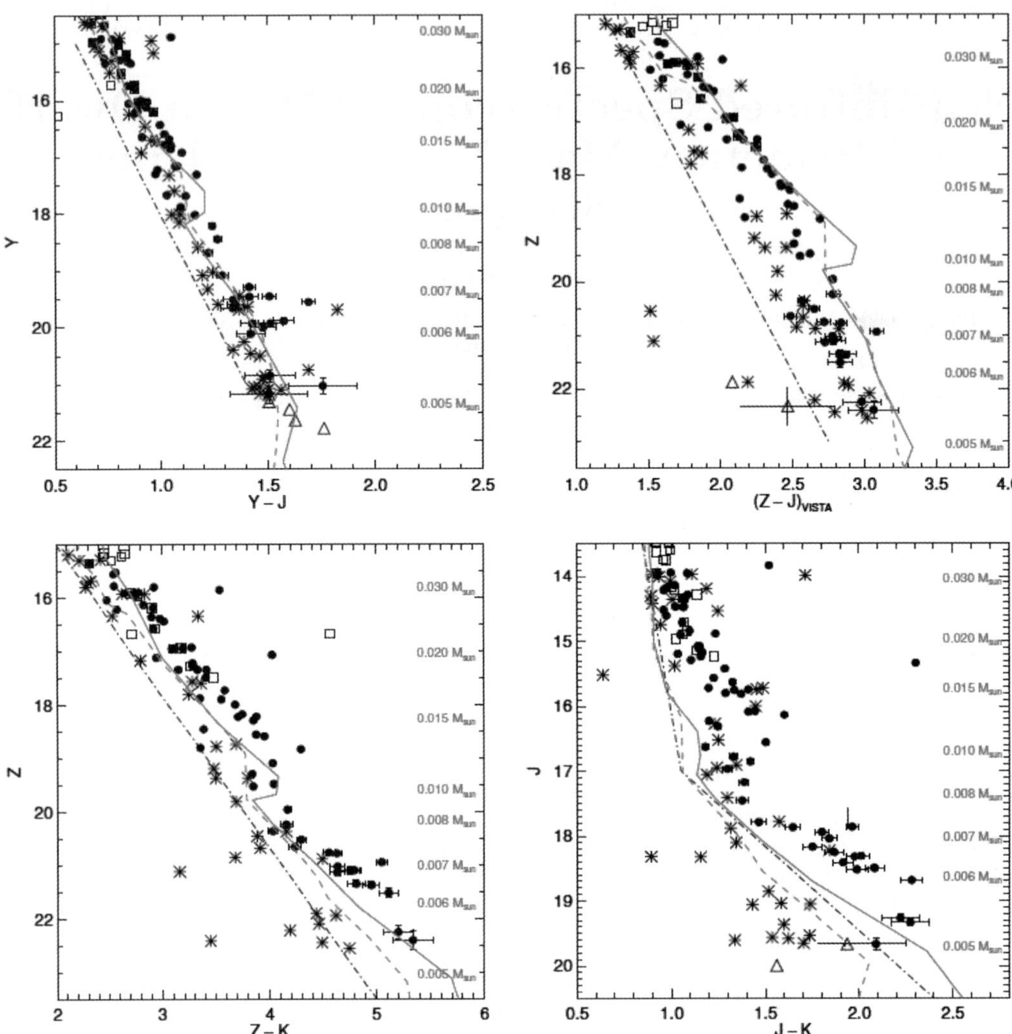

**Figure 1.** Colour-magnitude diagrams used to identify brown dwarf members of the USco association combining photometry from our deep VISTA *ZYJ* survey and public *HK* photometry from the UKIDSS Galactic Clusters Survey (Figure from Lodieu *et al.* 2013b).

## 3. VLT/X-shooter near-infrared spectroscopy

We obtained near-infrared (1.0–2.4 microns) intermediate-resolution (R~3900) spectroscopy with the X-shooter instrument on the Very Large Telescope (VLT) on 10–14 April 2015 in visitor mode. We observed 15 USco candidate members with $J = 17$–19.5 mag, equivalent to masses below the deuterium-burning limit. Moreover, half of the candidates with $J = 15$–17 mag (i.e. masses in the ~20–10 Jupiter mass range) identified in our deep VISTA survey are already confirmed members with optical spectral types (Luhman & Mamajek 2012). Those candidates have also VLT/X-shooter spectra available from the European Southern Observatory archive from an independent programme. All spectra show weak alkali lines and peaked *H*-band, signs of youth (Fig. 2). We also observe a diversity among the near-infrared spectra of these new members. We estimate preliminary spectral types in the L3–L7 range by comparison with known young members of nearby association and clusters (e.g. Lafrenière *et al.* 2010; Zapatero *et al.* 2014).

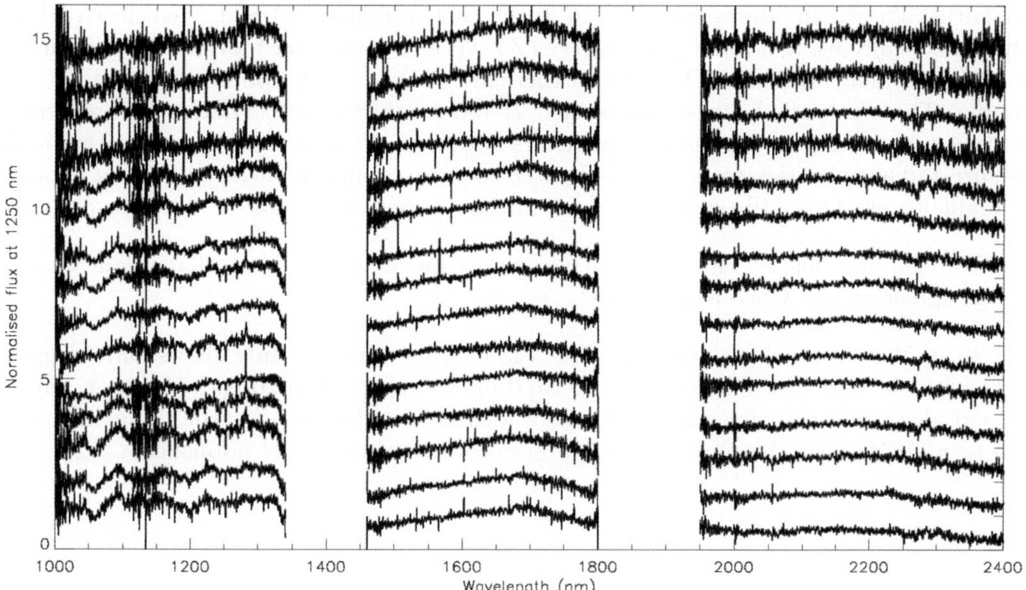

**Figure 2.** VLT/X-shooter near-infrared of 15 USco member candidates confirmed spectroscopically.

We compared our spectra to intermediate-age and very low gravity members of moving groups, suggesting that the latter might have dusty atmospheres or harbour discs. These results will be presented in more details in a future paper (Lodieu *et al.* 2015, in prep).

## References


de Bruijne *et al.* 1997, *ESA SP-402: Hipparcos - Venice '97*, p 575
Lafrenière, Jayawardhana, & van Kerkwijk 2010, *ApJ*, 719, 497
Lawrence, A., Warren, S. J., Almaini, O., *et al.* 2007, *MNRAS*, 379, 1599
Lodieu, N., Dobbie, P. D., & Hambly, N. C. 2011, *A&A*, 527, 24
Lodieu, N., Dobbie, P. D., Cross, N. J. G., *et al.* 2013b, *MNRAS*, 435, 2474
Lodieu, N. 2013a, *MNRAS*, 431, 3222
Luhman, K. L. & Mamajek, E. E. 2012, *ApJ*, 758, 31
Pecaut, M. J., Mamajek, E. E., & Bubar, E. J. 2012, *ApJ*, 746, 154
Preibisch, T. & Zinnecker, H. 1999, *AJ*, 117, 2381
Preibisch, T. & Zinnecker, H. 2002, *AJ*, 123, 1613
Song, I., Zuckerman, B., & Bessell, M. S. 2012, *AJ*, 144, 8
Zapatero Osorio *et al.* 2014, *A&A*, 568, 77


## Acknowledgements
My current contract is funded by the Ramón y Cajal fellowship number 08-303-01-02. I also acknowledge funding from an IAC internal project and another project financed by the Spanish Ministry of Economy and Competitiveness (MINECO).

## Discussion

L. MALO: You should measure radial velocities before claiming that these sources are members of the USco association

AUTHOR: These candidates are selected in the central region of USco. We selected them photometrically using 5-band photometry ($ZYJHK$) and proper motions. The sequence is well-separated from the field stars in this part of the association not affected by reddening. The weak alkali lines and peaked $H$-band confirm their youth. We are pretty confident that they are bona-fide members of USco. Nonetheless we will attempt to mesure radial velocities from our X-shooter spectra comparing with the latest models available to us. This will be presented in a future paper (Lodieu *et al.* 2015, in prep).

M. LIU: Do you observe any difference in the sequence of members that might be due to age spread in the USco association?

AUTHOR: We did not look into this. The VISTA survey covers only 13.5 square degrees in the central part of the association so we do not expect much spread. However, this is a good idea to check using our latest list of member candidates over the full association (Lodieu 2013a) once the map generated by Pecaut & Mamajek is published.

*Young Stars & Planets Near the Sun*
*Proceedings IAU Symposium No. 314, 2015*
*J. H. Kastner, B. Stelzer, & S. A. Metchev, eds.*

© International Astronomical Union 2016
doi:10.1017/S1743921315006328

# Rotation Profile of Kepler-63 from Planetary Transits

## Yuri Netto and Adriana Valio

Centro de Rádio Astronomia e Astrofísica Mackenzie (CRAAM), São Paulo, Brasil
email: `dirceuyuri@hotmail.com`

**Abstract.** Currently it is possible to estimate the rotation profile of a star that harbours a planet in an orbit such that it eclipses the star periodically. During one of these transits, the planet may occult a spot on the photosphere of the star, causing small variations in its light curve. By detecting the same spot in a later transit, it is possible to estimate the stellar rotation period. Here we present the results of this model for the case of the star Kepler-63, which has a planet in an orbit with high obliquity. This means that the planetary eclipse occults many latitude bands of the star, from near the equator to the poles. The results show that Kepler-63 has differential rotation of 0.133 rd/d and a relative differential rotation of 11.4 %.

**Keywords.** stars: rotation - starspots - planetary system - star: individual: Kepler-63

## 1. Introduction

A stellar rotation profile can be estimated if the star is eclipsed periodically by an orbiting planet. During one of these transits, the planet may occult a spot on the photosphere of the star, causing small variations in the transit light curve. By monitoring the positions of these spots in subsequent transits, it is possible to estimate the rotation period of a star (Silva-Valio 2008). Currently there are a total of over 1900 confirmed planets, and more than 1200 planets eclipse their host stars. Some stars with planets detected by CoRoT and Kepler satellites have been analysed and their rotation profiles obtained by this method (Silva-Valio & Lanza 2011).

## 2. Starspot model applied to Kepler-63

Kepler-63 is a young, solar-like star (age $\sim$ 210 Myr) (Sanchis-Ojeda *et al.* 2013) that displays high activity. It has mass of 0.984 $M_\odot$ and radius of 0.901 $R_\odot$. The planet Kepler-63b has approximately 0.4 Jupiter masses and an orbital period of 9.42 days. Interestingly, the planet presents a polar orbit (Sanchis-Ojeda *et al.* 2013). Figure 1 shows the configuration of the stellar inclination and the planet transit, according to Sanchis-Ojeda *et al.* (2013). The average rotation period of the star is obtained by analysing the Lomb-Scargle periodogram, which shows a strong peak at 5.4 days.

We applied a model that simulates planetary transits in front of a star (Silva 2003). The model assumes a synthetized image of a limb-darkened star with a planet on a circular orbit. The spots cause a slight increase in the intensity during a few minutes of the transit. The fit of the light curve by this model allows us to determine the physical characteristics of the spots, such as size, temperature and location.

This model was applied to the 150 observed transits of Kepler-63, using the stellar parameters reported in Sanchis-Ojeda *et al.* (2013). Some values had to be refined, such as the scaled semi-major axis and planet-to-star radius ratio, to better fit the transit

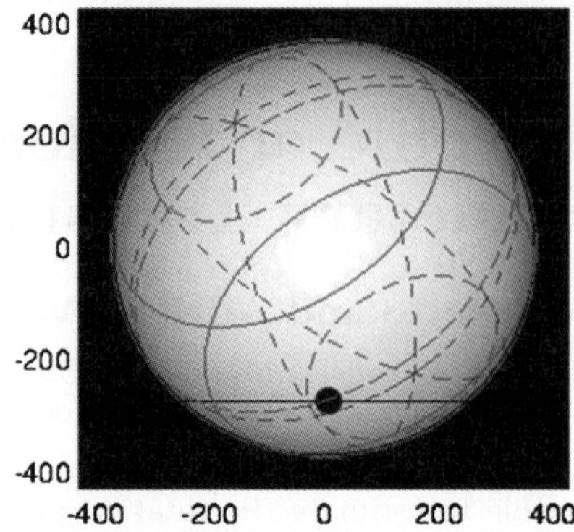

**Figure 1.** Planetary transit simulation for a planet on a polar orbit around Kepler-63.

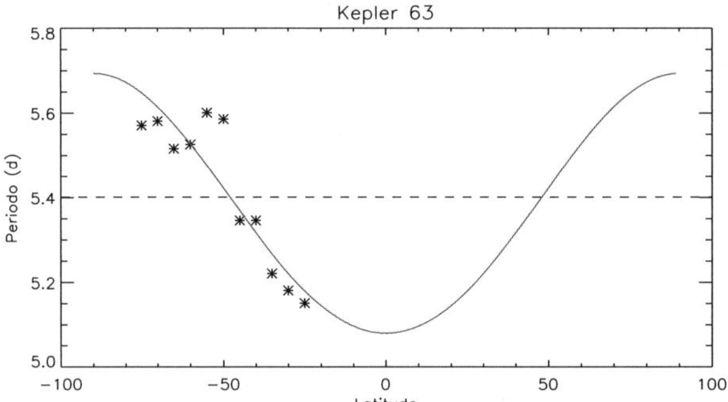

**Figure 2.** Kepler-63 rotation profile.

light curve. The results from the detection of 297 spots are listed in Table 1 in the form of mean values of the physical parameters.

| Parameters | Unit | Mean Value | Parameters | Unit | Mean Value |
|---|---|---|---|---|---|
| Radius | $(R_p)$ | $0.8 \pm 0.3$ | Radius | (km) | $(33 \pm 12) \cdot 10^3$ |
| Intensity | $(I_c)$ | $0.47 \pm 0.16$ | Temperature | (K) | $4700 \pm 400$ |

**Table 1.** Mean values of physical parameters of the star Kepler-63.

Due to the obliquity of the tilted stellar axis, the planet transit covers a range of stellar latitudes of Kepler-63 (Sanchis-Ojeda *et al.* 2013). To analyse the data obtained from the spots inferred by the light curve fits it was necessary to apply a rotation matrix to the position of the spots, since the model assumes orbits that are coplanar with the stellar equator. Thus, we obtained the spot position on the reference frame that rotates with the star.

The spots were separated into latitude bands of 5°. Using the model described in Silva-Valio & Lanza (2011) we determined the rotation period for each of these latitude bands. Assuming solar-like differential rotation ($\Omega = A - B \sin^2(lat)$) and the set of rotation period values obtained for different latitudes yields the rotation profile of Kepler-63 presented in Figure 2. Through these values, it is possible to calculate the differential rotation, $\Delta\Omega$, and relative differential rotation, $\Delta\Omega/\Omega_0(\%)$. For Kepler-63 we obtained $\Delta\Omega/\Omega_0 = 11.4\ \%$ and $\Delta\Omega = 0.133$ rd/d.

## 3. Summary and Conclusions

Assuming a differential rotation period like that of our Sun, we determined the differential rotation and relative differential rotation for Kepler-63.

The differential rotation of Sun ($\Delta\Omega = 0.050$ rd/d) and differential rotation of CoRoT-2 ($\Delta\Omega = 0.042$ rd/d), both stars of G spectral type (Valio 2013), present values more than twice lower than what we found for Kepler-63. Nevertheless, Kepler-63 presents a relative rotation period quite different from these stars. CoRoT-2, with an age between 0.13 and 0.5 Gyr, has a relative differential rotation of $\Delta\Omega/\Omega_0 = 3.0$ %, whereas the much older Sun ($\sim 4.6$ Gyr) presents a value of 22.1 % (Valio 2013), twice that of Kepler-63.

## References

Sanchis-Ojeda, R., Winn, J. A., Marcy, *et al.* 2013, *ApJ*, 775, 54
Silva, A. V. R. 2003, *ApJLett*, 585, 147
Silva-Valio, A. 2008, *ApJLett*, 683, 179
Silva-Valio, A. & Lanza, A. F. 2011, *A&A*, 529, A36
Valio, A. 2013, *New Quests in Stellar Astrophysics III: A Panchromatic View of Solar-Like Stars*, With and Without Planets, 472, 239

*Young Stars & Planets Near the Sun*
Proceedings IAU Symposium No. 314, 2015
J. H. Kastner, B. Stelzer, & S. A. Metchev, eds.

© International Astronomical Union 2016
doi:10.1017/S174392131500592X

# Star-Planet Interaction: The Curious Case of the Planet Spoon-feeding Its Host Star (and Other Amenities)

Ignazio Pillitteri[1,2] and S. J. Wolk[1] and A. Maggio[2] and T. Matsakos[3]

[1] SAO-Harvard CfA, 60 Garden St. Cambridge 02138 MA, USA
email: ipillitt@cfa.harvard.edu
[2] INAF-Osservatorio Astronomico di Palermo, Piazza del Parlamento 1, 90134 Palermo, IT
[3] Dept. Astronomy & Astrophysics, University of Chicago, 5640 S. Ellis Ave, Chicago, IL 60637, USA

**Abstract.** We report two cases of Star-Planet Interaction (SPI) in two systems with hot Jupiters: HD 189733 and HD 17156. We used HST-COS to study the FUV variability of HD 189733 after the planetary eclipse. With the support of MHD simulations, we evince that material is likely evaporating from the planet and accreting onto the parent star. This produces a hot spot on the stellar surface, co-moving with the planetary motion and responsible of the X-ray and FUV variability at peculiar planetary phases. In HD 17156, which hosts a hot Jupiter in an eccentric orbit, we observed an enhancement of the X-ray activity at the passage of its planet at the periastron. The origin can be due to magnetic reconnection between the planetary and stellar magnetic fields, or due to material tidally stripped from the planet and accreting onto the star.

---

**Keywords.** stars: activity, stars: coronae, stars: solar-type, (stars:) planetary systems, stars: individual (HD 189733, HD 17156)

## 1. Introduction

Hot Jupiters are close-in exoplanets inflated by the strong radiation from their host stars and lose their atmospheres. There is evidence that hot Jupiters can interact through magnetic fields and gravity with their parent stars, influencing the stellar activity at a detectable level. These phenomena constitute cases of Star-Planet Interaction (SPI).

## 2. HD 189733

In FUV observations of HD 189733 with HST COS during and after a planetary eclipse we found two episodes of strong variability of the line fluxes of ions of Si, C, and N that have not been observed in previous observations at planetary transits. The variability at the post planetary eclipse is analogous to the X-ray variability seen with *XMM-Newton* (Pillitteri *et al.* 2010, 2011, 2014) and points to stellar activity phased with the planetary motion. We interpret this activity as the signature of material detaching from the planet and falling onto the star (Fig. 1). This accretion creates a hot spot moving in phase with the planet and observed when it is emerging at the stellar limb (Pillitteri *et al.*, 2015; Matsakos *et al.*, 2015).

## 3. HD 17156

We have also studied the case of HD 17156, a system with a hot Jupiter in a very eccentric orbit. HD17156 is a G0 star with a hot Jupiter in a 21 days period orbit (Barbieri *et al.* 2007). The eccentricity of the orbit is 0.68, and the planet reaches a minimum separation of $\sim 7R_*$. We observed HD 17156 in September 2014 with *XMM-Newton* and HARPS-N when the planet was at

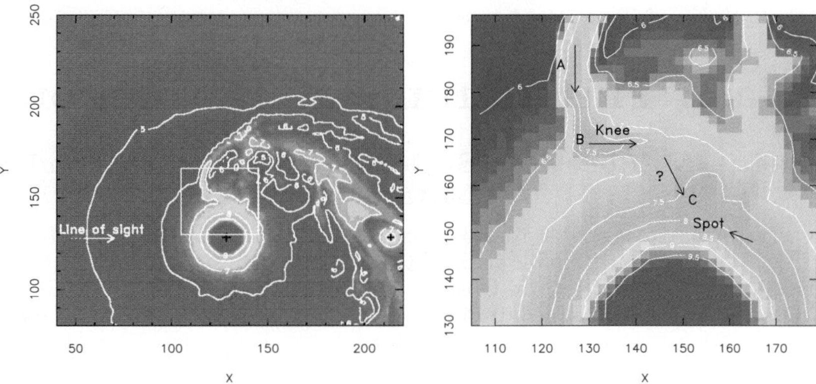

**Figure 1.** Particle density contours (units cm$^{-3}$) of an MHD simulation that models star–planet interactions (Matsakos *et al.*, 2015) between a hot Jupiter and its host (polar view). The star rotates counterclockwise, and the planet orbits the star along the same direction. The two "+" symbols shown on the left panel indicate the location of the star (red disk) and the planet (green disk). A close up of the impact region is depicted in the right panel, where the motion of the accreting plasma is marked with arrows. The shocked plasma is funnelled by the magnetized stellar wind in an almost radial trajectory close to the star (A), it forms a "knee" structure that consists of hot and dense plasma (B), and then accretes in a spot ahead of the orbital phase (C).

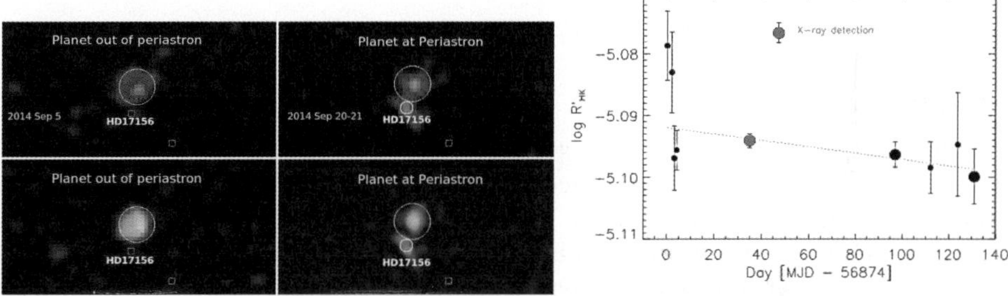

**Figure 2.** Left: XMM-EPIC count images (top panels) and *rgb* composite images (bottom panels) of HD 17156 during two *XMM-Newton* visits. Circles show the position of the detection of HD 17156, boxes are Simbad objects. An unknown X-ray source is also detected at $\sim 10$" from our target. Right: $\log R'_{HK}$ vs. MJD. Red solid symbols are the observations coordinated with *XMM-Newton* visits. Symbol sizes are proportional to the exposure times of the spectra.

the periastron and when it was farther out. We found an increase of coronal and chromospheric activity that suggests a systematic variability during periastron passages (Fig. 2). The cause might be either due a magnetic reconnection (i.e., flaring activity) when the planet was at its minimum separation, or due to material stripped from the planet and falling onto the star. The excess of X-rays has a soft spectrum and this favors the tidal stripping hypothesis.

## References

Barbieri, M., Alonso, R., Desidera, S., *et al.* 2009, *A&A*, 503, 601
Matsakos, T., Uribe, A., & König, A. 2015, *A&A* 578A, 6
Pillitteri, I., Maggio, A., Micela, G., *et al.* 2015, *ApJ* 805, 52
Pillitteri, I., Wolk, S. J., Lopez-Santiago, J., *et al.* 2014, *ApJ*, 785, 145
Pillitteri, I., Günther, H. M., Wolk, S. J., Kashyap, V. L., & Cohen, O. 2011, *ApJL*, 741, L18
Pillitteri, I., Wolk, S. J., Cohen, O., *et al.* 2010, *ApJ*, 722, 1216

*Young Stars & Planets Near the Sun*
*Proceedings IAU Symposium No. 314, 2015*
*J. H. Kastner, B. Stelzer, & S. A. Metchev, eds.*

© International Astronomical Union 2016
doi:10.1017/S1743921315006535

# Gaia: The Astrometry Revolution

**A. Sozzetti[1], M. Bonavita[2], S. Desidera[3], R. Gratton[3] and M. G. Lattanzi[1]**

[1] INAF - Osservatorio Astrofisico di Torino - Via Osservatorio 20, I-10025 Pino Torinese (Italy)
email: `sozzetti@oato.inaf.it`
[2] The University of Edinburgh, Royal Observatory, Blackford Hill, Edinburgh EH9 3HJ, UK
[3] INAF - Osservatorio Astronomico di Padova - Vicolo dell'Osservatorio 5, I-35122 Padova
(Italy)

**Abstract.** The power of micro-arcsecond ($\mu$as) astrometry is about to be unleashed. ESA's Gaia mission, now headed towards the end of the first year of routine science operations, will soon fulfil its promise for revolutionary science in countless aspects of Galactic astronomy and astrophysics. The potential of Gaia position measurements for important contributions to the astrophysics of planetary systems is huge. We focus here on the expectations for detection and improved characterization of 'young' planetary systems in the neighborhood of the Sun using a combination of Gaia $\mu$as astrometry and direct imaging techniques.

**Keywords.** planetary systems, astrometry, open clusters and associations: general, methods: numerical

---

## 1. Introduction

The study of nearby young moving groups (NYMGs) in the vicinity ($d \lesssim 100 - 200$ pc) of the Sun is of particular relevance to improve our understanding of many a key issue related to the early evolutionary stages of stars, circumstellar disks, and planetary systems. NYMGs are excellent laboratories indeed, as their precise and accurate identification, the determination of clean samples of bona-fide members (down to the sub-stellar regime), their origin, age, distance from Earth, and multiplicity properties (see, e.g., the review and contributed papers by Kastner, Mamajek, Pinsonneault, Elliott and Faherty, this volume) are keys to a) provide new insights into the early evolution of low- to intermediate-mass stars (see, e.g., the review and contributed papers by Feiden, Baraffe, Matt and Kraus, this volume), b) shed light on protoplanetary disk sculpting and dispersal (see, e.g., the review papers by Cieza, Birnstiel and Kospal, this volume), and c) globally understand giant and terrestrial planet formation, and the transition region between giant planets and brown dwarfs (see, e.g., the review and contributed papers by Mordasini, Chauvin, Marley, Allers and Youdin, this volume).

While data collected with a variety of techniques over the past two decades has allowed to make significant progress in the field thanks to the identification of hundreds of young nearby stars, the impact of new and future facilities on the study of NYMGs and their members is expected to be revolutionary. Quantum leaps in our knowledge (surprises included!) will come from the next generation of large-scale synoptic surveys in the visible (e.g., http://www.lsst.org/), sub-millimeter and milliter observations (e.g., Wilner, this volume), wide-field spectroscopic surveys at visible wavelengths (e.g., Martell, this volume), and direct imaging surveys in the visible and near-infrared (e.g., Marois, this volume). In this respect, the promise for huge progress in all the above mentioned key areas will soon be achieved when data will become available for ESA's billion star surveyor: Gaia.

## 2. Gaia: The Dawn of the Age of $\mu$as Astrometry

Gaia (http://www.cosmos.esa.int/gaia) is the first experiment set to demonstrate single-epoch measurement accuracies $\sigma_A \approx 20$ $\mu$as for bright stars (see, e.g. the review on the mission, including its payload, by de Bruijne *et al.* 2010). The mission is now entering its second year of science operations at L2, after its successful launch in December 2013. Gaia's exquisite astrometric sensitivity will allow to unravel the formation history, evolution, structure, and dynamics of the Milky Way, through measurements of the positions, motions, distances, and astrophysical parameters of the brightest 1,000 million stars in the sky (Perryman *et al.* 2001). At the end of an over six-months long commissioning phase, a number of issues arose, which pose a challenge to data reduction pipelines in order to demonstrate pre-launch performance estimates. First, significant stray light levels were identified. Second, the transmission of the optics slowly degrades with time, as a result of contamination by water ice. Finally, the intrinsic instability of the basic angle which separates the lines of sight of the two telescopes is larger than expected. While a number of additional calibration procedures are being put in place in order to cope with, mitigate, and possibly remove these undesired effects, the impact on the astrometric performance of Gaia is still under evaluation, but the hope is that such complications will still fit within the 20% margin included in pre-launch calculations (for a recent review of the issue, see de Bruijine *et al.* 2015).

## 3. Astrometric Planet Detection within the Gaia Pipeline

High-precision global astrometry with Gaia is poised to enable the detection of planetary-mass companions around stars in the Solar neighborhood. The determination of the astrometric orbits of extrasolar planets will be obtained as part of the non-single star (NSS) treatment within Coordination Unit 4 (Object Processing) of the Gaia Data Processing and Analysis Consortium. CU4 will tackle the NSS problem by attempting to derive, based on the available spectroscopic and photometric Gaia data, spectroscopic and/or eclipsing binary solutions. For astrometry, a cascade of increasingly more complex models will describe the data in terms of solutions containing derivatives of the stellar proper motion, accounting for variability induced motion, all the way to fully Keplerian astrometric orbital solutions, including where appropriate multiple companions. A Development Unit (DU437) has been specifically devoted to the modelling of the astrometric signals produced by planetary systems. The DU is composed of several tasks, which implement multiple robust procedures for (single and multiple) astrometric orbit fitting (such as Markov Chain Monte Carlo and genetic algorithms) and the determination of the degree of dynamical stability of multiple-component systems. This robust approach to orbit modeling is expected to allow coping with the complexities inherent to adjusting large, non-linear models to the data (e.g., Sozzetti 2005, 2014)

## 4. Planets Around Young Stars: The Gaia Potential

The size of the astrometric perturbation $\alpha$, espressed in arcsec, induced on the primary of mass $M_*$ by an orbiting planet of $M_p$ (both in $M_\odot$) and with a semi-major axis $a_p$ (in AU) scales with the distance $d$ (in pc), to the observer: $\alpha = (M_p/M_*) \times (a_p/d)$.

The sensitivity of Gaia astrometry to (single and multiple) giant planetary companions at intermediate separations around bright, nearby, solar-type dwarfs has been the objective of several works in the past (Lattanzi *et al.* 2000; Sozzetti *et al.* 2001; Casertano *et al.* 2008). Those early estimates have been recently revisited and complemented by Sozzetti

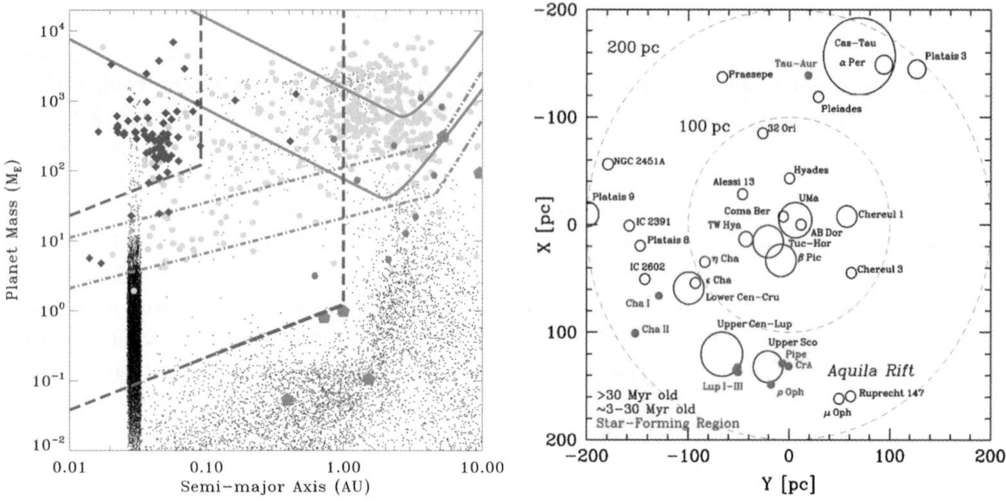

**Figure 1.** *Left:* Gaia exoplanets discovery space (solid curves) compared to that of Doppler (dashed-dotted lines) and transit (long-dashed curves) techniques. Detectability curves are defined on the basis of a 3-$\sigma$ criterion for signal detection (see Sozzetti 2010 for details). The upper and lower solid curves are for Gaia astrometry with $\sigma_A = 15$ $\mu$as, assuming a 1 $M_\odot$ primary at 200 pc and a 0.4 $M_\odot$ M dwarf at 25 pc, respectively, and survey duration set to 5 yr. The light-grey filled circles indicate the inventory of Doppler-detected exoplanets as of May 2010. Transiting systems are shown as dark-grey filled diamonds. Grey hexagons are planets detected by microlensing. Solar System planets are also shown (large grey pentagons). The small black dots represent a theoretical distribution of masses and final orbital semi-major axes (Ida & Lin 2008). *Right:* The closest ($d < 200$ pc) star forming regions and young stellar kinematic groups (image courtesy of E. Mamajek).

*et al.* (2014) and Perryman *et al.* (2014), who used improved (pre-commissioning) knowledge of the astrometric error budget and extended the studies to encompass a wider range of primary spectral types and limiting target magnitudes ($G = 20$ mag). The global figures on which all the above works converge speak of several thousands (possibly $\sim 10^4$) astrometrically detectable giant planets in the separation range $0.5 \leqslant a \leqslant 4.5$ AU from their parent stars (see Figure 1, left panel). The overall all-sky reservoir of stars around which Gaia will be sensitive to planetary-mass companions thus exceeds $10^6$.

As for the population of young stars near the Sun ($d < 200$ pc), there exist of order twenty or so nearby star-forming regions, young associations, open clusters and moving groups (see Figure 1, right panel) with ages in the approximate range $1 - 100$ Myr (Mamajek, this volume; see also Zuckerman & Song 2004, and references therein, and López-Santiago *et al.* 2006, and references therein). The likely number of bright ($V < 13 - 14$ mag) members is on the order of $10^3$ (Mamajek, private communication). All these stars will be observed by Gaia with enough astrometric sensitivity to massive giant planets ($M_p \gtrsim 2$ $M_J$) orbiting at $2 - 4$ AU. The possibility of detecting giant planets still forming in the protoplanetary disk would constitute fundamental observational evidence to validate the proposed theoretical models of giant planet formation. It will also probe the transition region between giant planets and brown dwarfs (e.g., Helled *et al.* 2014, and references therein) in a regime of orbital separation that would uniquely complement near- and mid-infrared imaging surveys (e.g., Burrows 2005, and references therein) for direct detection of young, bright, wide-separation ($a > 30 - 100$ AU) giant planets, such as those presently carried out by SPHERE (Beuzit *et al.* 2006) and GPI (Macintosh *et al.* 2014), and in the near future by JWST. The above is just one example of the broad range

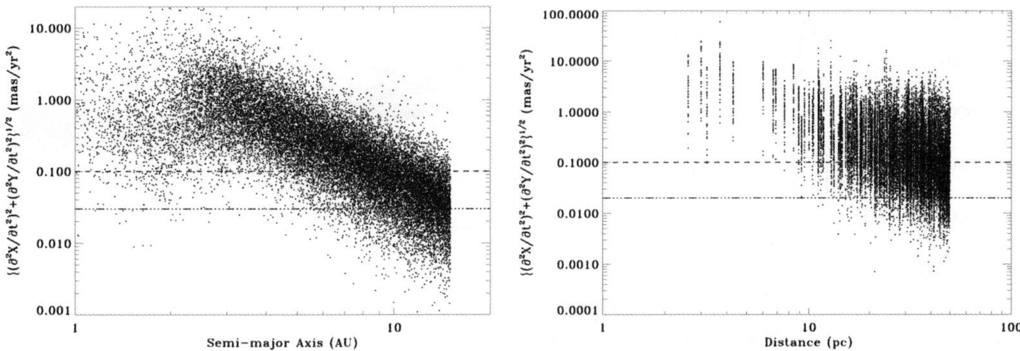

**Figure 2.** *Left:* Accelerations in the stellar motion induced by $1 - 70\ M_J$ orbiting companions at orbital separations $a < 15$ AU detected by SPHERE around a sample of $> 400$ targets of the GTO program with $V < 12$ mag and $d < 50$ pc. Dashed and dashed-dotted lines indicate 3-$\sigma_A$ detection limits with Gaia at mid-mission (2.5 yr) and at mission end (5 yr). Only 5% of the accelerations in Gaia astrometry go undetected at the end of the mission, with a high degree of completeness (99%) for $a < 7$ AU. *Right:* The same quantity expressed as a function of the distance of the systems from the Sun.

of applications to exoplanets science in which Gaia data will act as an ideal complement to (and in synergy with) many ongoing and future observing programs devoted to the indirect and direct detection and characterization of planetary systems, both from the ground and in space (Sozzetti 2015).

## 5. On the Synergy Between Gaia Astrometry and Direct Imaging

The new generation of high-contrast imaging devices (SPHERE, GPI) is primarily sensitive to young giant planets at wide separations. Mass estimates of any detected companion heavily rely on structural and evolutionary models, that carry in turn large uncertainties, particularly in the age estimates. Dynamical mass constraints are highly desirable, but difficult to obtain. Doppler techniques could help in principle, but a) the amplitudes of the radial velocity (RV) signals decrease with orbital period, and b) young stars are trouble (due to rotation and activity). For astrometry, the amplitude of the signature increases with $a$, while youth and stellar activity are not an issue for positional measurements with Gaia-like precision. It is thus worthwhile to investigate in detail the extents of the synergy between Gaia data and high-contrast imaging programs. We focus here in particular on the effectiveness of the combination of SPHERE/VLT direct detections of wide-separation giant planets with Gaia determinations of accelerations in the stellar motion due to the orbiting companions. The aim of this study, which details will be presented in a forthcoming publication (Bonavita *et al.* 2015, in preparation) is to provide improved constraints on the orbital architecture and mass, thereby helping in the modeling and interpretation of giant planets' phase functions and light curves.

### 5.1. *Simulation Setup*

The stellar sample used is composed of 439 $V < 12$ mag stars with an age cut-off at $t < 0.5$ Gyr and within 50 pc drawn from the SPHERE GTO sample. Synthetic planet populations (1 companion per star) are instead generated with mass and semi-major axis following the distributions from Cumming *et al.* (2008), extrapolated to 70 $M_J$ and 15 AU, and with the other orbital parameters randomly drawn from uniform distributions bracketed by their natural boundaries. Then, a large Monte Carlo simulation is run

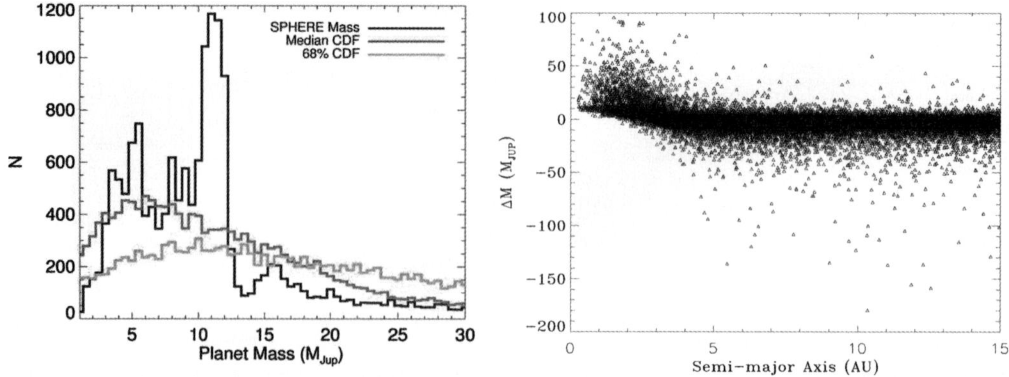

**Figure 3.** *Left:* Comparison between the distribution of the nominal companion masses based on SPHERE imaging and the median and 68.3% credibile intervals of the cumulative distribution function of the mass of the companions based on Gaia measurements of accelerations in the astrometry of the primaries and the angular separation at a single epoch from SPHERE. *Right:* Difference between median mass estimate from Gaia astrometry and the SPHERE value as a function of orbital separation.

using the MESS code (Bonavita *et al.* 2012) to identify a statistically significant sample of directly detected systems using up-to-date SPHERE detection limits (Zurlo *et al.* 2014). Next, Gaia-like simulations of the directly imaged systems are generated, using a setup described in Sozzetti *et al.* (2014). Astrometric detection of the companions is based on significant deviations from a five-parameter, single-star model, following which acceleration terms in the stellar motion are included in the model to account for curvature effects in the residuals. Finally, the methodology described in e.g. Torres (1999) is adopted to compute the cumulative distribution function (CDF) of companion masses inferred by Gaia astrometry that are compatible with the allowed range of orientations and eccentricities, and with separation estimates from SPHERE.

### 5.2. *Preliminary Results*

Figure 2 shows the magnitude of the acceleration terms (espressed in mas yr$^{-2}$) estimated by Gaia astrometry for the companions directly imaged by SPHERE as a function of the orbital separation (left panel) and system distance from the Sun (right panel). Only $\approx 10\%$ of the curvature signals are deemed not significant based on a time baseline spanning the nominal 5-yr duration of the Gaia mission. For $a < 7$ AU, the completeness levels are higher than 99%. Similar sensitivity limits apply to systems within $\sim 25$ pc and to companions with masses larger than $\approx 10$ M$_J$ (plot not shown).

The left panel of Figure 3 shows the comparison between the nominal value of the companion mass as derived from the SPHERE observations (based on inferences from theoretical models) and the median and $1 - \sigma$ confidence intervals of the mass CDF as obtained from the combination of detections of curvature in Gaia astrometry and a measurement of the projected separation from SPHERE (a model-independent estimate). Both the latter metrics tend to overestimate the mass of the companion with respect to the one inferred from the models. The systematic effect is likely due to the need (given the range of orbital separations) to fit the Gaia observations with improved descriptions of the curvature effects, such as time derivatives of the accelerations. In the right panel of Figure 3 we show instead how the effect reverses when orbital periods become comparable to the timespan of the observations: In these cases a full orbital model should instead be fitted to the observations.

## 6. Summary

The Gaia mission is bound to set the standards in high-precision astrometry for the next decade. Gaia's defining role in the exoplanet arena will be its ability to provide a large compilation of new, high-accuracy astrometric orbits of giant planets, unbiased across all spectral types, along with exquisitely precise parallaxes. There exists a huge synergy potential between Gaia and ongoing and planned exoplanet detection and (atmospheric) characterization programs, both from the ground and in space, for much improved understanding of many aspects of the formation, physical and dynamical evolution of planetary systems. We have started gauging the effectiveness of the combination of Gaia astrometry and SPHERE high-contrast imaging for constraining masses of directly-imaged companions around young stars in the solar neighborhood, as a means to reduce inherent degeneracies and uncertainties in model predictions. Our preliminary findings indicate that accelerations in Gaia astrometry for stars with companions detected by SPHERE will be easy to spot. We have identified a potential reference metric (the mass CDF) for the assessment of the quality of actual mass estimates without relying on model assumptions. Future work will focus on a) quantifying the uncertainties in companion mass estimates when other elements of information (e.g., the system age) are factored in, and b) determining the possible improvements in mass determinations when detection of orbital motion is obtained in both Gaia and SPHERE data.

## Acknowledgements

It is a great pleasure to acknowledge the SOC and LOC of IAU Symposium 314 for organizing a top-class event, spanning a wide range of topics connected to the astrophysics of young stars and their planets. This work has been funded in part by ASI under contract to INAF 2014-025-R.0 (Gaia Mission: The Italian Participation to DPAC).

## References

Beuzit, J.-L., *et al.* 2006, *The Messenger*, 125, 29
Bonavita, M., *et al.* 2012, *A&A*, 537, A67
Burrows, A. 2005, *Nature*, 433, 261
Casertano, S., Lattanzi, M. G., Sozzetti, A., *et al.* 2008, *A&A*, 482, 699
Cumming, A., *et al.* 2008, *PASP*, 120, 531
de Bruijne, J., Kohley, R., & Prusti, T. 2010, *Proc. SPIE*, 7731, id. 77311C
de Bruijne, J., Rygl, K., & Antoja, T. 2015, *EAS Pub. Ser.*, in press (arXiv:1502.00791)
Helled, R., *et al.* 2014, *Protostars and Planets VI*, University of Arizona Press, Tucson, 643
Lattanzi, M. G., Spagna, A., Sozzetti, A., *et al.* 2000, *MNRAS*, 317, 211
López-Santiago, J., *et al.* 2006, *ApJ*, 643, 1160
Macintosh, B., *et al.* 2014, *PNAS*, 111, 12661
Perryman, M. A. C., *et al.* 2001, *A&A*, 369, 339
Perryman, M. A. C., Hartman, J., Bakos, G. Á., *et al.* 2014, *ApJ*, 797, 14
Sozzetti, A., Casertano, S., Lattanzi, M. G., *et al.* 2001, *A&A*, 373, L21
Sozzetti, A. 2005, *PASP*, 117, 1021
Sozzetti, A. 2010, *EAS Pub. Ser.*, 42, 55
Sozzetti, A. 2014, *Mem. SAIt*, 85, 643
Sozzetti, A. 2015, *EAS Pub. Ser.*, in press (arXiv:1502.03575)
Sozzetti, A., Giacobbe, P., Lattanzi, M. G., *et al.* 2014, *MNRAS*, 437, 497
Torres, G. 1999, *PASP*, 111, 169
Zuckerman, B. & Song, I. 2004, *ARA&A*, 42, 685
Zurlo, A., *et al.* 2014, *A&A*, 572, A85

*Young Stars & Planets Near the Sun*
*Proceedings IAU Symposium No. 314, 2015*
*J. H. Kastner, B. Stelzer, & S. A. Metchev, eds.*

© International Astronomical Union 2016
doi:10.1017/S1743921315006080

# ALMA and the Future of Millimeter Imaging Observations

## David J. Wilner

Harvard-Smithsonian Center for Astrophysics
email: dwilner@cfa.harvard.edu

**Abstract.** The Nearby Young Moving Groups sample the critical age when primordial disks around stars complete their transformation into planetary systems with associated debris. Millimeter wavelengths provide direct access to cool material in these circumstellar disks. The high angular resolution of interferometry at these long wavelengths enables resolved observations of solids in an optically thin regime, as well as the thermal, chemical, and dynamical structure of gas, if present. In this contribution, I briefly review the evolving landscape of millimeter telescopes, with emphasis on the revolutionary capabilities of the new international Atacama Large Millimeter/submillimeter Array (ALMA) and describe pertinent early science results.

**Keywords.** circumstellar matter, planet-disk interactions, submillimeter: planetary systems

## 1. Introduction

It is not an exaggeration to say that high resolution imaging observations at millimeter wavelengths are undergoing a revolution, the direct result of a decades long international effort to develop and construct the Atacama Large Millimeter/submillimeter Array (ALMA). Several other contributons in this volume discuss ALMA results, notably from Session 3 of the conference on disks orbiting Nearby Young Moving Group members, including reviews of protoplanetary disk evolution (Cieza), transitional disk structure (van der Marel), disk chemistry (Öberg), and debris disks (Kospal). In this contribution, I take a step back to review the relevance of millimeter wavelength observations for studying circumstellar disks, and then I briefly discuss the landscape of millimeter telescopes to put ALMA into context, followed by a few selected examples from ALMA early science.

## 2. Millimeter Wavelength Observations of Disks

The "millimeter" regime spans about two orders of magnitude in wavelength, from about 300 microns, limited at shorter wavelengths by insufficient atmospheric transmission to the ground, to wavelengths of a few centimeters, melding into the domain of traditional radio astronomy. Figure 1 shows the full spectral energy distribution of TW Hya, representative of a nearby young star surrounded by a circumstellar disk. The spectral energy distribution is dominated by two components, the stellar photosphere that peaks in the optical, and a broad excess above the photosphere from thermal dust emission that extends from the infrared into the millimeter. It's very obvious from Figure 1 that emission at millimeter wavelengths is not significant in an energetic sense. However, the emission at millimeter wavelengths has useful characteristics that make observations of disks in this regime important. First, the emission becomes optically thin and therefore provides a direct tracer of mass (in solids). Second, millimeter emission allow unique access to cold material, 10's of K and below, thereby sampling the bulk of the disk material, including the midplane. Third, many common molecules have spectral lines in the

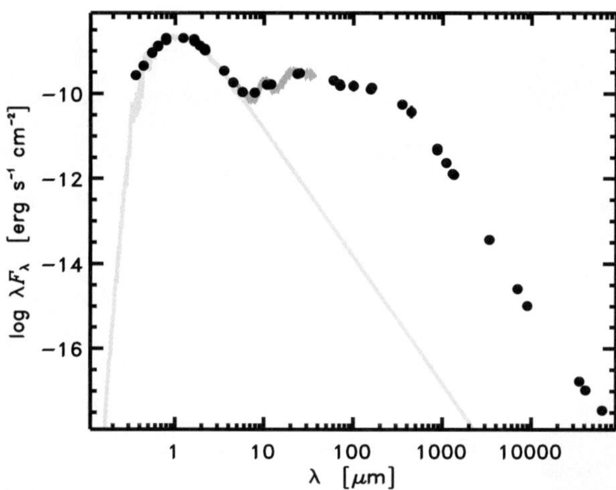

**Figure 1.** The spectral energy distribution of TW Hya (see Menu *et al.* 2014), showing a model stellar photosphere peaking in the optical (blue curve) and observations of dust emission in excess above the photosphere from the infrared through the millimeter.

millimeter, which offer probes of gas heating, cooling, and chemistry. With receivers using heterodyne techniques, the spectral resolution is naturally very high, $\Delta\lambda/\lambda > 10^6$, enabling detailed studies of gas motions. Finally, the stellar photosphere is relatively faint at these long wavelengths and contrast with disk material is not a problem, allowing straightforward access to planet-forming regions. These characteristics are all wonderful, but the catch is that emission from cold, optically thin material has very low brightness. As a consequence, very high sensitivity is required for imaging, especially at the high angular resolution needed for circumstellar disks. This is the key reason why the enormous sensitivity gain realized by ALMA is driving a revolution.

## 3. The Landscape of Millimeter Telescopes

Figure 2 shows a schematic view of telescopes currently operating in the 1.3 millimeter atmospheric window. The circles in this Figure are scaled by the physical diameter of the telescope antennas, which serves as a proxy for sensitivity (all else being equal). ALMA is clearly a very large telescope by the standards of existing facilities. Note that ALMA includes a main array of $50 \times 12$ meter antennas, and a compact array to obtain information on larger size scales with $12 \times 7$ meter antennas and four additional 12 meter antennas for total power observations. The 66 high precision (and movable) antennas that comprise ALMA are sited at 5000 meters elevation in the Chilean Andes, above much of the atmospheric water vapor that absorbs millimeter radiation. ALMA represents an improvement by one to two orders of magnitude in essentially every way over other millimeter telescopes. Besides the enormous collecting area (equivalent to a single $\sim 80$ meter antenna), the large number of antennas results in superior interferometric image fidelity, and the main array configurations can extend to 15 km to provide resolutions as high as 5 milliarcseconds at the shortest wavelengths. ALMA represents a fusion of international efforts under development since the 1980's, and is funded by an international partnership of North America (37.5%) Europe (37.5%), and East Asia (25%), in cooperation with the Republic of Chile. Up-to-date details about ALMA technical capabilities and management can be found at www.almaobservatory.org.

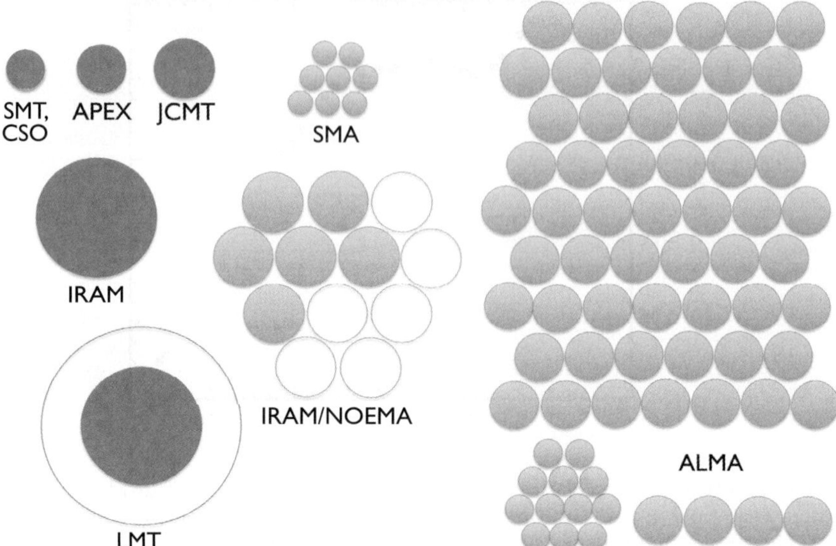


**Figure 2.** The landscape of millimeter telescopes. The circle size is proportional to antenna diameter. On the left, color coded in orange, are single dish telescopes; on the right, color coded in blue, are interferometers.

An improvement in observational capability this large comes along perhaps only once in the career of a typical observational astronomer. Community interest in using ALMA has been overwhelming. ALMA Cycle 0 early science began in October 2011 with just 16 antennas and angular resolutions no better than other arrays, but already the sensitivity gains were significant; 919 proposals were submitted and 113 projects selected (with time allocated according to partner share). The most recent Cycle 3 call, for operation with at least 36 antennas and baselines to 10 km, resulted in 1592 proposals submitted, the most ever for any observatory.

## 4. ALMA's Revolutionary Sensitivity and Imaging Speed

ALMA's power allows for observations that couldn't be done before in a reasonable time. For example, a few hour ALMA observation of a source at 50 pc distance is sensitive to dust masses less than a few tenths of a lunar mass (for standard dust opacities and disk temperatures). New statistical views of populations of sources become possible. Figure 3 shows a montage of ALMA Cycle 0 observations at 0.88 millimeters of 20 K and M-type stars surrounded by disks in the 5-10 Myr-old Upper Sco OB association, from Carpenter *et al.* 2014. Each source was observed in just a few minute snapshot, with lower noise *and* better imaging than could be obtained after integration times of hours using other arrays. These ALMA data were used by to measure the dust content in these disks, down to 0.3 Earth masses, with 13 detections (4 of which are resolved). The image of the transition disk surrounding the star J160421.7-213028 (second row, second column) is a particularly impressive (compare with the SMA observations of Mathews *et al.* 2012.) ALMA observations of this disk, which were also configured for the CO J=3-2 line, were further analyzed by Zhang *et al.* 2014 to show that 80% of the dust mass is concentrated in a 35 AU wide annulus, while the CO emission extends to both substantially larger and smaller radii. The strong radial concentration of millimeter dust emission in the gas disk suggests the presence of a pressure trap, perhaps created by a planet inside the cavity.

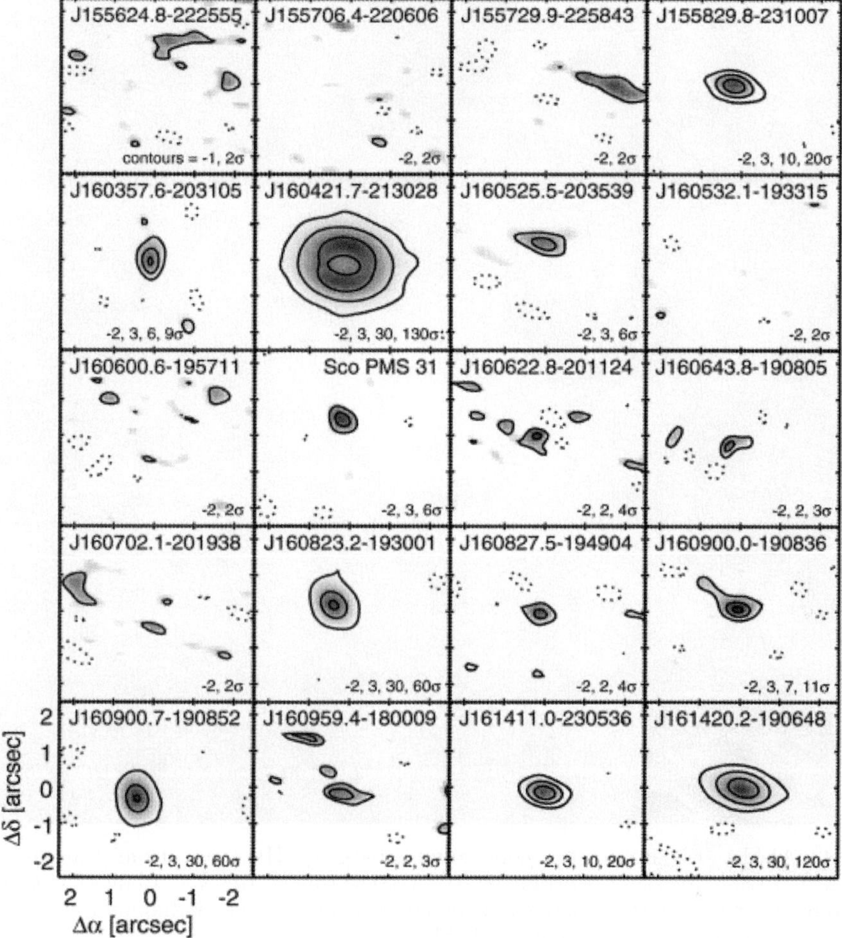

**Figure 3.** ALMA Cycle 0 0.88 millimeter snapshot images of 20 K-M type stars in Upper Sco, from Carpenter *et al.* 2014. The contour levels are indicated in the lower right of each panel; the $1\sigma$ sensitivity is typically 0.2 mJy beam$^{-1}$ (beam size $\sim 0.3$ arcsec).

A byproduct of ALMA observations of gas-rich circumstellar disks will be disk-based dynamical masses of their stellar hosts. Rosenfeld *et al.* 2012 present a proof of concept study of this technique using SMA observations of CO J=2-1 emission from the circumstellar disk orbiting the double-lined spectroscopic binary star V4046 Sgr, a member of the $\beta$ Pictoris moving group. Modeling the Keplerian velocity field of the disk around V4046 Sgr gives an estimate of the total mass of the central binary that is in excellent agreement with independent dynamical constraints obtained from monitoring the stellar radial velocities. This comparison in the V4046 Sgr system validates the absolute accuracy of this high precision (few percent) disk-based method for estimating stellar masses. With ALMA survey observations of gaseous disks around hundreds, if not thousands, of young stars in the coming years, disk-based dynamical masses will become an industry, and the results will provide critical tests of pre-main-sequence stellar evolution models. This SMA study also highlights the fact that the smaller millimeter telescopes can co-exist and thrive in the ALMA era, with a mix of innovative, highly focused programs

**Figure 4.** ALMA 1.3 millimeter continuum image of HL Tau, from the 05 November 2014 ALMA Press Release ("Revolutionary ALMA Image Reveals Planetary Genesis") at `www.almaobservatory.org/press-room/press-releases/`. The disk diameter is about 235 AU, and the beam size (35 milliarcseconds) is 5 AU.

and preparatory work (analogous to 2 meter class optical telescopes, or university based high performance computer clusters).

## 5. ALMA's Revolutionary Angular Resolution

The first results from the Long Baseline Campaign in fall 2014 as part of ALMA's Science Verification process are already remarkable, providing unprecedented angular resolution. Figure 4 shows the already iconic high resolution ALMA image of the disk surrounding the $< 1$ Myr-old star HL Tau at 1.3 millimeters (ALMA Partnership *et al.* 2015). The time on source was 4.5 hours, using 25 to 30 antennas. The beam size is 35 milliarcseconds, corresponding to 5 AU. Most striking is the pattern of concentric bright and dark rings in the disk, tracing zones of concentrated and depleted solid particles. Some, or perhaps all, of the grooves may be carved out by the gravitational influence of giant planets in formation. Given the young age of this system, which is still partly embedded in an envelope, the formation of a system of planets is perhaps surprising. However, other explanations for these nested rings have been proposed, such as magnetohydrodynamic zonal flows and condensation fronts of volatiles, which must be considered and tested. Interestingly, the innermost regions of the HL Tau disk in this image show brightness temperatures in excess of 200 K and are almost certainly optically thick. To probe dust substructure in the zone of terrestrial planet formation in this system will require high resolution observations at longer (centimeter) wavelengths, to penetrate the inner disk column; no existing radio telescope has this capability.

What substructure will disks around other young stars show? As experience with models of the formation of our Solar System demonstrate, it is dangerous to extrapolate from a single example to all other systems. We really don't know what to expect. Will other disks also be comprised of nested rings, as seen in HL Tau? Giant planets have long been predicted to open gaps in disk gas, and particle trapping in the pressure maxima outside these gaps can make these features very prominent in millimeter continuum emission. Will we see spiral waves in disks, expected to be launched by these giant planets? Or will we see stochastic clumps, as solids are concentrated in turbulent eddies throughout the disk? The stars in Nearby Young Moving Groups will be crucial proving grounds for testing theoretical ideas about planet formation and disk-planet interaction, because of their proximity. For the TW Hya disk, for example, ALMA can resolve low-amplitude features as small as 2 AU, sufficient to measure planetary gaps on scales compatible with our Solar System architecture, and to probe turbulent fluctuations with sizes comparable to the gas pressure scale height over most of the disk, thereby providing stringent constraints.

## 6. Concluding Remarks

The ALMA era in millimeter astronomy is well underway, with revolutionary sensitivity, image fidelity, and high angular resolution. The early science results from ALMA on disks around young stars are tantalizing, unexpected, and incredibly exciting. We have a lot to look forward to.

## References


Carpenter, J. M., Ricci, L., & Isella, A. 2014, *ApJ*, 787, 42
Mathews, G. S., Williams, J. P., & Ménard, F. 2012, *ApJ*, 753, 59
Menu, J., van Boekel, R., Henning, T., *et al.* 2014, *A&A*, 564, A93
Partnership, A., Brogan, C. L., Perez, L. M., *et al.* 2015, arXiv:1503.02649
Rosenfeld, K. A., Andrews, S. M., Wilner, D. J., & Stempels, H. C. 2012, *ApJ*, 759, 119
Zhang, K., Isella, A., Carpenter, J. M., & Blake, G. A. 2014, *ApJ*, 791, 42

*Young Stars & Planets Near the Sun*
*Proceedings IAU Symposium No. 314, 2015*
*J. H. Kastner, B. Stelzer, & S. A. Metchev, eds.*

© International Astronomical Union 2016
doi:10.1017/S1743921315006055

# Studying Young Stars with Large Spectroscopic Surveys

## Sarah L. Martell

School of Physics, University of New South Wales
email: s.martell@unsw.edu.au

**Abstract.** Galactic archaeology is the study of the history of star formation and chemical evolution in the Milky Way, based on present-day stellar populations. Studies of young stars are a key anchor point for Galactic archaeology, since quantities like the initial mass function and the star formation rate can be studied directly in young clusters and star forming regions. Conversely, massive spectroscopic Galactic archaeology surveys can be used as a data source for young star studies.

**Keywords.** stars: formation; Galaxy: evolution; Galaxy: stellar content

## 1. Introduction

The goals of Galactic archaeology, as laid out by Freeman & Bland-Hawthorn (2002), are to identify stars that formed together using their surface abundance patterns, and then to use those groups of co-formed stars to explore the history of star formation and chemical enrichment in the Milky Way. Since the typical mode for star formation is clustered (e.g., Lada & Lada 2003), and star-forming regions up to the masses of globular clusters ($\sim$ few $\times 10^5 M_\odot$) ought to be chemically well-mixed (e.g., Feng & Krumholz 2014), we expect that each star formation event should produce a large number of stars with matching abundance patterns.

During these stars' lifetimes they will drift, scatter and migrate through the Galaxy. This process begins with the evaporation of young clusters (e.g., Terlevich 1987; Murphy *et al.* 2010). However, the majority of a star's surface abundances will remain constant as it evolves, and it is this persistence of the initial abundance pattern that allows stars from those initial groups to be "chemically tagged" as belonging together (e.g., De Silva *et al.* 2011; 2013).

Identifying stars with matching abundance patterns is an interesting challenge - because some elements are produced in multiple nucleosynthetic sites, on different timescales, there is not a simple model for how stars should be arranged in the parameter space of chemical abundances ("C-space"). Simply matching as many abundances as can be measured is inefficient since there are a number of element groups that tend to be produced in the same astrophysical sites and enriched or depleted simultaneously. A number of ongoing and planned survey projects will be determining a large number of elemental abundances per star for many stars; quantifying the most efficient and informative ways to search for groups within that abundance data will be an important challenge for those survey teams.

## 2. Spectroscopic Galactic archaeology surveys

Chemical tagging can be carried out at different levels with the data from many large survey projects, whether it is focused on searching for unusual stars (e.g., Martell &

Grebel 2010; Martig *et al.* 2014), or on identifying groups of stars that formed together (Bland-Hawthorn *et al.* 2010; Mitschang *et al.* 2013). The abundance precision needed to select co-formed groups of stars is quite high, estimated as 0.05 to 0.1 dex in [X/H]†. This is a difficult but achievable level of precision, even for large survey projects with automated data reduction and analysis processes.

Recent theoretical work evaluating the prospects of identifying co-formed stars based solely on abundance information (Ting *et al.* 2015) has found that it may be quite difficult, depending on the intrinsic differences between the abundance patterns of independent groups of stars. If there is not much separation between the groups in C-space, then it will naturally be quite difficult to distinguish them. However, the addition of age information (for example, requiring that stars with similar abundance patterns must also fall near a single isochrone, as is done in Mitschang *et al.* 2014) is a powerful tool in confirming that a group of stars did form at the same time. Even in the case when individual co-formed groups cannot be identified, the form and strength of structure in C-space still carries information about the star formation rate, the initial cluster mass function, and the minimum mass for star-forming events.

Since Galactic archaeology is so interested in the mode and rate of star formation, studies of present-day young stars have the potential to be quite important to our understanding of Galactic history. As one example, direct measurements of the stellar initial mass function in a young cluster or moving group are far more concrete than distributions inferred from much older populations. In addition, detailed abundance determinations for young stars can be used to test whether star-forming regions are as well-mixed as predicted, and how well the abundance patterns of co-forming stars match.

Ironically, the level of detail available to studies of present-day young stars can make it difficult to combine them with old samples. The ages for old stars typically carry error bars of $\pm 1$ Gyr, and studies based on old stars would not be able to distinguish substructures within a star-formation region where the stars formed $10 - 100$ Myr apart. Studies of pre-main-sequence stars can determine ages to a much finer resolution, and as a result the membership criteria for young associations and moving groups (e.g., Malo *et al.* 2013) can be quite complex. The processes and algorithms for group-finding in Galactic archaeology data will need to recognise the fact that star formation is a distributed and asynchronous process.

## 3. Opportunities for young star studies in existing and future survey data sets

The three large high-resolution spectroscopic Galactic archaeology surveys currently in operation (GALAH, de Silva *et al.* 2015; APOGEE, Holtzman *et al.* 2015; Gaia-ESO, Gilmore *et al.* 2012) all take different approaches to including young stars in their sample. Gaia-ESO specifically observed a number of young associations, which have produced some of their first scientific results. Spina *et al.* (2014) find that the metallicity in the Chamaeleon 1 star-forming region is mildly sub-Solar, similar to other star-forming regions in the Solar neighbourhood. Jeffries *et al.* (2014) find that the $\gamma$ Velorum cluster comprises two independent groups along the same line of sight, with slightly different systemic velocities and ages.

† By convention, elemental abundance ratios written in square brackets are expressed on a logarithmic scale normalised to the Sun. For an element X in a star, $[X/H]= \log(N_X/N_H)_{star} - \log(N_X/N_H)_\odot$

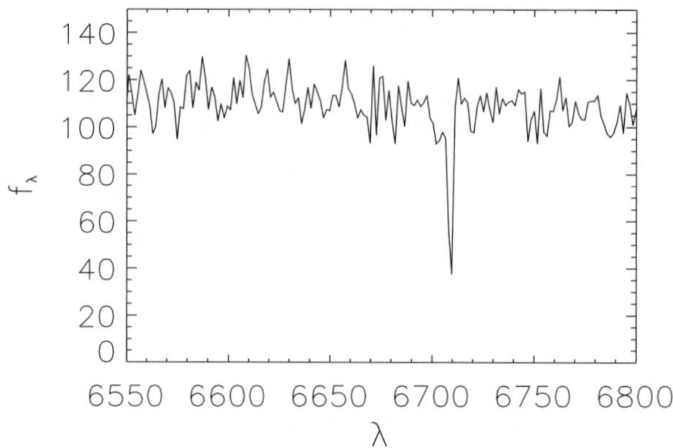

**Figure 1.** Portion of a LAMOST spectrum of a previously unknown Li-rich M0 star. Further data mining should uncover more stars like this.

APOGEE devotes a fraction of its observing time to ancillary science, and the IN-SYNC ancillary program (Cottaar *et al.* 2014) has observed nearly 3500 young ($\leqslant 10$ Myr) stars across IC 348 and the Pleiades. They find that there is a range in radius at fixed stellar mass in IC 348, indicating an age range of 25%. GALAH has a small number of specifically targeted fields, including the Kepler-2 (K2, Howell *et al.* 2014) Campaign 2 field, which covers the Scorpius-Centaurus association. The targeting in GALAH is aimed at red giants to be observed by K2, and not specifically at young stars, but nevertheless a useful fraction of the over 2000 stars reported as candidate young stars by Rizzuto *et al.* (2015) have been observed by GALAH. Given the very thorough sky coverage of GALAH, it is quite likely that other star-forming regions, young associations and young clusters will also be part of the GALAH sample.

Finally, there are other spectroscopic resources that bear exploration and data-mining. The WEAVE (Dalton *et al.* 2012) and 4MOST (de Jong *et al.* 2012) surveys, which are both due to start observing in the next few years, will collect massive data sets and could involve targeted observations of young stars, with involvement from the community. The Large Sky Area Multi-Object Fibre Spectroscopic Telescope (LAMOST, Luo *et al.* 2015) survey has been acquiring low-resolution (R$\sim$ 1800) spectra since 2011, and its first data release includes spectra and stellar parameters for over one million A, F, G and K stars. The first LAMOST data release also includes over 120,000 stars classified as M dwarfs based on their spectra, providing a rich data set in which to search for spectroscopic youth indicators like chromospheric activity and strong Li 6708Å absorption. Figure 1 shows an example of a previously unreported Li-rich M star in LAMOST. This spectrum was identified during a cursory examination of the public LAMOST data, using the Li 6708Å absorption index defined in Martell & Shetrone (2013). It is classified by LAMOST as an M0 star.

## References

Bland-Hawthorn, J., Krumholz, M. R., & Freeman, K. 2010, *ApJ*, 713, 166
Cottaar, M., Covey, K. R., Meyer, M. R. *et al.* 2014, *ApJ*, 794, 125
Dalton, G., Trager, S. C., Abrams, D. C. *et al.* 2012, *Proc. SPIE*, 8446, 84460P
de Jong, R. S., Bellido-Tirado, O., Chiappini, C. *et al.* 2012, *Proc. SPIE*, 8446, 84460T
De Silva, G. M., Freeman, K. C., Bland-Hawthorn, J. *et al.* 2011, *MNRAS*, 415, 563
De Silva, G. M., D'Orazi, V., Melo, C. *et al.* 2013, *MNRAS*, 431, 1005
De Silva, G. M., Freeman, K. C., Bland-Hawthorn, J. *et al.* 2015, *MNRAS*, 449, 2604

Feng, Y. & Krumholz, M. R. 2014, *Nature*, 513, 523
Freeman, K. C. & Bland-Hawthorn, J. 2002, *ARA&A*, 40, 487
Gilmore, G., Randich, S., Asplund, M. *et al.* 2012, *ESO Messenger*, 147, 25
Holtzman, J. A., Shetrone, M., Johnson, J. A. *et al.* 2015, *arXiv:1501.04110*
Howell, S. B., Sobeck, C., Haas, M. *et al.* 2014, *PASP*, 126, 398
Jeffries, R. D., Jackson, R. J., Cottaar, M. *et al.* 2014, *A&A*, 563, 94
Lada, C. J. & Lada, E. A. 2003, *ARA&A*, 41, 57
Luo, A.-L., Zhao, Y.-H., Deng, L.-C. *et al.* 2015, *arXiv:1505.01570*
Malo, L., Doyon, R., Lafreniere, D. *et al.* 2013, *ApJ*, 762, 88
Martell, S. L. & Grebel, E. K. 2010, *A&A*, 519, 14
Martell, S. L. & Shetrone, M. D. 2013, *MNRAS*, 430, 611
Martig, M., Rix, H.-W., Silva Aguirre, V. *et al.* 2014, *MNRAS*, 451, 2230
Mitschang, A. W., de Silva, G., Sharma, S., & Zucker, D. B. 2013, *MNRAS*, 428, 2321
Mitschang, A. W., de Silva, G., Zucker, D. B. *et al.* 2014, *MNRAS*, 438, 2753
Murphy, S. J., Lawson, W. A., & Bessell, M. S. 2010, *MNRAS*, 406, L50
Rizzuto, A. C., Ireland, M. J., & Kraus, A. L. 2015, *MNRAS*, 448, 2737
Spina, L., Randich, S., Palla, F. *et al.* 2014, *A&A*, 568, 2
Terlevich, E. 1987, *MNRAS*, 224, 193
Ting, Y.-S., Conroy, C., & Goodman, A. 2015, *ApJ*, 807, 104

## Discussion

PROF. TRIMBLE: Are there any likely candidates for stars that co-formed with the Sun?

DR. MARTELL: The open cluster M67 has been suggested as a place where the Sun may have formed, but mainly based on age and metallicity. You would need to match the abundance patterns across more elemental abundances to be more certain. Trying to match the Sun to the stars it co-formed with based on Galactic orbits is less certain than chemical tagging, since possibility of scattering by spiral arms or giant molecular clouds in the time since the Sun formed is not negligible.

PROF. MAMAJEK: Is anyone already searching the LAMOST data for active stars?

DR. MARTELL: Not that I know of.

*Young Stars & Planets Near the Sun*
*Proceedings IAU Symposium No. 314, 2015*
*J. H. Kastner, B. Stelzer, & S. A. Metchev, eds.*

© International Astronomical Union 2016
doi:10.1017/S1743921315005943

# eROSITA - Nearby Young Stars in X-rays

## J. Robrade

Hamburger Sternwarte, Germany
email: jrobrade@hs.uni-hamburg.de

**Abstract.** X-ray surveys are well suited to detect, identify and study young stars based on their high levels of magnetic activity and thus X-ray brightness. The eROSITA instrument onboard the Spectrum-Roentgen-Gamma (SRG) satellite will perform an X-ray all-sky survey that surpasses existing data by a sensitivity increase of more than an order of magnitude. The 4 yr survey is expected to detect more than half a million stars and stellar systems in X-rays.

**Keywords.** X-rays: stars, stars: activity, stars: coronae

---

## 1. Introduction

The X-ray emission from cool stars originates from magnetic coronae that contain hot plasma at MK temperatures and volume complete X-ray surveys have shown that coronae are ubiquitous around all types of cool stars (Schmitt, 1997). Magnetic activity is driven by stellar dynamos with activity levels covering a range of $\log L_{\rm X}/L_{\rm bol} \approx -3 \ldots -7$. Activity is particularly strong in young stars due to their fast rotation and declines as stars spin down during their lifetime through the loss of angular momentum via magnetized winds (see S. Matt, this Issue). The efficiency of a stellar dynamo and thus the generation of magnetic activity depend not only on rotation, but also on the convective turnover time, thereby introducing a mass/spectral type dependent property with higher efficiencies at lower masses. When combining both parameters in the Rossby number $Ro = P_{\rm rot}/\tau_c$, the dynamo efficiency can be uniformly described for solar and late-type stars. The resulting stellar activity-rotation relation that has been studied in detail in X-ray astronomy, see e.g. Pizzolato *et al.* (2003), Wright *et al.* (2011). Key features are a saturated regime with a constant activity level around $\log L_{\rm X}/L_{\rm bol} \approx -3$, an unsaturated regime with a power-law decline, $L_{\rm X}/L_{\rm bol} \propto Ro^{\beta}$ and a threshold at $Ro \approx 0.1 - 0.15$. Summarized, at fast rotation X-ray brightness depends only on stellar luminosity and beyond a mass dependent rotation period it decreases with slower rotation.

The strong decline of coronal X-ray emission by several magnitudes over the stellar lifetime allows for an identification of stellar youth. In addition, younger and more active stars are not only X-ray brighter, but also have hotter coronae and harder X-ray spectra. Observations at multiple wavelengths have shown that in general emission from hotter plasma declines more steeply, leading to a pronounced contrast in X-ray observations. For example, the 'Sun in Time' project showed that early G stars are typically saturated for about 100 Myr and faint by about two orders of magnitude at relevant X-ray energies already during the first Gyr of their life, see e.g. Güdel *et al.* (1997), Ribas *et al.* (2005). Not only the average stellar activity level depends on age, but also the average age at which a star of a given spectral type leaves the saturated X-ray regime. Thereby a quite robust age-dating via X-ray observations is possible for ensembles of stars at ages above roughly 100 Myr, but saturation effects, variability and intrinsic scatter, especially of rotation periods at stellar youth, considerably restrict the applicability of this method

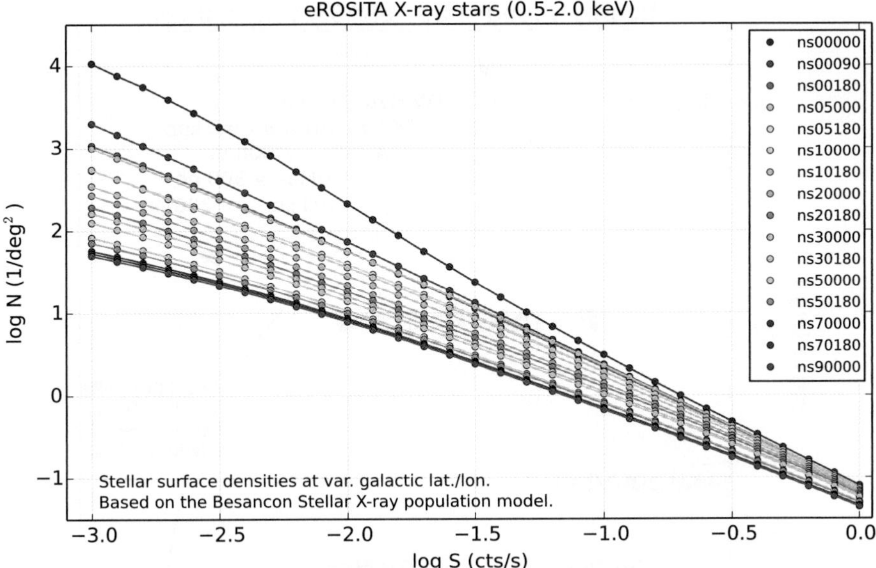


**Figure 1.** Simulated stellar densities in the eROSITA All-Sky Survey towards various galactic directions, the typical survey sensitivity is at $\log S \approx -1.9 \, \mathrm{cts\,s^{-1}}$.

for individual objects. Nevertheless, a classification as 'young' often suffices, e.g. when a distinction between nearby young moving group members and older field stars matters.

## 2. The eROSITA All-Sky Survey

eROSITA (extended Roentgen Survey with an Imaging Telescope Array) is the primary instrument onboard the Spectrum-Roentgen-Gamma (SRG) satellite which will be launched around end 2016 and placed in an L2 orbit. It has been developed under the leadership of the Max Plank Institute for Extraterrestrial Physics (MPE, PI institute) and consists of seven co-aligned X-ray telescopes, each equipped with a CCD-type detector and a field of view (FOV) with a diameter of 1.03 deg. eROSITA will perform a 4 yr imaging all-sky survey at low to medium X-ray energies ($0.3-10.0$ keV). The eRASS (eROSITA All-Sky Survey) is unprecedented, given its sensitivity, energy range, spectral and angular resolution as well as its eightfold sky coverage.

Important technical parameter of the eROSITA instrument are (on-axis/FOV average) a HEW of 15/28 arcsec at 1.5 keV, an eff. area of 2400/1400 cm$^2$ at 1.0 keV and an energy resolution (FWHM) of 60 eV at 0.3 keV and of 140 eV at 6 keV. The all-sky survey scan paths follow mainly an ecliptical geometry and accumulate in total an average sky exposure depth of about 2.5 ks. eROSITA/SRG is a joint German/Russian project and eROSITA data will be shared in equal parts between the partners; the respective priority areas are implemented as a sky-spit through the galactic center, the German side is the 'Galactic West' (gal. lon. $180-360$ deg). A description of the mission and its science objectives are given in the eROSITA Science Book, see Merloni *et al.* (2012).

With an average survey sensitivity of $F_{\mathrm{X}} \approx 1 \times 10^{-14}$ erg cm$^{-2}$ s$^{-1}$ it is expected to detect about $0.7 \times 10^6$ stars, using simulations based on the Besancon Stellar X-ray population model (Guillout *et al.*, 1996). As shown in Fig. 1, the typical stellar surface density is a strong function of galactic viewing direction; here the on average younger and X-ray brighter disk population further enhances the latitude contrast. Separating the

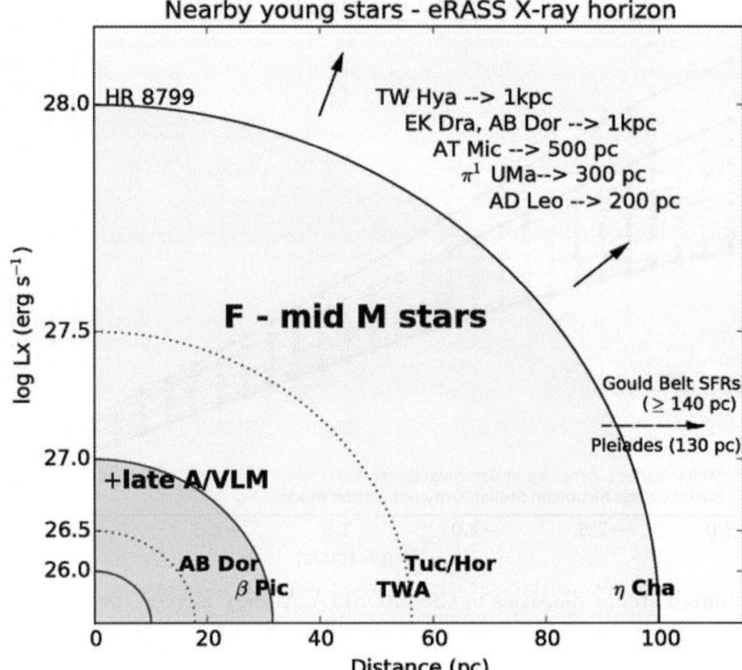



**Figure 2.** Sketch of the X-ray horizon of the eROSITA All-Sky Survey, average sky-sensitivity.

X-ray stars by mass/spectral type, late-type K and M dwarfs dominate given their X-ray brightness and large volume densities; separating by age, young stars (age $\lesssim$ 200 Myr) contribute in total at a fraction of about 40 % and start to dominate at low galactic latitudes. Naturally, they include the members of nearby young stellar populations, moving groups, clusters and associations.

The sensitivity increase of the eRASS by a factor of roughly 20 compared to the RASS (ROSAT All-Sky Survey) will help to overcome many issues related to the brightness limits of the existing data and significantly enlarge the volume surveyed with sufficient depth at X-ray energies. The deepened X-ray view of our galactic neighborhood and the multiple synergies have a high potential to set new levels in a variety of studies within the field of stellar astrophysics.

## 3. Nearby young stars in the eRASS

The 'local' X-ray sky is dominated by coronal emission from various types of cool stars and stellar systems, ranging from solar-type stars to very low mass stars and from pre-main sequence stars to older disk populations. Besides discovering new nearby young stars with the eRASS, their high X-ray fluxes also enable the investigation of coronal properties, variability and other characteristics. X-ray emission provides important diagnostics on magnetic activity phenomena and their evolution throughout the stellar lifetime, further it influences circumstellar disks and planetary systems. Overall, nearby young stellar populations are especially suited for X-ray studies.

The sketch shown in Fig. 2 depicts the typical eRASS sensitivity, when following the rough scheme of this symposium by focussing on objects within 100 pc around the Sun that are younger than 200 Myr. This region includes well studied moving groups and young associations like $\beta$ Pic, AB Dor, TWA, Tuc/Hor or $\eta$ Cha. For young stars within this volume the eRASS provides a virtually complete X-ray coverage of stellar targets with spectral types in the range of early F to mid M. The eRASS sensitivity translates into a detection limit of $L_X \approx 1 \times 10^{24} \times d^2 (\mathrm{pc})$ erg s$^{-1}$ and the X-ray horizons for exemplary young stars are indicated in the upper right corner. Taking again the 'young suns' example, stars like EK Dra (G 1.5, 100 Myr, $P_{\mathrm{rot}} = 2.7$ d, $L_X \approx 10^{30}$ erg s$^{-1}$) can be detected up to a distance of about 1 kpc and stars like $\pi^1$ UMa (G 1.5, 300 Myr, $P_{\mathrm{rot}} = 4.7$ d, $L_X \approx 10^{29}$ erg s$^{-1}$) up to roughly 300 pc.

In terms of activity levels, the flux limit transforms at a distance of 100 pc to sensitivities of $\log L_X/L_{\mathrm{bol}} \gtrsim -6.3$ at spectral type early F (5 L$_\odot$) and $\log L_X/L_{\mathrm{bol}} \gtrsim -3.3$ at mid M (0.005 L$_\odot$). This allows for the detection of virtually all so far unrecognized nearby young stars. In addition, the sensitivity of the eRASS is sufficient to detect a large fraction of the members of moderately older (UMa cluster, Hyades) or more distant (Pleiades) groups, opening the way to a more complete sampling of many well characterized, coeval stellar populations at ages of up to 1 Gyr. By doing so, the eRASS sample broadens the stellar parameter space suitable for a systematic investigation of the magnetic activity evolution, the activity-rotation-age relation and coronal properties, utilizing well defined stellar ensembles at different ages. eROSITA will also detect X-ray emission from late A and late M stars, but towards higher masses the increasingly thinner convective zones create weaker dynamos (low $L_X/L_{\mathrm{bol}}$) and towards lower masses stars become fainter (low $L_{\mathrm{bol}}$). Therefore stars at the hot and at the cool end of the magnetic activity range are intrinsically X-ray fainter and correspondingly their X-ray horizon will be closer.

In the regime of intermediate mass stars, the young HAeBe stars (Stelzer *et al.*, 2009) and the magnetic ApBp stars (Robrade & Schmitt, 2011) are prime targets for X-ray surveys. Both types of stars show diverse X-ray phenomena and a broad range of X-ray luminosities was observed in previously studied objects. The eRASS is promising to shed light on the general X-ray properties of the respective class.

Going to sub-stellar objects, the most massive, youngest and hottest Brown Dwarfs (Williams *et al.*, 2014) are the most promising ones and X-ray detections will likely be limited to M-type Brown Dwarfs that are either very nearby or in star forming regions.

## 4. Star forming regions

Active star formation is absent in the 100 pc solar neighborhood. However, several nearby groups like the TW Hya Association ($d \approx 50$ pc, age $\sim 10$ Myr) or the $\eta$ Cha cluster ($d \approx 97$ pc, age $6-8$ Myr) still contain a few classical T Tauri stars (CTTS), i.e. young stars showing strong accretion signatures and a circumstellar disk. Significant amounts of molecular gas or dust associated to these groups is virtually absent and many of these older groups have already started to disperse, underlining the value of wide-field X-ray surveys for identifying and studying possible members.

The TWA and $\eta$ Cha cluster are thought to belong to the Sco-Cen OB association, which includes also active star forming regions like Chamaeleon, Lupus or Rho Oph. These are part of the Gould Belt, a nearby starburst region that contains further low-mass SFRs (star forming region) like Taurus-Aurigae and with Orion it harbors the closest active massive SFR. Massive stars are a different topic in X-ray studies, but of strong interest due to their importance for galactic evolution and chemical enrichment.

All classes of pre-main sequence stars are known to be X-ray bright and SFRs will be of prime interest to study the youngest stellar populations. In contrast to the situation for young field stars, larger amounts of interstellar and circumstellar material are typically present and X-ray absorption becomes important. The coverage of X-rays up to 10 keV, providing better sensitivity for embedded sources and the good positional accuracy are

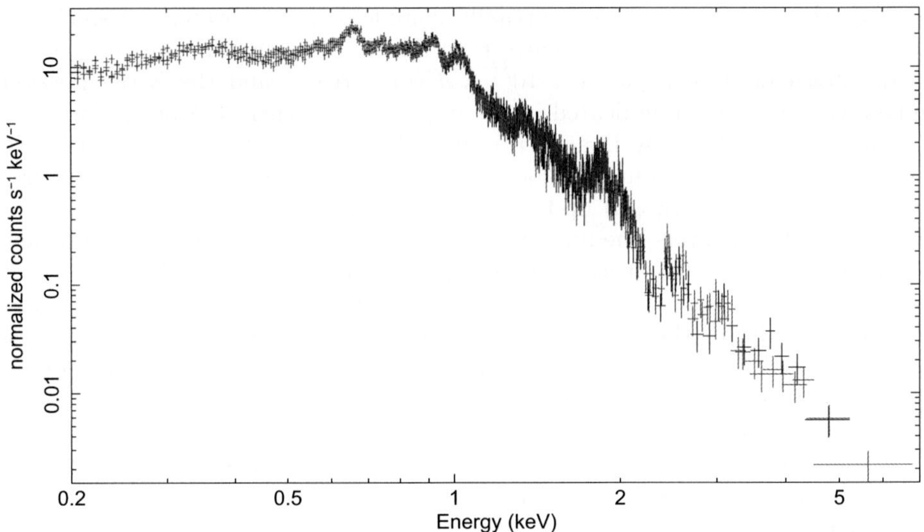

**Figure 3.** Simulated X-ray spectrum of the young M dwarf binary AT Mic, 2 ks survey exposure (eRASS/3 simulation runs).

here further advantages of the eRASS data. Depending on distance and line-of-sight absorption, the evolutionary sequence from young CTTS over the virtually diskless WTTS (weak-line T Tauri stars) to post T Tauri/ZAMS (zero-age main sequence) stars will be covered for a large range of stellar masses in many of these regions. In star formation studies, the eRASS complements very well to deeper pointed observations with Chandra or XMM-Newton, that typically focus on the cores of SFRs.

## 5. Beyond detections

Beside X-ray detections and fluxes, the eRASS will obtain spectra and light curves for all X-ray brighter sources and many of these will be nearby young stars. Several thousands of stars have more than 1000 detected X-ray photons and several ten thousand more than 200 photons. Fig. 3 shows a simulated eRASS spectrum of AT Mic, a mid M-dwarf binary that belongs to the $\beta$ Pic moving group. The spectral quality is with about 25000 counts high enough to study the coronal plasma temperature distribution, elemental abundances or line-of-sight absorption; even time-resolved spectroscopy becomes possible. If sample size is more important, lower quality data can be used to perform simple spectral modelling or derive the spectral hardness of the detected photons, thereby putting constraints on coronal parameters or the classification of the respective object.

Variability is another important issue in stellar X-ray astronomy and many stars show distinct brightness variations. The eRASS allows to investigate the time domain for many targets in greater detail and naturally provides X-ray measurements on several timescales. During the build up of the 4 yr survey, each source is visited at least eight times with a 0.5 yr cadence; in addition the scanning law of the satellite provides several FOV passages during each visit with a cadence of 4 h. Thereby the eRASS provides for the first time large sets of short- and long-term X-ray time series with a single instrument.

Among the science themes to be addressed with time resolved data are the characterization of typical X-ray activity levels and their long-term stability, coronal variability and its amplitudes, flare statistics and X-ray activity cycles. Further, strong X-ray flares are quite common in a variety of stars, enabling the detection of additional sources with too low quiescent flux. The X-ray spectra allow to address, among others, topics related to coronal physics, magnetic activity evolution or the high-energy irradiance of circumstellar disks, exoplanets and their atmospheres.

## 6. Synergies

While the eROSITA data alone is already impressive, its scientific value is greatly enhanced when combining the X-ray sources and their properties with other suitable datasets, most importantly to obtain identifications and distances as well as classifications and characterizations of their counterparts. In this respect the optical brightness of the stellar content within the eROSITA survey is a great advantage. Most eRASS stars will have magnitudes of $V \lesssim 15$ mag and Gaia (see A. Sozetti, this Issue) will provide accurate positions, distances, 3-D space motions and basic stellar parameters; the remaining fraction typically has at least partial or less accurate Gaia parameters.

In addition, a wealth of very useful multi-wavelength data from IR to UV wavelength exists, often with large area or all-sky coverage and several future wide-area surveys are planned. Besides the numerous surveys in the optical regime, both photometric and spectroscopic (see K. Covey; S. Martell, this Issue), a virtually complete all-sky coverage is e.g provided by the NIR/IR surveys 2MASS and WISE. Infrared observations are also an important supplement to optical data, when identifying very red or high-extinction sources. In X-rays itself, the RASS and several mission archives enable the study of X-ray variability on timescales of many years up to decades.

Finally, large catalogs and data bases with rotation measurements or related activity indicators can also be combined with the X-ray data and dedicated follow-up observations for outstanding sources are envisaged. Likewise nearby star searches like RECONS or SUPERBLINK and object class, planet host or young nearby moving group membership catalogs are just a few examples of valuable complementary data to put the X-ray stars in a broader astrophysical context.

### Acknowledgements

I acknowledge support from DLR under the grant 50QR0803.

### References

Güdel, M. Guinan, E., & Skinner, S. 1997, *ApJ*, 483, 947
Guillout, P., Haywood, M., Motch, C., & Robin, A. C. 1996, *A&A*, 316, 89
Merloni, A., Predehl, P., Becker, W. *et al.* 2012, *arXiv1209.3114M*
Pizzolato, N., Maggio, A., Micela, G., Sciortino, S., & Ventura, P. 2003, *A&A*, 397, 147
Ribas, I., Guinan, E., Güdel, M., & Audard, M. 2005, *ApJ*, 622, 680
Robrade, J. & Schmitt, J. H. M.. M. 2011, *A&A*, 531A, 58
Schmitt, J. H. M. M. 1997, *A&A*, 318, 215
Stelzer, B. Robrade, J., Schmitt, J. H. M. M., & Bouvier, J. 2009, *A&A*, 493, 1109
Williams, P. K. G., Cook, B. A., & Berger, E. 2014, *ApJ*, 785, 9
Wright, N. J., Drake, J. J., Mamajek, E. E., & Henry, G. W. 2011, *ApJ*, 743, 48

### Discussion

G. Hussain: Will there be public data releases?

J. Robrade: There will be publicly available eRASS data products/catalogs. Speaking for the German side, it is planned to have a final as well as intermediate data releases after respective proprietary periods of about two years. If you are interested in collaborating, give us a note; the Hamburg Observatory is leading the eROSITA Stars WG.

*Young Stars & Planets Near the Sun*
Proceedings IAU Symposium No. 314, 2015
J. H. Kastner, B. Stelzer, & S. A. Metchev, eds.

© International Astronomical Union 2016
doi:10.1017/S1743921315006468

# Precise Near-Infrared Radial Velocities

Peter Plavchan[1], Peter Gao[2], Jonathan Gagne[3], Elise Furlan[4],
Carolyn Brinkworth[5], Michael Bottom[2], Angelle Tanner[6], Guillem
Anglada-Escude[7], Russel White[8], Cassy Davison[8], Sean Mills[9], Chas
Beichman[4], John Asher Johnson[10], David Ciardi[4], Kent Wallace[11],
Bertrand Mennesson[11], Gautam Vasisht[11], Lisa Prato[12], Stephen
Kane[13], Sam Crawford[11], Tim Crawford[11], Keeyoon Sung[11], Brian
Drouin[11], Sean Lin[11], Stephanie Leifer[11], Joe Catanzarite[14], Todd
Henry[15], Kaspar von Braun[12], Bernie Walp[16], Claire Geneser[1], Nick
Ogden[1], Andrew Stufflebeam[1], Garrett Pohl[1] and Joe Regan[1]

[1]Missouri State University
email: plavchan@missouristate.edu
[2]Caltech, [3]University of Montreal, [4]NASA Exoplanet Science Institute, [5]National Center for
Atmospheric Research/ University Corporation for Atmospheric Research, [6]Mississippi State
University,[7]University College London,[8]Georgia State University,[9]University of
Chicago,[10]Harvard,[11]NASA Jet Propulsion Laboratory,[12]Lowell Observatory,[13]San Francisco
State University,[14]SETI Institute,[15]Georgia State University,[16]NASA Ames

**Abstract.** We present the results of two 2.3 $\mu$m near-infrared (NIR) radial velocity (RV) surveys to detect exoplanets around 36 nearby and young M dwarfs. We use the CSHELL spectrograph ($R \sim 46{,}000$) at the NASA InfraRed Telescope Facility (IRTF), combined with an isotopic methane absorption gas cell for common optical path relative wavelength calibration. We have developed a sophisticated RV forward modeling code that accounts for fringing and other instrumental artifacts present in the spectra. With a spectral grasp of only 5 nm, we are able to reach long-term radial velocity dispersions of $\sim$20–30 m s$^{-1}$ on our survey targets.

**Keywords.** instrumentation: spectrographs, techniques: radial velocities

## 1. Introduction

We are carrying out an exoplanet search to exploit the "M dwarf opportunity." Approximately 70% of main sequence stars are M dwarfs, spanning a factor of 5 in mass/radius and $\sim 10^3$ in luminosity. Due to their smaller masses and radii and cooler temperatures compared to FGK stars, transiting exoplanets produced deeper eclipses of M dwarfs, exoplanet habitable zones are closer, and exoplanet radial velocity amplitude reflex motions are larger. More recently, a new "M dwarf opportunity" for finding exoplanets has been identified with the Kepler mission: not only are planets smaller than 4 $R_\oplus$ more common than Jovian planets, they are predominantly found around lower mass stars (Howard *et al.* 2012, Dressing & Charbonneau 2013).

There are several factors spoiling the "M dwarf opportunity." First, M dwarfs are red and faint, with $V - K > 3.5$ mag and only 4 M4 or later dwarfs with $V < 12$ mag. Additionally, small planetary RV signals are comparable in amplitude and time-scale to numerous periodic perturbing stellar jitter signals such as rotational spot modulation & magnetic activity cycles (Vanderberg *et al.* 2015, Robertson *et al.* 2015). One solution to both of these challenges is to observe in the NIR, where M dwarfs put out much of their bolometric luminosity, and the amplitude of stellar jitter is diminished (Reiners *et al.* 2010, Plavchan *et al.* 2013a,b,2015, Anglada-Escude *et al.* 2012).

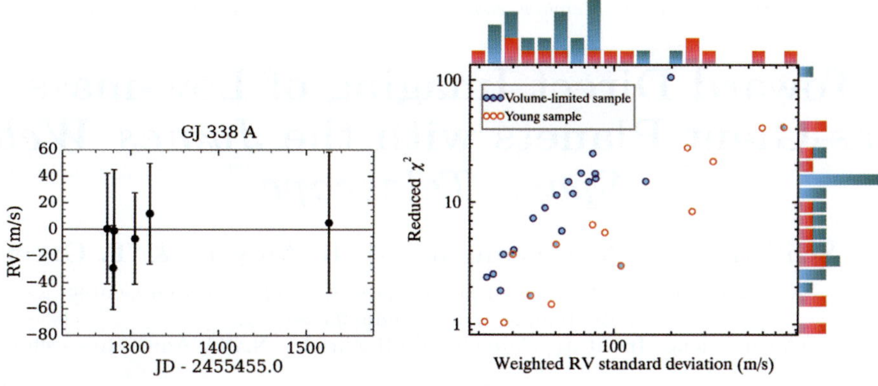

**Figure 1.** Left: CSHELL NIR RV time-series for GJ 338A ($N_{obs}$=6, rms=22 m s$^{-1}$, $\chi_r^2 = 2.4$). Right: CSHELL NIR RV survey results, plotted as a function of RV rms and reduced chi-squared for both young and nearby M dwarfs in red and blue respectively.

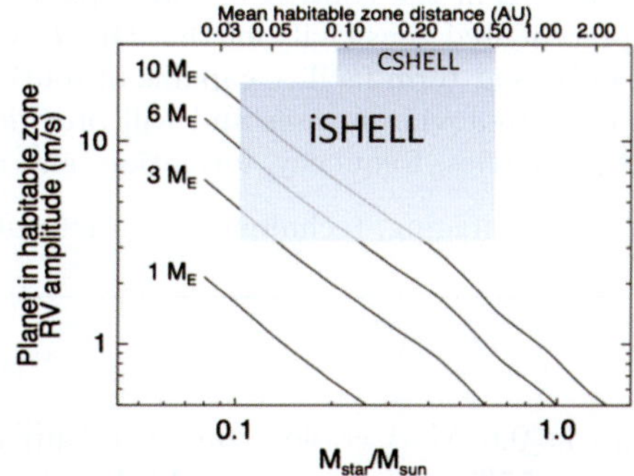

**Figure 2.** A comparison of the potential exoplanet sensitivity of CSHELL and iSHELL. The solid black lines correspond to the RV semi-amplitude of Habitable Zone (HZ) exoplanets with the indicated masses as a function of stellar mass.

## 2. Results and Conclusions

We have obtained our goal precision of $\sim$30–60 m s$^{-1}$ in our young M dwarf survey from 2010–2012, and $\sim$20–30 m s$^{-1}$ in 2014 at higher SNR for nearby M dwarfs not previously surveyed by visible RV teams. We present a sample RV curve and our survey results in Figure 1. We are in the process of publishing our detailed survey results.

Looking towards the future, we have made the gas cells for iSHELL ($R \sim$ 75,000, spectral grasp of 250 nm), which will replace CSHELL in 2016. We expect to get to $<$ 5 m s$^{-1}$ precision with iSHELL due to the increased spectral grasp alone (Figure 2). The increased efficiency, resolution, no fringing, and smaller number of bad pixels will provide further improvements.

## References

Anglada-Escude, G., *et al.*, 2012, *PASP*, 124, 586

Dressing, C. & Charbonneau, D., 2013, *ApJ*, 767, 95

Howard, A., *et al.* 2012, *ApJS*, 201, 15

Reiners, A., *et al.*, 2010, *ApJ*, 710, 432

Plavchan, P., *et al.*, 2015, *NASA ExoPAG report*, arXiv:1503.01770

Plavchan, P., *et al.*, 2013a, *SPIE*, 8864, 0G

Plavchan, P., *et al.*, 2013b, *SPIE* 8864, 1J

Robertson, P., *et al.*2015, *ApJ* 801, 79

Vanderberg, A., *et al.*, 2015, in prep.

*Young Stars & Planets Near the Sun*
Proceedings IAU Symposium No. 314, 2015
J. H. Kastner, B. Stelzer, & S. A. Metchev, eds.



# Toward Direct Imaging of Low-mass Gas-Giant Planets with the *James Webb Space Telescope*

J. E. Schlieder[1], C. A. Beichman[2], M. R. Meyer[3], & T. Greene[4]

[1]NASA Postdoctoral Program Fellow, NASA Ames Research Center
email: joshua.e.schlieder@nasa.gov
[2]NASA Exoplanet Science Institute/Caltech, [3]ETH Zürich, [4]NASA Ames Research Center

**Abstract.** In preparation for observations with the *James Webb Space Telescope* (*JWST*), we have identified new members of the nearby, young M dwarf sample and compiled an up to date list of these stars. Here we summarize our efforts to identify young M dwarfs, describe the current sample, and detail its demographics in the context of direct planet imaging. We also describe our investigations of the unprecedented sensitivity of the *JWST* when imaging nearby, young M dwarfs. The *JWST* is the only near term facility capable of routinely pushing direct imaging capabilities around M dwarfs to sub-Jovian masses and will provide key insight into questions regarding low-mass gas-giant properties, frequency, formation, and architectures.

**Keywords.** stars: low-mass, stars: imaging, techniques: high-resolution

## 1. Introduction

M dwarfs, low-mass stars ($\leqslant$0.6 $M_\odot$) cooler than the Sun ($\leqslant$3900 K), are the most common stars in the Galaxy ($\sim$75%). The youngest M dwarfs in the Solar neighborhood (d$\leqslant$100 pc) are members of loose associations having common Galactic kinematics and ages $\sim$10–130 Myr; young moving groups (YMGs, Zuckerman & Song 2004). Due to their proximity, youth, and intrinsically low-luminosities, M dwarf YMG members are prime targets for direct exoplanet imaging. Additionally, a sizable population of $\sim$Jupiter to $\sim$Neptune mass planets is detected at moderate separations ($\sim$1–10 AU) around M dwarfs by gravitational micro-lensing surveys (Cassan *et al.* 2012). For these reasons, many dedicated searches to identify the missing M dwarf members of YMGs have been undertaken, e.g. Kraus *et al.* (2014), and several surveys to directly image their planets are ongoing (Bowler *et al.* 2015). Despite the identification of several hundred likely new M dwarf YMG members and high-contrast imaging observations of a sizable fraction of that sample, YMGs still lack the large numbers of M dwarfs expected (Kraus *et al.* 2014) and only a few planetary-mass companions on wide orbits have been imaged. To probe the known population of low-mass gas-giants around M dwarfs via direct imaging, more members of the nearby young sample must be identified and facilities with greater sensitivity are required.

## 2. A Search For Nearby Young M Dwarfs and the Current Sample

We have identified nearly 400 candidate M dwarf YMG members using a selection algorithm that hinges on proper motion, photometry, and activity (see Schlieder *et al.* 2012). For the last 3 years we have pursued an all-sky, spectroscopic, follow-up program to identify true YMG members in the sample that we call *CASTOFFS* (Schlieder *et al.* 2015). We have observed $\sim\frac{3}{4}$ of our candidates using high-resolution optical (MPG 2.2m/FEROS,

CAHA 2.2m/CAFE) or medium-resolution near-IR (IRTF 3.0m/SpeX) spectroscopy and our analyses continue. Our high-resolution optical spectra have so far revealed more than 50 new M dwarf YMG members, several isolated young field M dwarfs, and a dozen spectroscopic binaries. Our analyses of the near-IR spectra are just beginning and young M dwarfs with low surface-gravity features are already apparent in the data.

We have compiled an up to date list of young, low-mass stars within 100 pc from the literature and added to it our new identifications from *CASTOFFS*. We primarily include stars having both spectroscopic confirmation of youth and at least partial kinematics (proper motion and RV) consistent with YMG membership. A few well characterized, young field stars are also included. From more than 30 literature references over the last $\sim$15 years and our new results, we find 440 $\sim$K5–M9 systems (338 known or presumed single, 102 multiples) with ages $\sim$10–400 Myr at distances $\lesssim$100 pc. This is nearly a factor of five increase over the M dwarf YMG sample from Torres *et al.* 2008. Only $\sim\frac{1}{3}$ of the sample has measured parallaxes, the remaining distances are kinematic or photometric estimates. There are fewer stars within 25 pc than anticipated and the population falls off very quickly beyond $\sim$M4 type and distances $\gtrsim$50 pc, indicating the potential for more discoveries with ongoing, dedicated searches.

## 3. *James Webb Space Telescope* Survey Simulations and Future Work

The *James Webb Space Telescope* will provide unprecedented imaging sensitivity. Observing at $\sim$4.5 $\mu$m with a coronagraph, the NIRCam instrument is expected to provide a contrast of $10^{-5}$ at $1''$ separations with a very low background limit. To explore these capabilities in the context of M dwarfs, we have performed imaging survey simulations on our young M dwarf sample using the Monte Carlo methods described in Beichman *et al.* (2010). Young planet magnitudes were taken from extended COND03 models (Baraffe *et al.* 2003) covering 0.1–5 $M_{\rm Jup}$. The simulated survey results reveal routine imaging sensitivity to planets $<$0.5 $M_{\rm Jup}$ at $>$50 AU separations. In the best cases, the simulations predict *JWST*/NIRCam will detect $\sim$0.1 $M_{\rm Jup}$ ($\sim$2 $M_{\rm Nep}$) planets at $<$10 AU, directly probing the known micro-lensing population.

Our preliminary simulations indicate that *JWST*/NIRCam imaging can routinely detect sub-Jovian mass planets around nearby, young M dwarfs. However, further work is warranted. New constraints on the M dwarf planet population are available and new planet evolution and atmosphere models will provide better predictions of young, low-mass planet luminosities. We plan to include these new results in our future simulations. Our continuing *CASTOFFS* survey and improved survey simulations will provide high priority targets for early *JWST* GTO and GO proposals and pave the way for the first direct images of low-mass gas-giant planets.

## References

Baraffe, I., *et al.* 2003, *A&A*, 402, 701
Beichman, C. A., *et al.* 2010, *PASP*, 122, 162
Bowler, B., *et al.* 2015, *ApJS*, 216, 7
Cassan, A., *et al.* 2012, *Nature*, 481, 167
Kraus, A. L., *et al.* 2014, *ApJ*, 147, 146
Schlieder, J. E., *et al.* 2012, *AJ*, 143, 80
Schlieder, J. E., *et al.* 2015, *Proceedings of CS18*, Editors: G. van Belle and H.C. Harris, 919
Torres, C. A. O., *et al.* 2008, *HSFR*, Editor: Bo Reipurth, 2, 757
Zuckerman, B. & Song, I. 2004, *ARA&A*, 42, 645

*Young Stars & Planets Near the Sun*
*Proceedings IAU Symposium No. 314, 2015*
*J. H. Kastner, B. Stelzer, & S. A. Metchev, eds.*

© International Astronomical Union 2016
doi:10.1017/S1743921315006274

# Young Stars & Planets Near the Sun in 2015: Five Takeaways and Five Predictions

## Michael C. Liu

Institute for Astronomy, University of Hawaii, 2680 Woodlawn Drive, Honolulu, HI 96822, USA

**Abstract.** I present a highly biased and skewed summary of IAU Symposium 314, "Young Stars and Planets Near the Sun," held in May 2015. This summary includes some takeaway thoughts about the rapidly evolving state of the field, as well as some crowd-sourced predictions for progress over the next ~10 years. We predict the elimination of 1–2 of the currently recognized young moving groups, the addition of 3 or more new moving groups within 100 pc, the continued lack of a predictive theory of stellar mass, robust measurements of the gas and dust content of circumstellar disks, and an ongoing struggle to achieve a consensus definition for a planet.

**Keywords.** stars, brown dwarfs, planetary systems

---

## 1. Five Takeaways from 2015

> *"Every great cause begins as a movement, becomes a business, and*
> *eventually degenerates into a racket."* — Eric Hoffer

IAU Symposium 314 has been an incredible testament to the progress made in studying young stars near Earth since the last dedicated meeting on this topic almost 15 years ago, in 2001. Thanks the diligence of conference participants Matthew Kenworthy and Rahul Patel, short summaries of every conference talk are available on Facebook (`https://goo.gl/pvHUHd`). So like many other things in today's world, the traditional role of a conference summary talk has been superseded thanks to social media. Thus instead of trying to recapitulate the content of this week-long meeting, instead I provide here some personal impressions about the state of the field.

1. **Learn to live with ambiguity.** Stellar characterization is currently experiencing a renaissance. For young stars, there has been ample progress in key topics such as stellar age-dating, assessment of moving group membership, and spectroscopic characterization. However, we should remember that ambiguity is inherent to this field. Many of the key physical properties that we seek may simply be inaccessible.

The star formation process in molecular clouds shows us that most stars form in bound clusters. The dissolution of such clusters is inevitable over time, and stars are lost from these primordial groups into the field (talk by Lada). Thus, when we find young stars and brown dwarfs in the field, we should accept that the birth history for a significant fraction of them will be unrecoverable. In short, not every young star that you find should belong to a moving group.

As the census of moving groups has grown, so have the methods used to assign membership, with the current pinnacle being the quantitative BANYAN method (talks by Malo & Gagne). Nevertheless, the known groups overlap in both velocity ($UVW$) and space ($XYZ$). Thus, spatial and kinematic information may not be sufficient, on its own, to ever assign some young stars to a specific group. Similarly, the challenges of stellar age dating can be expected to persist, rendering an inherent fuzziness in our ability to assign ages to young stars in groups and in the field.

Finally, even objects that have a large suite of high quality measurements can present a challenge when deriving physical properties. A very timely example is the recent activity in determining ages and masses for young brown dwarfs, from the stellar/substellar boundary all the way into the planetary mass regime. Their young ages are reflected in their spectra through gravity-dependent absorption features. But spectral features alone are remarkably degenerate for young objects of different masses and ages — similar spectra (and colors and magnitudes) between objects do not necessarily correspond to the similar masses and ages (talk by Allers).

2. **Everything gets more complicated the closer you look.** Fields of astronomical research often have a natural progression, as the quantity and quality of data grow. In the early phases, data are sparse and thus simple paradigms of how things work can be remarkably successful. This is followed by a natural compulsion to deepen and expand our data, which then sweeps away much of the initial paradigm. At this IAU meeting, we saw many examples of the growth of complexity. For instance, there is the long-standing picture of the nearest OB association, Sco-Cen, whereby the history of its star formation is thought to have sequentially migrated across the region. This simple concept now needs to be revisited given the more detailed maps of its stellar distribution, which show a patchwork of different ages sewn together on the sky (talk by Mamajek). More broadly, stellar activity modeling is now building an integrated theoretical framework that combines interior modeling, angular momentum evolution, mass loss, and magnetic fields (talk by Matt). And yet we only need to look at the work being done on our Sun by the solar physics community to see how far we have to go even given a plethora of observations.

3. **Initial conditions matter.** Stellar evolution theory has made significant advances in modeling the range of initial conditions relevant to the earliest stages of star formation, namely magnetic fields, accretion, rotation, and photospheric spots (talks by Feiden, Baraffe, and Somers). This is encouraging, given the challenges of modeling these physical effects. However, such efforts also bring to light the large role that initial conditions play in the final observational properties, even many tens to hundreds of millions of years after formation. Since initial conditions are largely inaccessible to observations, and many are fundamentally unknowable (e.g. the early accretion history of a star), we made be heading towards a situation where the failings between models and data are naturally attributed to "initial conditions." Thus, in the same way that historical astronomers invoked epicycles to explain the discrepancies between the observed motion of the planets and their theoretical model (namely the Earth-centric universe), we may have found our own modern epicycles in magnetic fields, accretion, and rotation.

4. **You can't always get what you want.** A prime driver for the identification of young stars near Earth is the benefit for directly imaging exoplanets and circumstellar disks. Tremendous gains have been made in this arena, both through the identification of the best targets (i.e., the youngest, nearest stars) and the development of high-contrast adaptive optics systems to distinguish the planets from their bright host stars. This two-pronged approach has been enormously successful, yielding $\approx 10$ directly imaged planets to date and with more expected to come imminently from Gemini/GPI, VLT/SPHERE, and SPHERE/HiCIAO. However, during the same time that target identification and planet imaging capabilities have been improving, concurrent theoretical efforts in exoplanet formation and evolution have revealed the intense degeneracies of different formation models (cold/warm/hot start) when manifested into the available observables (luminosity, age, magnitudes, temperature, etc.). Thus, like a small child who desires many presents on Christmas but then cannot open the shrink-wrapped boxes to get to

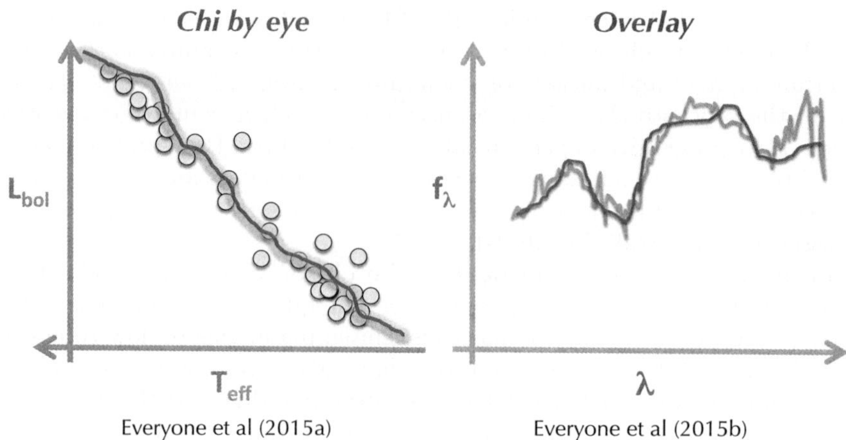


**Figure 1.** Representation of common practices for the interplay of models and data. We need to do better.

the toys, we may soon achieve our long-sought goal of having a rich census of directly imaged planets, and yet understanding of how they form and evolve may be elusive.

5. **Stop comparing models to data and start actually testing models with data.** "Testing models with data" is a near-ubiquitous statement in proposals seeking telescope time and funding support. This IAU Symposium has amply shown the advances both in the theoretical models and the observational data. And yet it also shows the significant gap in our efforts to integrate the two aspects. Many results shown here, and many more results published in the literature, are not truly tests of the models (Figure 1). For stellar/substellar evolution, our very rich data sets are paired with state-of-the-art evolutionary and atmospheric models simply by placing the two on the same plot — this "chi-by-eye" approach neglects the very rich information available in the numbers, distributions, and outliers of the data, as well as the more subtle features of the models. Similarly, in determining physical parameters for young brown dwarfs and exoplanets, an oft-used technique is to derive temperatures and gravities by overlaying models onto photometric and spectroscopic data (likewise with higher-order quantities such as clouds and chemistry, as well as "uncertainties" on all these quantities). This "overlay" method is inherently an open loop process, with no actual validation of the models used to derive the physical parameters. We should endeavor to do better in order to yield the full fruit of our efforts.

## 2. Five Predictions for 2025

The future is very bright for study of young stars and planets near Earth given the wide range of upcoming astronomical facilities on the horizon, e.g., ALMA, Pan-STARRS, GAIA, eROSITA, JWST, LSST, and a host of ELTs. We can easily see the tremendous growth in this field by comparing the advances from the 2001 meeting and our 2015 gathering. What does the next decade hold?

In an attempt to forecast the future (and to provide some legacy value for this conference proceedings), the audience† was presented with 5 key scientific questions that will be addressed in the next 10 years. Since prognostication is a difficult task for any one individual, we invoked the method of the "wisdom of the masses." A common example

† This refers to the audience that was actually present for the final talk of the conference, which amounted to 68% of the registered participants.

**Figure 2.** The wisdom of the masses is a common method to discern the answer to questions that are too difficult for any individual to answer, such as the number of jellybeans in a jar.

of this method is counting jellybeans in a jar: it is hard for any one person to guess the right number, but the median of guesses from a large group of people will yield an accurate result (Figure 2). In this spirit, the audience voted on three possible outcomes for each of the key science questions. The results are summarized here. (Note that the vote-counting was done in real-time from the podium, so it may not be exactly accurate. Not everyone voted on every question, so the total number of votes for each question is not the same.)

1. **How many of the currently known young moving groups will be eliminated?** Some of the known YMGs are quite secure in terms of their legitimacy, while the reality of others is currently being debated. For reference, we consider the 7 groups adopted in the BANYAN model (TWA, $\beta$ Pic, AB Dor, Tuc-Hor, Columba, Argus & Carina) as the current census. *CHOICES: None (7 votes), 1–2 (82 votes), >2 (3 votes).*

2. **How many new young groups will be found within 100 pc?** With the revolutionary astrometric datasets from GAIA and LSST, as well as complementary massive spectroscopic surveys and the new all-sky X-ray mission eROSITA, we can anticipate significant progress in finding/defining new moving groups. *CHOICES: None (2 votes), 1–2 (25 votes), >2 (65 votes).*

3. **Will we have a complete predictive theory of how stars get their mass?** While observational studies have made great advances in mapping the physical conditions and resulting distributions (in mass, age, position, etc.) of forming stars and their natal molecular clouds, a predictive theory of the stellar initial mass function remains an outstanding challenge. *CHOICES: Yes (1 vote – Chabrier), No (91 votes).*

4. **Will we be able to robustly measure the gas and dust masses of circumstellar disks?** We are at the start of a great observational revolution in sub-millimeter astronomy thanks to ALMA. To advance studies of disks and the associated planet formation process, an accurate inventory of the gas and dust components is needed. *CHOICES: Yes (48 votes), No (27 votes).*

5. **Will we have a consensus definition of a "planet"?** The diversity of objects that might warrant the prized label of a planet has expanded now that observational facilities are capable of directly detecting objects down to a few Jupiter masses, both free-

**Figure 3.** A selection of acronyms presented at this meeting. *Top row:* PALMS (B. Bowler), LACEWING (A. Riedel), HAZMAT (E. Shkolnik). *Middle row:* SACY (P. Elliott), SPHERE (G. Chauvin), DALI (N. van der Marel). *Bottom row:* BASS (J. Gagne), BANYAN (L. Malo), JASON (S. Murphy).

floating and as companions around higher mass objects. The 2003 definition produced by the International Astronomical Union, using the deuterium-burning limit ($\approx$13 $M_{\rm Jup}$) as the dividing line between planets and brown dwarfs, has amply been shown to be an inadequate description of nature. Will we have a clear view on this in 10 years? *CHOICES: Yes (14 votes), No (78 votes).*